Active Radar Cross Section Reduction

Theory and Applications

Hema Singh

Rakesh Mohan Jha

CAMBRIDGE
UNIVERSITY PRESS

CAMBRIDGE
UNIVERSITY PRESS

Cambridge House, 4381/4 Ansari Road, Daryaganj, Delhi 110002, India

Cambridge Univerisity Press is part of the University of Cambridge.

It furthers the University's mission by disseminating knowledge in the pursuit of education, learning and research at the highest international levels of excellence.

www.cambridge.org
Information on this title: www.cambridge.org/9781107092617

First published 2015

Printed in India by Akash Press, New Delhi 110 020

A catalogue record for this publication is available from the British Library

Library of Congress Cataloging-in-Publication Data
Jha, R. M. (Rakesh Mohan), 1959-
Active radar cross section reduction : theory and applications / Rakesh Mohan Jha, Hema Singh.
pages cm
Includes bibliographical references and index.
Summary: "Provides a comprehensive discussion on active and passive radar cross section (RCS) estimation and techniques to achieve the low observable aerospace platforms"-- Provided by publisher.
ISBN 978-1-107-09261-7 (hardback)
1. Radar cross sections. 2. Adaptive antennas. 3. Antenna arrays. 4. Radar--Interference. 5. Echo suppression (Telecommunication) 6. Stealth aircraft. I. Singh, Hema. II. Title.
TK6580.J478 2015
629.1—dc23
2014034055

ISBN 978-1-107-09261-7 Hardback

To
Professor R. Narasimha

CONTENTS

List of Tables

List of Figures

List of Abbreviations

ACM	Array correlation matrix
AMC	Artificial magnetic conductors
AWGN	Additive white gaussian noise
CFRP	Carbon fibre reinforced polymer
CLAS	Conformal load-bearing antenna structure
CNT	Carbon nanotube
DC	Direct current
DDD	Direct data domain
DF–GSC	Decision feedback sidelobe canceller
DNG	Double negative
DOA	Direction-of-arrival
DOF	Degree of freedom
DPS	Double positive
EBG	Electronic band gap
ECM	Equivalent currents method
EM	Electromagnetic
EMC	Electromagnetic compatibility
EMI	Electromagnetic interference
ENG	Epsilon negative
ERAKO	Electronic radar with conformal array antenna
ESPRIT	Estimation of signal parameters via rotational invariance techniques
FDMA	Frequency division multiple access
FDTD	Finite difference time domain
FEM	Finite element method
FHC	Filled-hole-compression
FHR	Fraunhofer institute for high frequency physics and radar techniques
FPGA	Field programmable gate array
FSS	Frequency selective surface
GA	Genetic algorithm
GA-CG	Genetic algorithm and conjugate gradient
GFRP	Glass fibre reinforced polymer
GO	Geometrical optics
GSC	Generalised sidelobe canceller
GTD	Geometrical theory of diffraction
HF	High frequency
INR	Interference to noise ratio
LCMP	Linearly constrained minimum power
LCMV	Linearly constrained minimum variance
LHM	Left handed material
LMS	Least mean square
LP	Linear programming
LS	Least squares
MMSE	Minimum mean square error
MNG	Mu negative
MoM	Method of moments
MSE	Mean square error
MUSIC	Multiple signal classification
NLMS	Normalised least mean square
NU	Non-uniform
NURBS	Non-uniform rational B-spline

OHC	Open-hole-compression
OML	Outer mould line
PEC	Perfect electric conductor
PMC	Perfect magnetic conductor
PO	Physical optics
PSLL	Peak sidelobe level
PSO	Particle swarm optimisation
QFRP	Quartz fibre reinforced polymer
QPSK	Quadrature phase-shift keying
QRD	QR decomposition
RAM	Radar absorbing material
RAS	Radar absorbing structure
RCS	Radar cross section
RCSR	Radar cross section reduction
RF	Radio frequency
RHM	Right handed material
RLS	Recursive least square
RMIM	Receiving-mutual impedance method
SINR	Signal-to-noise-interference-ratio
SLL	Sidelobe level
SMI	Sample matrix inversion
SMILE	Scheme for spatial multiplexing of local elements
SNR	Signal to noise ratio
TDMA	Time division multiple access
TE	Transverse electric
TM	Transverse magnetic
UAV	Unmanned aerial vehicle
UHF	Ultra high frequency
UTD	Uniform theory of diffraction
UWB	Ultra wideband

Preface

Logic will get you from A to Z; imagination will take you everywhere.

- Anonymous

Evading detection by radar has been one of the fascinating topics in aerospace engineering. Initial intuitive attempts towards achieving low-observable platforms, such as fighter aircraft, unmanned aircraft, missiles and even battle ships, came from the application of radar absorbing materials (RAM) and shaping to reduce the radar cross section (RCS) of the platform.

These efforts on RAM design have continued since 1940s. Likewise, the efforts on shaping have run parallel to the developments in the theory of electromagnetic (EM) scattering and diffraction. The edge-diffraction EM formulations of 1960s resulted in the ongoing efforts in 1970s towards eventual realisation of the Lockheed F-117 *Nighthawk*; this fighter aircraft was characterised by the faceted planar exterior for RCS reduction. Likewise, the subsequent standardisation of the EM curved surface-diffraction formulations reflected in the blended Northrop Grumman B-2 *Spirit* bomber aircraft structure in 1980s.

However, it must be understood that both the shaping and RAM were in the domain of passive RCS reduction (RCSR).Once these concepts were formalised, it became apparent that these essentially catered to the narrowband RCSR. The radar had to merely switch or scan over a wider frequency range to overcome the stealth strategies of the low observables.

Thus there is a need for broadbanding the electromagnetic design for airborne and naval structures. The desire is to evade detection for any incoming radar frequency, polarisation and direction of arrival. Indeed, the fervent wish list is to "some-how" sense and generate an anti-wave for cancellation. This is the essence of active radar cross section reduction (RCSR).

Active RCSR towards realisation of low-observable platforms hinges on the integration of (i) active antenna elements, (ii) onboard/airborne antenna analysis, and (iii) conformal EM analysis.

The onboard/ airborne antenna analysis formulations have matured over the last three decades. The conformal antennas provide the advantage of least obscuration and conflict with competing aerodynamic requirements. Conformal antenna studies have been actively carried out for last four decades. However, over the last ten years, the modern conformal antenna theory appears to have been systematised.

The active antenna elements have been extensively studied over the last two decades. Indeed the phased antenna arrays provide one, albeit not the only, route to generate adaptive patterns towards active RCSR. Thus, it is our contention that active RCSR is an idea whose time has come!

The Centre for Electromagnetics (CEM), to which these authors are affiliated, has an active commitment towards RCS studies. Since RAM was once considered as a "highly classified area", we at the CEM took the initiative to demystify this topic in the form of a book:

> Vinoy, K.J. and R.M. Jha, *Radar Absorbing Materials: From Theory to Design and Characterisation.* Kluwer Academic Publishers, Norwell, Boston, USA, ISBN: 0792 397 533, 209 p., 1996.

The book in your hand is the second in this trilogy.

For the reasons discussed above, the emphasis in this book is on the phased antenna array analysis and algorithms as applied to active RCSR. All aspects, including mutual coupling, which takes one towards viable technological realisation of active RCSR, are discussed in details.

However, the end goal of this book is the low-observable platforms. Hence apart for the phased antenna arrays, advancement in the RAM design, plasma stealth, active FSS elements and metamaterial designs are also discussed as a parallel stream of concept.

Hema Singh
Rakesh Mohan Jha

Acknowledgements

We would like to sincerely thank Mr Shyam Chetty, Director, CSIR-National Aerospace Laboratories (NAL), Bangalore for his continued support for the various activities of the Centre for Electromagnetics, and for the formal permission to write this book.

Thanks are also due to Dr U. N. Sinha, Distinguished Scientist, CSIR, for his constant encouragement to take up this endeavour and complete the manuscript at the earliest.

The authors are also affiliated to the Academy of Scientific and Innovative Research (AcSIR), New Delhi, a deemed university. It is a pleasu re to acknowledge the numerous interactions with Prof. Nagesh Iyer, Acting Director, AcSIR, who has often stressed that lateral thinking and innovation should be the corner stone for CSIR scientists and academicians.

The cautionary notes, not necessarily pessimistic, were often rung by Prof. P. R. Mahapatra (radar signal processing) and Prof. Dipankar Banerjee (advanced material technologies), two of the stalwarts from Indian Institute of Science (IISc) Bangalore. We assure them that every suggestion they made was considered carefully while developing this novel idea of active RCSR.

The Centre for Electromagnetics (CEM) has been fortunate to have several scientists with varied expertise. We take particular pleasure in acknowledging constant professional interactions and technical discussions with Dr R. U. Nair (radomes & EM material characterisation), Dr Shiv Narayan (FSS and metamaterials) and Dr Balamati Choudhury (metamaterials and EM soft computing).We would also like to mention the experimental support received from Mr K. S. Venu, Technical Officer, CEM.

We also take this opportunity to thank our CEM collaborator Mr Jason P. Bommer, Manager, Boeing R&T Seattle, WA, USA. Our collaborative work on aircraft RF simulations, which has been extensively published, finds a perspective in this book in the context of absorbers and metamaterials. We would also like to thank Mr Jason Bommer for facilitating interesting and stimulating discussions with some of the pioneers in metamaterials at Boeing R&T, Seattle.

Writing a voluminous book is never easy. We would like to thank Ms Arya Menon, currently at the CEM, CSIR-NAL for going through parts of the manuscript to help us reduce the syntax errors.

It is a pleasure to acknowledge the encouragement and support received from Mr Manish Choudhary, Commissioning Editor, Cambridge University Press, India.

Hema Singh would like to thank her daughter Ishita and parents for their constant cooperation and encouragement during the preparation of the book. She would also like to acknowledge her brothers Rajeev, Sanjeev, Sandeep, and their spouses for their immense support and belief in completing this book.

R. M. Jha would like to thank his wife Renu and daughter Vishnupriya for their constant understanding and support during the writing of this book. He would also like to acknowledge his daughter Kanupriya Vazandar and her spouse Vishal Vazandar for the sheer excitement and enthusiasm that they exuded towards this endeavour.

Hema Singh
Rakesh Mohan Jha

1 Introduction to Radar Cross Section Reduction

1.1 Introduction

The concept of stealth or radar cross section (RCS) reduction and control has been a topic of interest since World War II. Attempts were initially made to reduce the detectability of the aircraft by employing wood and other composites as aircraft materials since they were less reflective to the radar waves than a metal. Following the initial systematisation, one realised that shaping and coating [by radar-absorbing materials (RAMs)] emerged as the primary techniques for the RCS reduction (RCSR).

RCSR through shaping is readily apparent in case of stealth fighter aircraft F-22. The edges at principal and drooping ends of wings and rear end of the aircraft have similar angular sweep. Further, the fuselage and canopy are smooth-surfaced with slopes at sides. The shapes of the surface interfaces, such as the doors at bay and the seam of the canopy, are saw-wave-type. The vertical airfoil of aircraft tails is slant. The front side of its engine is obliterated and includes a serpentine-shaped engine duct. Finally, all the weapons are stored within the aircraft itself. These alterations in the conventional shape of the aircraft resulted in considerable RCSR of the aircraft.

In contrast, RAM coatings have been used since 1950s to achieve low-RCS aircraft design. RAM was also useful in mitigating the coupling effect and cross talk between the antennas mounted on the surface of the aircraft. The reconnaissance airplane Lockheed U-2 and the fighter aircraft F-117 are few examples where RAM has been used for RCSR.

Sufficient knowledge base was created over time regarding the scattering behaviour of aircraft structures. The parameters that played a significant role in overall scattering characteristics of these structures were identified. For example, flat plates and cavities were observed to result in large radar return at normal incidence. Similarly, the inlet and exhaust systems of the fighter aircraft were identified as significant contributors towards the aircraft RCS in front-on and rear-on angles, while its vertical tail dominated the radar signature from other angles at the sides.

Numerical techniques were developed over years for the quantitative estimation of scattering from different parts of the aircraft structure. This facilitated the balanced design of aircraft with optimum RCS. Such aircraft include the F-117A- and B-2-type stealth aircraft.

The frontal RCS can be reduced by avoiding shapes and angles of high radar return. Multiple reflections are one of the important factors apart from orientation of the shape and polarisation of the impinging wave. If the wave enters into a long, closed perfect electric conductor enclosure, it undergoes multiple bounces and may result in large scattered field towards the radar source. The field associated with radar return can be reduced by coating the inner surface of the enclosure with RAM or redesigning the shape of the enclosure. For example, a curved duct can be useful in increasing the reflections significantly, thereby attenuating the incident energy without any adverse effect on its aerodynamic performance. Such a cavity, in particular, should have large cross-sectional aspect ratio. The SR-71 engine duct inlet is an example of such multiple bounce low-RCS design.

A low-frontal RCS is important in aerospace vehicles. However, it is a real challenge to achieve this when a large radar antenna exists within the nose radome. Attempts have been made to reduce the frontal RCS by mounting the radar antenna at an angle offset to the nose, but it does not lead to low observability. Another option is to redesign the slender nose as done in the cases of YF-23 and Boeing X-32 aircraft. The main aim of this is to block the signals from hostile radar sources to enter into the nose cone radome. This can be done by diverting the impinging radar waves or by reflecting them back from a flat plate-like antenna. Alternately, one can adapt the antenna pattern and place nulls towards the hostile radars; this may be considered as an example of *active RCSR*.

The RAMs and alignment of inlet surfaces require intelligent design. Moreover, the frequency range over which low observability is required is large and covers two to three orders of magnitude frequently. It is also known that the constitutive parameters of coatings depend on frequency and temperature. RAMs are discussed further in Section 1.3.2. Within the engine nozzle of the vehicle, air passes with a large velocity thereby rising the temperatures very high. Thus, the RAM coatings are expected to be functional at high temperatures. Russian researchers have developed coatings and techniques in the stealth design that can reduce the head-on RCS of a Sukhoi Su-35 fighter aircraft by a factor of 10, thereby halving the radar range for the target detection. Moreover, the Su-35 aircraft consists of a treated cockpit canopy that reflects the impinging radar waves and conceals the RCS contribution from metallic components.

Such electromagnetic (EM) designs attempt to alter the aircraft surface characteristics with an objective to make the aircraft 'invisible'. Essentially, the effective area of the aircraft reflecting EM waves that are detected by the radar is reduced. This concept is known as stealth.

The stealth techniques initially employed were frequency dependent and, thus, limited in their overall effectiveness. The RCS estimation and control involves many researchers and engineers of various disciplines; the attributes pertaining to EM, signal processing, materials, structural aspects, aerodynamics, etc. have to be considered simultaneously.

1.2 The concept of target signatures

Although EM signatures are the primary concept that come to our mind in the context of the detection of aerospace vehicles, other signatures are also considered for this purpose. These include acoustic (noise), optical (visible), infrared (thermal) and radar signatures.

Acoustic signatures: The acoustic signatures of an aircraft (2 cm–16 m wavelength) are due to the aerodynamic noise from its vortices, wings, rotors, propellers and engines. The intensity of noise is directly proportional to the wingspan loading and speed. Reduction of such signatures contributes towards acoustic stealth. Electric motors are less noisy than turbo and piston engines, but are limited to short-endurance applications. However, small mass and low aerodynamic drag contribute significantly to noise reduction. Keeping the distance constant, the sound attenuation is inversely proportional to the wavelength square. This makes low-frequency noise an important factor for acoustic stealth. This is the reason for mounting turbo engines above the wings in large tactical aircraft, which facilitates shielding of the compressor noise and the efflux noise from the ground.

The type of engines is also an important factor for the generation of acoustic signatures. A four-cycle piston engine has more fuel consumption efficiency than a two-cycle one; the frequency of noise is comparatively lower in the two-cycle piston engine. This leads to lower noise attenuation in four-cycle piston engines and hence make them less preferred from the acoustic stealth point of view.

Such combustion noise can be reduced by coating the engine with sound-absorptive materials. The increase in weight due to extra coatings can be mitigated by covering only those areas that emit.

Optical signatures: The size and shape of a vehicle are important factors for detection by optical signatures. The contrast against the background is also crucial in fixing the threshold of detection. The background luminance depends upon the atmospheric conditions and the target position with respect to the sun. The surface texture of the vehicle and the atmosphere-reflected illumination are other factors that decide the strength of optical signatures (0.4–0.7 μm).

Thermal signatures: The thermal or infrared signatures (0.75 μm–1 mm) are due to the heat generated by the aircraft engine jets, propellers and rotors. The exhaust heat from the aircraft can be prevented by travelling towards the ground. This path diversion enables reducing the detection probability of the aircraft by the detectors at the ground base. Moreover, low-emissive materials (e.g. Ag, Al) can be used to avoid radiation. The engine exhausts should be screened by other airframe components to deflect the thermal radiation away from the ground.

Radar signatures: Radar (3 mm–30 cm wavelength) signatures are related to the *radio frequency* (RF) emissions from the aircraft. These are primarily reflected RF signals. They can be reduced by either applying RAM coatings or shaping the aircraft. Special attention may be given to the hot spots including the edges and corners of the aircraft. Care must also be taken so that no surface of the aircraft is illuminated by radar signal at normal incidence. Likewise, for the signals incident at smaller angles, vertical surfaces such as fins become significant contributors to radar signatures. It is also desirable to avoid corner-reflectors-type geometry, since surfaces meeting at right angles give rise to strong radar returns. The most significant contribution comes from antennas/sensors mounted over the vehicle. These sensors and antennas might add to the RF signatures of the vehicle.

1.3 Radar cross section of an aircraft

RCS is an estimate of observability of a target, which in turn, depends on its external features and EM properties. The RCS essentially relates the EM energy of the receiver reflected from the target to the incident EM energy (Knott et al. 2004). When EM wave is incident on a body, part of the energy is absorbed. The remaining energy is accounted by the phenomena of reflection and diffraction. An important characteristic that explains the EM scattering phenomenon is the *electrical dimension* of the scatterer. The point of concern is that the radar signal returning from shapes other than spherical ones depends on the polarisation of incident wave. The polarisation of EM waves may vary with scattering. Surface ray propagating on a general surface has finite torsion and its path of propagation cannot be restricted to a plane. Thus, its direction changes continuously, that leads to a change in polarisation.

The RCS of an object has an apparent size as seen by the radar. It is essentially a coherent summation of the contributions from various scattering centres of the target once illuminated by the radar. In other words, the target structure including the various hot spots would re-radiate the EM energy impinging on the target. These individually scattered fields with the associated amplitude and phase add up to the resultant scattered field. The total scattered field includes reflection in specular directions, diffraction at sharp edges, corners, multiple scattering, surface waves, creeping waves, shadowing effect, etc. Thus, the shape and size of the target, and hence the scattering centres, decide the extent and scintillation of the RCS with respect to the aspect angle and frequency.

An aircraft, for example, would return the nose-on incidence mainly due to the engine inlets. If one moves away from the nose-on angles, the principal wing edge becomes one of the major contributors to the overall RCS. For the angle of incidence beyond this limit and up to 70°, the scattering is primarily from the forward fuselage and engine nacelles. Beyond this and up to normal incidence, scattering in the broadside direction has significant contribution from the fuselage and vertical stabilizers. Similarly, when the aspect angles are near ±180°, the engine exhausts dominate the scattering. Thus, the overall structural RCS of the aircraft will be coherent summations of all such returns.

Further, the aircraft structure has numerous sensors and antennas mounted over it for various applications. This may enhance the scattering cross section of the aircraft considerably. The scattering from the antenna or antenna array system is due to its structure and feed network. When the feed network is matched, the scattering cross section is termed as antenna structural RCS (Shrestha et al. 2008). This is due to the fact that in such a condition there will be no reflections from the feed and the antenna RCS will be due to its structure only. In other words, the antenna structural RCS is a function of currents induced over the surface of the antenna by an incident wave (Jenn 1995). It must be borne in mind that although other definitions based on varied interpretations are prevalent for antenna structural RCS in an open domain, the definition mentioned above shall be used in this book.

When the feed network is not matched, the antenna RCS will be due to its structure as well as the reflections from the feed network. This is referred to as *antenna mode scattering* (Yuan et al. 2008). Thus, the antenna RCS is the sum of the antenna structural RCS and antenna mode scattering.

The structural RCS of the antenna not only depends on the antenna structure but also on the platform (aircraft) over which it is mounted (Perez et al. 1997). The aircraft may be considered as a set of wedges and facets, or as a hybrid of parametric surfaces (Wang et al. 2001), or even the Non-Uniform Rational B-Spline (Domingo et al. 1995). This is followed by the use of numerical electromagnetics techniques such as uniform theory of diffraction (UTD), physical optics (PO), Electronic Counter Measures (ECM), Method of Moments (MoM), Finite Difference Time Domain (FDTD), etc. for estimating the total scattering to arrive at the total structural RCS of the aircraft. Since the asymptotic high-frequency techniques are often involved, its prerequisite, viz. the ray tracing is discussed in Section 1.3.1.

1.3.1 Ray-tracing techniques

The ray tracing, in principle, determines every possible ray-path between the source and the observation point. The computational cost of ray tracing, being a geometric method, does not depend on electrical dimensions of the structure. The ray-paths are based on generalised Fermat's principle. The direct ray, and all the reflected, double reflected, diffracted, reflected-diffracted, surface wave, and creeping waves are taken into account.

In free space, the principles of *geometrical optics* (GO) are employed. A complex field represents the amplitude and the direction of the wave. Ray tracing is essentially the determination of exact location of the *reflection*, *launching* and *shedding points* on the surface. The field associated with the ray at the receiving point is determined by coherent summation of each individual contribution of these ray-paths.

In RCS estimation, ray tracing is essentially used with high-frequency asymptotic methods when the electrical dimensions are greater than the wavelength of the impinging wave. These asymptotic methods include geometrical optics (GO), geometrical theory of diffraction (GTD), physical optics (PO), physical theory of diffraction (PTD), uniform theory of diffraction (UTD), etc.

In fact, for electrically large objects, low-frequency approaches, *viz.* the ent method, finite element method (FEM) are often infeasible. Despite the availability of high-speed, large memory computers, the size of the object that can be handled *via* such methods is too small for any use in actual cases. In contrast, the high-frequency methods, although simple, provide more accurate results. The simplicity of these methods lies in the assumption that each part of the target scatters energy independent of all the other parts. Therefore, the fields induced on a section of the target are only due to the wave impinging upon it, and not on the scattered waves from the other parts. This makes it relatively simple to estimate the induced fields and to integrate them over the body to obtain the RCS.

A ray at a given surface point may be defined in three possible ways: (a) finite-length (point-to-point) rays, (b) semi-infinite-length (point-to-direction) rays and (c) infinite-length rays. There can be two symmetric sub-configurations, *viz.* for (a) near-field to far-field transformation and (b) far-field to near-field transformation. Such analysis may be handled on the basis of the reversibility of the ray-path. The last configuration, infinite-length rays, may be either monostatic or bistatic.

Most critical step in ray-tracing method is to obtain the intersection of rays with the surface. Several methods have been introduced in the open domain to determine the point of intersection. The planar/non-planar surface at which the ray hits is divided in different ways to determine the point of reflection. One such method involves recursive subdivision within bounded volume (Whitted 1980). Here the surface is divided and the bounded volume is produced for each sub-surface. This process continues till the ray does not intersect the surface. The bounded volume may be considered as a sphere or a closed box (Whitted 1980, Pharr and Humphreys 2010). Another method is to divide the surface into triangular facets (Kaijiya 1982, Snyder and Barr 1987). However, the faceted surfaces lead to high computational cost.

This problem of high computational cost can be circumvented algebraically by converting the parametric surfaces into implicit formulation (Manocha and Demmel 1994). This results in a problem of intersection of two planes in a parametric space. The solution of the system of equations provides the curves formed by the plane–surface intersection. The numerical technique based methods, *viz.* Laguerre's method (Kajiya 1982), recursive Newton method (Martin et al. 2000) may be used for the solution. Alternatively, the implicit equations can be solved by expanding them as a high-order matrix determinant (Manocha and Demmel 1994). However, these methods are limited by their increase in computational complexity for higher degree surface. Some researchers approximate the surfaces as a plane surface and determine the initial point from the ray intersection with bounded volumes (Martin et al. 2000, Sturzlinger 1998). Optimisation algorithms like quasi-Newton iteration and conjugate gradient method can also be used for estimating the point of intersection (Joy and Murthy 1986).

If there is no intersection, the ray will travel towards the receiver. This is referred to as the direct ray. The field associated with the direct ray at a distance s is expressed as (Pathak and Kouyoumjian 1974, Pathak et al. 2013)

$$E_{Direct}(s) = E_i(0).A(s).e^{-jks}, \tag{1.1}$$

where e^{-jks} is the phase of the ray-field and $k = \dfrac{2\pi}{\lambda}$.

$A(s)$ is the amplitude variation given by

$$A(s) = \sqrt{\frac{\rho_1 \rho_2}{(\rho_1 + s)(\rho_2 + s)}}, \tag{1.2}$$

where ρ_1 and ρ_2 represent the principal radii of curvature of the wave front at a given surface point.

When a ray hits the surface, its propagation depends on the surface characteristics. For a transparent or semi-transparent surface, the rays may get reflected from the surface or transmitted through it. The attenuation in the ray-field depends on the constitutive parameters of the surface. The transmitted field at the distance s is given by

$$E_T(s) = E_i \,.\, A^t.\, T.\, e^{-jks}, \tag{1.3}$$

where the transmission coefficient depends on the polarisation of the incident ray (Jordan and Balmain 1976). The amplitude variation A^t depends on the radii of curvature of the surface and that of the incident wave, given by (Kouyoumjian 1965)

$$\frac{1}{\rho_1^t} = \frac{1}{2}\left(\frac{1}{\rho_1^i} + \frac{1}{\rho_2^i}\right) + \frac{1}{f_1};\; \frac{1}{\rho_2^t} = \frac{1}{2}\left(\frac{1}{\rho_1^i} + \frac{1}{\rho_2^i}\right) + \frac{1}{f_2}, \tag{1.4}$$

(f_1, f_2) being the focal distances and (ρ_1, ρ_2) being negative (positive) for the concave (convex) surface.

The reflected field on the surface is given by

$$E_R(s) = E_i \,.\, A^r.\, R\,.e^{-jks}, \tag{1.5}$$

where the reflection coefficient, R, depends on the nature of polarisation (parallel and perpendicular).

Another phenomenon that takes place when a ray hits a corner or an edge is diffraction. Rays that hit normal to the surface generates the waves of cylindrical wavefront. In contrast, for obliquely incident rays, diffracted wave will be conical. The diffracted field on the surface is given by (Kouyoumjian 1965)

$$E_D(s) = E_i \,.\, A^d.\, D\,.e^{-jks}, \tag{1.6}$$

where D is the diffraction coefficient that depends on polarisation.

When a ray hits the surface tangentially, it travels along the local geodesic of the surface and leaves the surface tangentially. This is referred to as the *creeping wave*. These ray trajectories are determined using the generalised Fermat's principle. Further, a ray may undergo multiple interactions, e.g. reflection followed by an edge diffraction and then again by transmission. The drawback of ray tracing is that the computational complexity increases due to such multiple propagation phenomena. Therefore, acceleration and optimisation procedures are required towards efficient and fast solutions.

1.4 RCS reduction

Presently, the extensive knowledge base of passive techniques can be employed for controlling the EM scattering. These passive techniques frequently involve either shaping or applying RAM. The effectiveness of such methods depends on the frequency, incident angle and polarisation of the impinging wave.

The main objective behind shaping the structure is to minimize the amount of energy that is backscattered towards the radar. This type of RCS control has been found to be effective for the monostatic radars. If, for an aerospace vehicle, e.g. aircraft, missile or UAV, the profile of the structure is designed so that only a small angular range is available to the radar, then stealth may be achieved by considering that angular region. However, it is important to keep in mind that RCSR in one aspect angle is frequently accompanied by an increase in the RCS at another aspect angle. Both the approaches of shaping and RAM are often considered simultaneously to achieve the acceptable low observability over the operational frequency band.

Apart from the above-mentioned passive techniques, other methods such as usage of artificial magnetic conductors (AMC) (Paquay et al. 2007), frequency selective surfaces (FSS) and active RCSR have also been proposed towards RCSR. The AMC structure (Yeo and Kim 2009) scatters the incident energy towards offset directions thereby reducing the specular reflection considerably.

The basic difference between the passive and active techniques of RCSR is that in passive techniques, the scattered wave from one part of the target cancels the same from the other part due to amplitude and phase difference while the active techniques involve the cancellation of the incoming wave through destructive interference with the scattered field within the array or sensor-based system. The active cancellation of the impinging waves makes the platform 'invisible' to the probing radar sources.

1.4.1 RCS reduction by shaping

RCSR through shaping is a high-frequency technique based on geometric optics. If the object is electrically large, the incident wave will be reflected mainly towards the specular direction. A cylindrical surface, for example, would produce specular reflections along its length when observed sidewise. A spherical surface would reflect from any point independent of its orientation. Reflections from other directions become significant only when the specular reflections are suppressed or eliminated. The structure is shaped to reduce the edges, surface discontinuities and corners (e.g. dihedrals, trihedrals). The main intention is to redirect the reflected waves in non-specular directions, thereby minimising the backscattering (Lynch 2004).

The shaping must comply with the aerodynamic requirements of the vehicle (e.g. aircraft, missiles, ships, etc.). If the nose of the missile could be made round instead of pointed, the specular reflections would be reduced considerably. Another example is the engine inlets. If the shape of the intake duct is made curved, then reflections from the inside walls of the inlet and the engine can be reduced. Furthermore, the recessing of inlets inside the fuselage would hide the engine opening from the radar. The stealth aircraft F-117 has shaped wings and fuselage designed for minimum reflection towards the radar. The RAS design is based on the principle of reflection. If the angle of incidence is varied from end-fire to the broadside, there is an increase in reflection towards the source.

The backscattered fields due to the edge-diffracted waves are also important contributors towards overall RCS. These diffracted waves are coherent for straight discontinuity and perpendicular incidence (Ufimtsev 1996). Such diffraction takes place at trailing edges of wings of an aircraft, gap between wing and flaps or rudders, edges of cargo doors, etc. Such backscattering can be reduced by either indenting the edges, replacing the surface with electromagnetically soft surface or applying RAM over the surface.

1.4.2 RCS reduction by RAM

Essentially, RAM absorbs the incident EM energy and converts it into heat, thereby reducing the scattered energy towards the radar. RAM is known to be quite effective in controlling the backscattering than forward scattering (Hiatt et al. 1960). RAMs have relatively high values of imaginary part of permittivity and permeability. Such coatings result in change in polarisation of the scattered waves.

Narrowband RAM coatings, such as the Salisbury screen and Dallenbach layer, have been used since 1950s. However, modern radar systems span a wide range of frequency. Hence, the need for wideband RAMs is apparent. A typical RAM employed on aircraft could be a ferrite-based paint or a composite. However, there are significant implications of using RAM. Firstly, most of them are toxic. Secondly, RAM coatings require precise application techniques, as the coating thickness and smoothness must be uniform across the platform.

Ideally a RAM should not impose weight penalty due to speed and pay load considerations. It should possess high mechanical strength and should be anticorrosive, chemically stable and should not get charged at high temperature. It must have a wideband RCSR. Lastly, it should be effective in all directions (Vinoy and Jha 1996). The RAM application process typically involves robotic sprayers that can accurately control the coating thickness. Furthermore, these coatings require strict constitutive parameter tolerances as well as uniformity in order to achieve the desired result. Therefore, the cost of implementation of RAM is often too high. Another issue is that RAM also increases the weight of the platform. This may have notable impact on the vehicle performance aerodynamically.

For different platforms, RAM coatings have been developed with appropriate combination of rubber, cotton-glass, epoxy and mica. Other possibilities are graphite fibres, Kevlar and ferrites. The materials can be of different forms such as sheets, honeycombs, laminates, etc. Ferrite materials in forms of flakes, wires or microspheres can be loaded into glass–epoxy or silicon rubber. The inks and coatings can be applied on kapton film or epoxy honeycombs.

Radar-absorbing paints are also coated over the surface of the vehicles. These paints consist of small ferrite particles that are polarised towards the impinging wave. Such paints are prepared by mixing solid iron oxides with various polymer resins, such as epoxy and plastics. The constitutive parameters including thickness of the paint, fix the resonance frequency for maximum absorption.

1.4.3 Active RCS reduction

For an aerospace vehicle, it is not only the structure of the vehicle that contributes to the radar signatures, but the antennas and sensors mounted on the vehicle are also of concern. Even if the structures are designed efficiently for stealth via shaping and RAM coating, there is a significant contribution from these sensors. The antenna RCS, which has two components, *viz.* the structural mode RCS and antenna mode RCS, is another important issue to be dealt with.

There can be situations when the RCS of the antenna mounted on the platform dominates the RCS of the platform. In such a case, the radiation pattern of the phased antenna array, for example, can be controlled through adaptive weight estimation. This feature relates to active RCS control and, hence, RCSR.

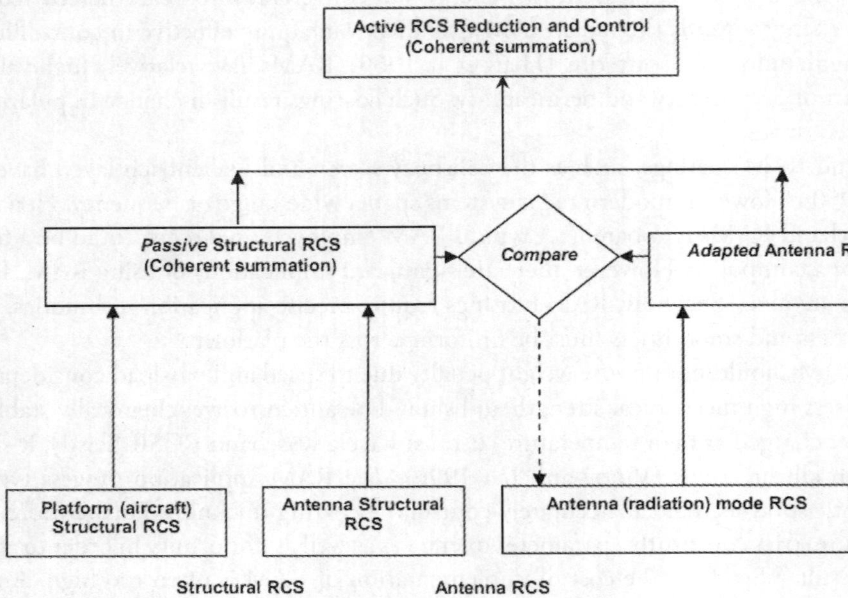

Figure 1.1 The concept of active RCS reduction and control

The active RCSR of a vehicle relates to the adaptive array, digital beam forming and field programmable gate array (FPGA) based system. The phased array along with an efficient weight adaptation generates the pattern nullifying each probing radar and simultaneously maintaining sufficient gain in the desired directions. The FPGA system does signal analysis, database search, and waveform generation and control. The structural RCS and the antenna RCS of the vehicle are determined before hand and stored in database, which is integrated with the active RCS so as to eventually generate low RCS for the aircraft. This is done within the signal processing and control module of the phased-array system. The module analyzes the measured radar signal parameters and thereafter searches the corresponding target echo data in the RCS database, and accordingly makes a real-time adjustment of the coherent echo amplitude and phase parameters.

The active RCSR is essentially a combination of both software and hardware realisation, enabled by utilizing high-speed microelectronic devices, phased-array antenna techniques and signal processing methods. The active RCSR through phased arrays, once coherently integrated, can cancel out the structural RCS of the platform over which the array is mounted, and hence contributes towards a low-observable platform.

The concept of overall active RCSR of an aerospace vehicle is shown in Figure 1.1. The idea is to achieve significant RCSR and a control for arbitrary practical scenarios. It is apparent that the RCS estimation and control of phased arrays is an important milestone towards the overall RCSR of the platform. This involves estimation of the antenna array RCS as well as the development and implementation of an efficient algorithm towards a desired adapted pattern in real-time situations.

1.5 Organisation of the book

The RCS estimation and control of aircraft-like structure involve inter-disciplinary expertise. The scattering characteristic of a body is not only an EM problem since it also involves materials and structural issues besides highly efficient computational capabilities. The RCS of the phased array mounted on an aerospace vehicle is one of the main contributors to overall signature of the vehicle. It is therefore essential to have accurate estimation of antenna RCS including its feed network and mutual coupling effects. This book covers the analytical estimation of RCS and the mutual coupling effect in phased arrays.

The book begins with the RAM analysis for a low-observable platform. Chapter 2 outlines the EM propagation in classical multilayered dielectric–dielectric and dielectric-metamaterial media. The reflection and transmission coefficients for semi-infinite and finite dielectric layers are discussed. These coefficients are also presented when the dielectric layers are replaced with metamaterials. The novel dielectric-metamaterial media characteristics can be exploited towards the design of anti-reflection coatings. The reflection/transmission behaviour of such coatings is discussed.

The scattering of signals by a phased array depends on the geometrical configuration of the array, its frequency of operation and the employed feed network. For a phased array operating with a frequency equal to that of the radar, the effect of the array geometry and feed network become prominent. In general, the antenna feed consists of orderly arranged radiators, phase shifters and couplers. As all these devices may not possess identical terminal impedances, significant mismatches might exist at each level of the feed network. This results in the reflection of the incident signal, which propagates from the radiators towards the receive port. These individual scattered fields add up coherently to high RCS under certain scenarios.

Since the antenna RCS is an important contributor to the overall RCS of the vehicle, it is described in Chapter 3. The analytical formulation of in-band scattering of phased array for both the series and parallel feed network is discussed in detail. The antenna elements are considered to be isotropic radiators. The role of design parameters of the phased array, such as number of antenna elements, inter-element spacing, beam scan angle, electrical length between the couplers and the coupling coefficients, in total antenna RCS is explained.

Once the antenna RCS is known, one can able to relate it to the active RCS of the phased array, which, in turn, is related to the adapted antenna pattern in a given signal environment. The signal scenario may have multiple desired sources along with the probing ones. These sources can be narrowband as well as wideband. The phased array should be able to cater to such signal environments even in the case of multipath effect. If the phased array does not transmit any energy towards any of the probing sources, it would be, in principle, 'invisible' towards them, thereby drastically reducing the observability of the aircraft. This capability of the phased array depends on an efficient adaptive algorithm for its optimum weights towards the generation of the adapted antenna pattern.

Chapter 4 presents the active RCSR in a phased array. Various types of adaptive algorithms, viz. least mean square algorithm and its forms, recursive least square algorithm, weighted least square algorithm and standard matrix inversion algorithm are discussed. The probe suppression in phased arrays is presented for different signal environments including the multipath effect.

In practical situations, the coupling between antenna elements plays a significant role in the performance of the phased array. Chapter 5 describes the mutual coupling effect in dipole arrays with different configurations. The dependence of output signal-to-noise ratio (SNR) on the mutual coupling effect is presented for different dipole arrays. The effect of parameters such as inter-element spacing, dipole length, geometric arrangement (side-by-side, parallel-in-echelon), etc. on mutual impedance and SNR is discussed. This mutual coupling must be included into the RCS estimation of a dipole array.

Chapter 6 describes the formulation of RCS of a dipole array with series and parallel feed network including mutual coupling effect. The impedance at different levels of the feed network, *viz.* at the terminals of dipole elements, phase shifters and coupler levels is derived. The signal is followed as it enters into the array aperture and travels through the phase shifters and couplers towards the terminal load. The scattering at different levels of the array systems is determined in terms of reflections and transmissions due to impedance mismatch. The parametric analysis towards the optimisation of array RCS is also discussed. The role of dipole parameters, aperture distribution, array configuration, terminal load impedance and mutual coupling effect is analysed.

Phased arrays have applications in sidelobe cancellers. The ability to suppress the impinging interferences in these cancellers can be utilised for active RCSR. The added advantage of such sidelobe cancellers is their robustness towards direction-of-arrival (DOA) mismatch. The enhanced performance of such cancellers is due to the inclusion of blind equalisation and implementation of different constraints in adaptive array processing. Chapter 7 describes the performance of such sidelobe cancellers in active RCSR. Schemes, such as generalised sidelobe cancellers (GSC), GSC with decision feedback (DF–GSC), DF–GSC with blind and semi-blind adaptation are discussed. The active cancellation of probing sources is presented in such schemes.

Several novel and exciting concepts have recently emerged in the area of RCSR. These ideas have been effective in reducing the observability over frequency band. Chapter 8 discusses a few of these emerging trends in the RCSR techniques. The idea of stealth was further reinforced with embedded antennas and conformal load-bearing antennas replacing the conventional antennas protruding out of the surface. Moreover, *frequency selective surface* (FSS) facilitated the bandwidth enhancement and RCS control of the structures. FSS-based design became popular in antenna array design, low-RCS radomes, and low-observable structures. Metamaterials with novel propagation characteristics too can be used for RCS control. Another area of interest is plasma-based RCSR. The appropriate combination of both passive and active RCSR techniques is expected to achieve an efficient stealth design of structures, be it a missile, aircraft or an unmanned air vehicle.

1.6 Conclusion

RCSR is often achieved by using the principle of scattering or absorption. Any radar can detect a target only if the EM wave scattered from the target reaches the radar. This fact is exploited to reduce the detection probability either by diverting the impinging waves in the directions away from the receiver via proper shaping or by absorbing the incident wave using RAM. These passive techniques involve appropriate EM design with respect to either the shape or material of the target. However, such passive techniques (coating and shaping) are often frequency dependent. The detectability can be further minimised with the help of the state-of-the-art metamaterial, FSS and plasma-based technologies. The FSS or metamaterial structures can be embedded in the aircraft structure for low-RCS design.

Active RCSR on the other hand covers dynamic scenario where the number of radar sources may vary for an aircraft. A phased array mounted on an aircraft caters to the situation by adapting its radiation pattern such that no energy is transmitted towards the probing radar, thereby making it 'invisible' to the radar. Such an active RCSR has great potential towards achieving low observability. An appropriate combination of passive and active RCSR towards a stealth aircraft can be achieved through a concurrent engineering approach.

References

Domingo, M., F. Rivas, J. Perez, R. P. Torres, and M. F. Catedra. 1995. 'Computation of the RCS of complex bodies modeled using NURBS surfaces.' *IEEE Antennas and Propagation Magazine* 37(6): 36–47.

Hiatt, R. E., K. M. Siegel, and H. Weil. 1960. 'Forward scattering by coated objects illuminated by short wavelength radar.' *Proceedings of IRE* 48: 1630–35.

Jenn, D. C. 1995. *Radar and Laser Cross Section Engineering*. AIAA Education Series. Washington, DC. 476.

Jordan, E. D. and K. C. Balmain. 1976. *Electromagnetic Waves and Radiating Systems.* Prentice-Hall of India. 753.

Joy, K. I. and N. B. Murthy. 1986. 'Ray tracing parametric surface patches utilising numerical techniques and ray coherence.' *ACM Siggraph Computer Graphics* 20(4): 279–85.

Kajiya, J. T. 1982. 'Ray tracing parametric patches.' *Computer Graphics* 16(3): 245–54.

Knott, E. F., J. Shaeffer, and M. T. Tuley. 2004. *Radar Cross Section*. Second edition. Raleigh, NC: Scitech Publishers, 611.

Kouyoumjian, R. G. 1965. 'Asymptotic high-frequency methods.' *Proceedings of IEEE* 53: 864–76.

Lynch, D. A. 2004. *Introduction to RF Stealth.* Raleigh, NC: Scitech/Peter Peregrinus, 575.

Manocha, D. and J. Demmel. 1994. 'Algorithms for intersecting parametric and algebraic curves I: Simple intersections.' *ACM Transactions on Graphics* 13(1): 73–100.

Martin, W., E. Cohen, R. Fish, and P. Shirley. 2000. 'Practical ray tracing of trimmed NURBS surfaces.' *Journal of Graphics Tools* 5(1): 27–52.

Paquay, M., J. C. Iriarte, P. Gonzalo, and P. D. Maagt. 2007. 'Thin AMC structure for radar cross-section reduction.' *IEEE Transactions on Antennas and Propagation* 55: 3630–38.

Pathak, P. H., G. Carluccio, and M. Albani. 2013. 'The uniform geometrical theory of diffraction and some of its applications.' *IEEE Antennas and Propagation Magazine* 55(4): 41–69.

Pathak P. H. and R. G. Kouyoumjian. 1974. 'A uniform geometrical theory of diffraction for an edge in a perfectly conducting surface.' *Proceedings of IEEE* 62: 1448–61.

Perez, J., J. A. Saiz, O. M. Conde, R. P. Torres, and M. F. Catedra. 1997. 'Analysis of antennas on board arbitrary structures modeled by NURBS surfaces.' *IEEE Transaction on Antennas and Propagation* 45: 1045–53.

Pharr, M. and G. Humphreys. 2010. *Physically Based Rendering: From Theory to Implementation*. Burlington, MA: Elsevier, 1167.

Shrestha, S., M. D. Balachandran, M. Agarwal, L. -H. Zou, and K. Varahramyan. 2008. 'A method to measure radar cross section parameters of antennas.' *IEEE Transactions on Antennas and Propagation* 56: 3494–500.

Snyder, J. M. and A. H. Barr. 1987. 'Ray tracing complex models containing surface tessellations.' *ACM Siggraph Computer Graphics* 21(4): 119–28.

Sturzlinger, W. 1998. 'Ray tracing triangular trimmed free-form surfaces.' *IEEE Transactions on Visualisation and Computer Graphics* 4(3): 202–14.

Ufimtsev, P. Y. 1996. 'Comments on diffraction principles and limitations of RCS reduction techniques.' *Proceedings of IEEE* 84: 1828–51.

Vinoy, K. J. and R. M. Jha. 1996. *Radar Absorbing Materials: From Theory to Design and Characterisation*. Norwell, Boston: Kluwer Academic Publishers, 209.

Wang, S., Z. -C. Shih, and R. -C. Chang. 2001. 'An efficient and stable ray tracing algorithm for parametric surfaces.' *Journal of Information Science and Engineering* 18: 541–61.

Whitted, T. 1980. 'An improved illumination model for shaded display.' *ACM Graphics and Image Processing Magazine* 23: 343–49.

Yeo, J. and D. Kim. 2009. 'Novel tapered AMC structures for backscattered RCS reduction.' *Journal of Electromagnetic Waves and Applications* 23: 697–709.

Yuan, H. W., X. Wang, S. X. Gong, and P. Zhang. 2008. 'A numerical method for analyzing the RCS of a printed dipole antenna with feed structure.' *Journal of Electromagnetic Waves and Applications* 22: 1661–70.

2

RAM Analysis for Low-Observable Platforms

2.1 Introduction

The electromagnetic (EM) wave propagation within a given material mainly depends on its constitutive parameters, *viz.* permittivity (ε), permeability (μ) and thickness of material. In macroscopic sense, these parameters describe the effects of induced electric and magnetic polarisations within the medium. These constitutive parameters can have either negative or positive values. The negative permittivity and permeability are known to be associated with the metamaterials (Veselago 1968). A metamaterial can be epsilon negative index (ENG) or mu negative index (MNG) or double negative index (DNG) materials (Shelby et al. 2001). Various analytical methods have been proposed for the plane wave analysis of EM propagation within the multilayered structures (Ziolkowski and Heyman 2001), (Kong 2002), (Cory and Zach 2004). The EM propagation within such materials exhibits negative refraction and other interesting propagation phenomena. The characteristic reflection/transmission behaviours of dielectric-metamaterial coatings have strategic applications towards the control of radar signatures of aerospace structures and, hence, low-observable platforms.

In this chapter, a systematic description of EM propagation through layered media is presented. These multilayered media include both dielectric–dielectric and dielectric-metamaterial media. The dielectric media is taken to be homogeneous. For a plane EM wave incident on planar layers having different dielectric and magnetic parameters, the reflected and transmitted fields are determined. The EM propagation within the layers is explained in terms of reflection/transmission from different types of media. This includes semi-infinite media, dielectric layers with finite thickness and lossy dielectric layers. The presence of absorption in a lossy medium is taken into account by the complex nature of wave numbers. The reflective behaviour of the multilayered media with respect to the frequency of the incident EM wave, thickness and constitutive parameters of the layers is discussed. A proper choice of design parameters would facilitate the design and development of radar-absorbing material (RAM) coatings.

2.2 EM propagation in classical multilayered media

In this section, the basic expressions for the impedance of multilayered medium, the Fresnel's coefficients for reflection and the transmission for parallel and perpendicular polarisations are discussed. The EM propagation is expressed in terms of these coefficients. These media can be semi-infinite, dielectric layers with finite thickness or lossy dielectric layers (Brekhovskikh et al. 1965). The presence of absorption in a lossy medium is accounted for by the complex nature of wave numbers. The dependence of the reflection coefficient on the thickness of the dielectric layer is analysed.

Reflection and transmission of EM waves: A plane EM wave is expressed as

$$\vec{E} = \vec{E}_o e^{i(\vec{k}.\vec{r}-\omega t)} \tag{2.1}$$

where $\vec{E}_o$ is the plane wave amplitude. The wave vector k is given by

$$k = \frac{\omega}{c}\sqrt{\varepsilon\mu} \tag{2.2}$$

where $\omega = 2\pi f$ with f being the frequency of the plane wave travelling in the medium.

Correspondingly, the magnetic field $\vec{H}$ associated with the plane wave is given by

$$\vec{H} = \frac{c}{\mu\omega}\vec{k}\times\vec{E} \tag{2.3}$$

where $c = 3 \times 10^8$ m/sec, $\mu = \mu_0\mu_r$, is the permeability of the medium. $\mu_0 = 4\pi \times 10^{-7}$ H/m, μ_r is the relative permeability of the medium.

The characteristic impedance of a medium η is given by

$$\eta = \sqrt{\frac{\mu}{\varepsilon}} \tag{2.4}$$

where $\varepsilon = \varepsilon_0\varepsilon_r$ is the permittivity of the medium, $\varepsilon_0 = 8.854 \times 10^{-12}$ F/m and ε_r is the relative permittivity of the medium.

For a wave incident on a plane boundary, the normal impedance Z is defined as

$$Z = \frac{E_t}{H_t} \tag{2.5}$$

If a plane wave impinges on the boundary at an incident angle θ (Figure 2.1), it will be partially reflected and partially transmitted, depending upon the material properties, the incident angle and the frequency.

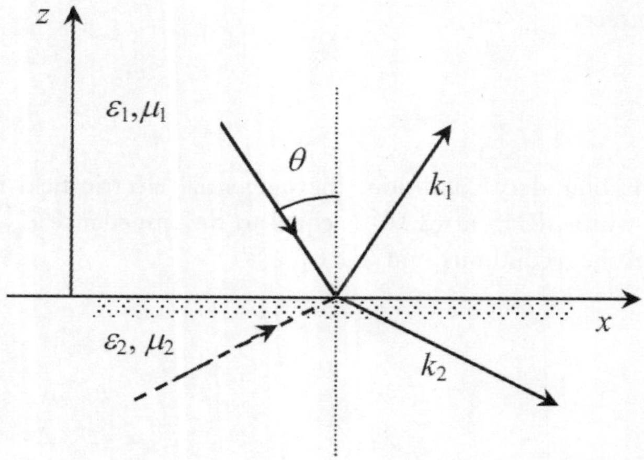

Figure 2.1 Reflection and refraction of a plane electromagnetic wave at a plane interface

Since $E_t = E, H_t = H\cos\theta$, for perpendicular polarisation, i.e. when E vector is perpendicular to the plane of incidence, from (2.5) one gets

$$Z_1 = \frac{E}{H\cos\theta} = \frac{\eta}{\cos\theta} \tag{2.6}$$

where η is the characteristic impedance of the medium.
Likewise, $E_t = E\cos\theta$ and $H_t = H$ for parallel polarisation, so one gets

$$Z_1 = \frac{E\cos\theta}{H} = \eta\cos\theta \tag{2.7}$$

For the reflected wave,

$$\frac{E_t}{H_t} = -Z_1 \tag{2.8}$$

In case of normal incidence (θ = 0°), the absolute value of the normal impedance and the characteristic impedance of the medium are same. If (2.1) represents the incident electric field at the interface, the total electric field in the first medium (ε_1,μ_1) for perpendicular polarisation can be expressed as the sum of the incident field and the reflected field, i.e.

$$E_T = E_o \exp(-jky\cos\theta)\exp(jkx\sin\theta) + E_o \exp(jky\cos\theta)R_\perp \exp(jkx\sin\theta)$$

$$E_T = E_o\left[\exp(-jky\cos\theta) + R_\perp \exp(jky\cos\theta)\right]\exp(jkx\sin\theta) \tag{2.9}$$

where $R_\perp$ is the reflection coefficient for perpendicular polarisation. Substituting y = 0 in (2.9), the tangential electric and magnetic fields at the boundary are given by

$$E_t = E_o(1+R_\perp)\exp(jkx\sin\theta);\ H_t = E_o\frac{1}{Z_1}(1-R_\perp)\exp(jkx\sin\theta)$$

Thus, at the interface,

$$\frac{E_t}{H_t} = Z_1 \frac{1 + R_\perp}{1 - R_\perp} \tag{2.10}$$

According to the boundary condition, the tangential electric field is continuous at the interface. In other words, RHS of (2.10) is equal to the impedance Z_2, a ratio of tangential field components in the second medium (ε_2, μ_2),

$$Z_1 \frac{1 + R_\perp}{1 - R_\perp} = Z_2 \tag{2.11}$$

or $$R_\perp = \frac{Z_2 - Z_1}{Z_2 + Z_1} \tag{2.12}$$

The reflection coefficient, $R_{||}$ for parallel polarisation is given by the same expression; however, the value of the impedance Z_1 will be given by (2.7) instead of (2.6). Thus, a generalised expression for reflection coefficient can be written as

$$R = \frac{Z_2 - Z_1}{Z_2 + Z_1} \tag{2.13}$$

The transmission coefficient will be $T = 1 + R$;

$$T = \frac{2Z_1}{Z_1 + Z_2} \tag{2.14}$$

The constitutive parameters of a medium such as permittivity, permeability are complex quantities. Thus, the reflection/transmission coefficients will also be complex. In other words, these coefficients have both amplitude and phase associated with them.

2.2.1 Semi-infinite media

The Fresnel reflection coefficient depends on the electrical parameters of the medium. If a plane EM wave of frequency ω impinges at the interface, separating two semi-infinite media (ε_1, μ_1; ε_2, μ_2), as shown in Figure 2.2, the transmission and reflection coefficients (Knott et al. 1985) are given by

$$R_{||,\perp} = \mp \frac{Z_1 - Z_2}{Z_1 + Z_2} \tag{2.15}$$

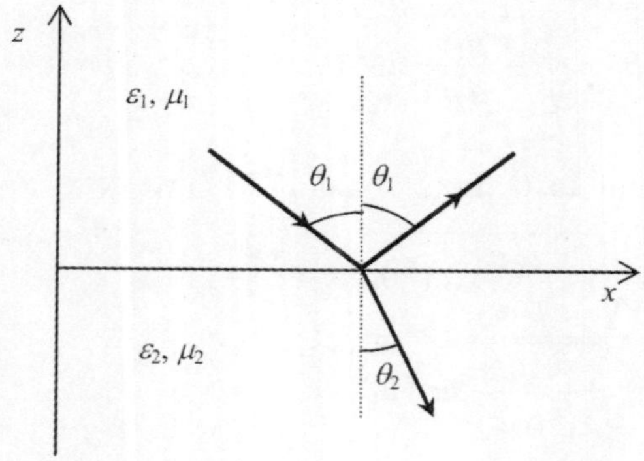

Figure 2.2 A plane EM wave incident on an interface between two semi-infinite media of different material constants

Using (2.6) and (2.7), the reflection coefficients for parallel polarisation (Jordan and Balmain 1976) can be expressed as

$$R_{\|} = \frac{\eta_2 \cos\theta_2 - \eta_1 \cos\theta_1}{\eta_2 \cos\theta_2 + \eta_1 \cos\theta_1} = -\frac{\eta_1 \cos\theta_1 - \eta_2 \cos\theta_2}{\eta_1 \cos\theta_1 + \eta_2 \cos\theta_2} = -\frac{\frac{\eta_1}{\eta_2}\cos\theta_1 - \cos\theta_2}{\frac{\eta_1}{\eta_2}\cos\theta_1 + \cos\theta_2}$$

$$R_{\|} = -\frac{\cos\theta_1 \sqrt{\frac{\mu_1 \varepsilon_2}{\varepsilon_1 \mu_2}} - \cos\theta_2}{\cos\theta_1 \sqrt{\frac{\mu_1 \varepsilon_2}{\varepsilon_1 \mu_2}} + \cos\theta_2} \tag{2.16}$$

Likewise, for perpendicular polarisation,

$$R_{\perp} = \frac{\frac{\eta_2}{\cos\theta_2} - \frac{\eta_1}{\cos\theta_1}}{\frac{\eta_2}{\cos\theta_2} + \frac{\eta_1}{\cos\theta_1}} = \frac{\eta_2 \cos\theta_1 - \eta_1 \cos\theta_2}{\eta_2 \cos\theta_1 + \eta_1 \cos\theta_2} = \frac{\frac{\eta_2}{\eta_1}\cos\theta_1 - \cos\theta_2}{\frac{\eta_2}{\eta_1}\cos\theta_1 + \cos\theta_{21}}$$

$$R_{\perp} = \frac{\sqrt{\frac{\mu_2 \varepsilon_1}{\varepsilon_2 \mu_1}}\cos\theta_1 - \cos\theta_2}{\sqrt{\frac{\mu_2 \varepsilon_1}{\varepsilon_2 \mu_1}}\cos\theta_1 + \cos\theta_2} \tag{2.17}$$

Applying Snell's law of refraction at the interface,

$k_1 \sin\theta_1 = k_2 \sin\theta_2$, where $k_i = \omega\sqrt{\mu_i \varepsilon_i}$

$$\cos\theta_2 = \sqrt{1 - \frac{k_1^2}{k_2^2}\sin^2\theta_1} = \sqrt{1 - \frac{\mu_1\varepsilon_1}{\mu_2\varepsilon_2}\sin^2\theta_1} \tag{2.18}$$

Substituting the value of $\cos\theta_2$ in (2.16) and (2.17) we get,

$$R_{\|} = -\frac{\cos\theta_1\sqrt{\frac{\mu_1\varepsilon_2}{\varepsilon_1\mu_2}} - \sqrt{1 - \frac{\mu_1\varepsilon_1}{\mu_2\varepsilon_2}\sin^2\theta_1}}{\cos\theta_1\sqrt{\frac{\mu_1\varepsilon_2}{\varepsilon_1\mu_2}} + \sqrt{1 - \frac{\mu_1\varepsilon_1}{\mu_2\varepsilon_2}\sin^2\theta_1}} \tag{2.19}$$

$$R_{\perp} = \frac{\sqrt{\frac{\mu_2\varepsilon_1}{\varepsilon_2\mu_1}}\cos\theta_1 - \sqrt{1 - \frac{\mu_1\varepsilon_1}{\mu_2\varepsilon_2}\sin^2\theta_1}}{\sqrt{\frac{\mu_2\varepsilon_1}{\varepsilon_2\mu_1}}\cos\theta_1 + \sqrt{1 - \frac{\mu_1\varepsilon_1}{\mu_2\varepsilon_2}\sin^2\theta_1}} \tag{2.20}$$

If $\mu_1 = \mu_2 = \mu_0$, the expressions (2.19) and (2.20) become

$$R_{\|} = -\frac{\sqrt{\frac{\varepsilon_2}{\varepsilon_1}}\cos\theta_1 - \sqrt{1 - \frac{\varepsilon_1}{\varepsilon_2}\sin^2\theta_1}}{\sqrt{\frac{\varepsilon_2}{\varepsilon_1}}\cos\theta_1 + \sqrt{1 - \frac{\varepsilon_1}{\varepsilon_2}\sin^2\theta_1}} \tag{2.21}$$

$$R_{\perp} = \frac{\sqrt{\frac{\varepsilon_1}{\varepsilon_2}}\cos\theta_1 - \sqrt{1 - \frac{\varepsilon_1}{\varepsilon_2}\sin^2\theta_1}}{\sqrt{\frac{\varepsilon_1}{\varepsilon_2}}\cos\theta_1 + \sqrt{1 - \frac{\varepsilon_1}{\varepsilon_2}\sin^2\theta_1}} \tag{2.22}$$

The dependence of reflection coefficients on the incident angle for different lossless semi-infinite media is presented in Figure 2.3. It is apparent that $R_{\|}$ is zero for a particular incident angle called Brewster angle. This angle depends on the ratio $\varepsilon_2/\varepsilon_1$ (Ruck et al. 1970). Moreover, at normal incidence ($\theta = 0°$), the magnitude of the reflection coefficients approaching both from the parallel and perpendicular polarisations are equal.

The reflection and transmission coefficients for a lossless dielectric are inter-related as

$$|R|^2 + |T|^2 = 1 \tag{2.23}$$

Therefore, the transmission coefficient may also be written as

$$|T| = \sqrt{1 - |R|^2} \tag{2.24}$$

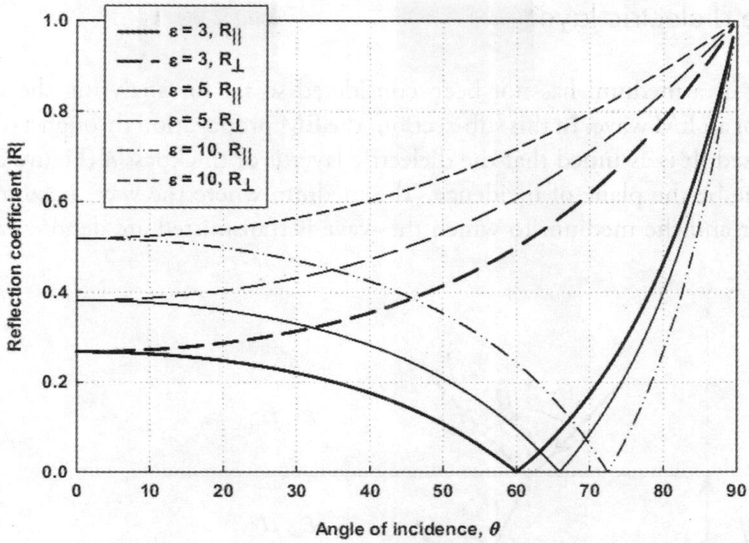

Figure 2.3 Variation of reflection coefficient with respect to the angle of incidence at an interface between two semi-infinite lossless media

Figure 2.4 shows the dependence of transmission coefficient on the incident angle for a plane EM wave incident at an interface of two lossless semi-infinite media.

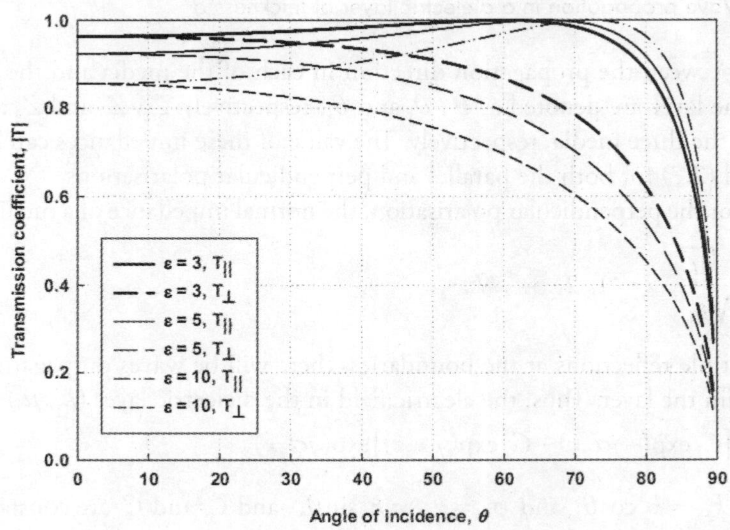

Figure 2.4 Variation of transmission coefficient with respect to the angle of incidence at an interface between two semi-infinite lossless media

2.2.2 Plane dielectric layer

The thickness of a medium has not been considered so far for analysing the reflection and transmission of an EM wave. In this sub-section, the EM propagation through a plane dielectric layer is discussed. It is assumed that the dielectric layer is of thickness d (Figure 2.5).

Let xz-plane be the plane of incidence. The medium where the wave is coming from, the dielectric layer and the medium to which the wave is transmitted are denoted as 1, 2 and 3, respectively.

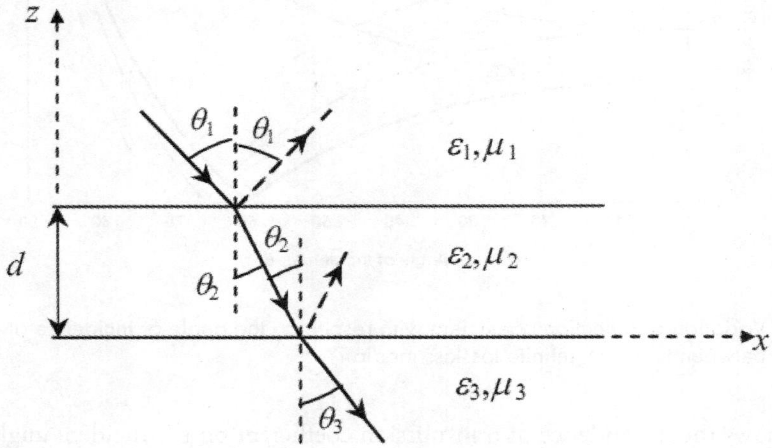

Figure 2.5 Wave propagation in a dielectric layer of thickness d

The angles between the propagation direction in each of the media and the normal to the boundary of the layer are denoted as θ_1, θ_2 and θ_3, respectively. Z_1, Z_2 and Z_3 are the normal impedances of the three media, respectively. The value of these impedances can be determined using (2.6) and (2.7) for both the parallel and perpendicular polarisations.

In general, for the perpendicular polarisation, the normal impedance of a medium is given by

$$Z_i = \frac{1}{\cos\theta_i}\sqrt{\frac{\mu_i}{\varepsilon_i}}, \quad i = 1, 2, 3\ldots N$$

Due to multiple reflections at the boundaries, there will be waves propagating in opposite directions within the layer. Thus, the electric field in the dielectric layer (ε_2, μ_2) is expressed as

$$E_2 = E_{2y} = [C_1 \exp(-j\alpha_2 z) + C_2 \exp(j\alpha_2 z)]\exp(j\sigma_2 x) \tag{2.25}$$

where $\alpha_2 = k_{2z} = k_2 \cos\theta_2$ and $\sigma_2 = k_{2x} = k_2 \sin\theta_2$, and C_1 and C_2 are constants.

The corresponding tangential component of the magnetic field can be written as

$$H_{2x} = \frac{1}{Z_2}[C_1 \exp(-j\alpha_2 z) - C_2 \exp(j\alpha_2 z)]\exp(j\sigma_2 x) \tag{2.26}$$

At the boundary $z = 0$, the ratio of E_{2y} / H_{2x} must be equal to the impedance of Medium 3, i.e. Z_3.

Thus, at $z = 0$, dividing (2.25) by (2.26), one gets

$$Z_2 \frac{C_1 + C_2}{C_1 - C_2} = Z_3$$

or $$\frac{C_2}{C_1} = \frac{Z_3 - Z_2}{Z_3 + Z_2} \qquad (2.27)$$

At the interface of the dielectric layer and Medium 1, i.e. at $z = d$, one has

$$Z_{in} = \left(\frac{E_{2y}}{H_{2x}}\right)_{z=d} = \frac{C_1 \exp(-j\alpha_2 d) + C_2 \exp(j\alpha_2 d)}{C_1 \exp(-j\alpha_2 d) - C_2 \exp(j\alpha_2 d)} Z_2$$

$$= \frac{\exp(-j\alpha_2 d) + {}^{C_2}\!/_{C_1} \exp(j\alpha_2 d)}{\exp(-j\alpha_2 d) - {}^{C_2}\!/_{C_1} \exp(j\alpha_2 d)} Z_2$$

Substituting ${}^{C_2}\!/_{C_1}$ from (2.27), one gets

$$Z_{in} = \frac{\exp(-j\alpha_2 d) + \left(\frac{Z_3 - Z_2}{Z_3 + Z_2}\right) \exp(j\alpha_2 d)}{\exp(-j\alpha_2 d) - \left(\frac{Z_3 - Z_2}{Z_3 + Z_2}\right) \exp(j\alpha_2 d)} Z_2$$

$$= \frac{(Z_3 + Z_2)\exp(-j\alpha_2 d) + (Z_3 - Z_2)\exp(j\alpha_2 d)}{(Z_3 + Z_2)\exp(-j\alpha_2 d) - (Z_3 - Z_2)\exp(j\alpha_2 d)} Z_2$$

$$= \frac{Z_3\left(\exp(j\alpha_2 d) + \exp(-j\alpha_2 d)\right) - Z_2\left(\exp(j\alpha_2 d) - \exp(-j\alpha_2 d)\right)}{Z_2\left(\exp(j\alpha_2 d) + \exp(-j\alpha_2 d)\right) - Z_3\left(\exp(j\alpha_2 d) - \exp(-j\alpha_2 d)\right)} Z_2$$

$$Z_{in} = \frac{Z_3 - jZ_2 \tan\alpha_2 d}{Z_2 - jZ_3 \tan\alpha_2 d} Z_2 \qquad (2.28)$$

If the input impedance of a dielectric layer is known, the wave reflection from a multilayered medium can be analysed in terms of the layer's input impedance instead of the impedance of the reflecting medium.

In Medium 1, the electric and magnetic fields are given as

$$E_{1y} = \left[C_3 \exp\{-j\alpha_1 (z - d)\} + C_4 \exp\{j\alpha_1 (z - d)\}\right] \exp(j\sigma_1 x) \qquad (2.29)$$

$$H_{1x} = \frac{1}{Z_1}\left[C_3 \exp\{-j\alpha_1(z-d)\} - C_4 \exp\{j\alpha_1(z-d)\}\right]\exp(j\sigma_1 x) \tag{2.30}$$

At $z = d$, the ratio of E_{1y} to H_{1x} must be equal to the input impedance of the dielectric layer. Thus, at $z = d$, dividing (2.29) by (2.30), one gets

$$\left(\frac{E_{1y}}{H_{1x}}\right)_{z=d} = Z_{in} = Z_1 \frac{C_3 + C_4}{C_3 - C_4}$$

$$\frac{C_4}{C_3} = \frac{Z_{in} - Z_1}{Z_{in} + Z_1} \tag{2.31}$$

This gives the reflection coefficient as

$$R = \frac{Z_{in} - Z_1}{Z_{in} + Z_1} \tag{2.32}$$

Substituting the value of Z_{in} from (2.28), one gets

$$R = \frac{\dfrac{Z_3 - jZ_2 \tan\alpha_2 d}{Z_3 - jZ_1 \tan\alpha_2 d} Z_2 - Z_1}{\dfrac{Z_3 - jZ_2 \tan\alpha_2 d}{Z_2 - jZ_3 \tan\alpha_2 d} Z_2 + Z_1} = \frac{Z_3 Z_2 - Z_2 Z_1 - \left(Z_2^2 - Z_3 Z_1\right) j \tan\alpha_2 d}{Z_3 Z_2 + Z_2 Z_1 - \left(Z_2^2 + Z_3 Z_1\right) j \tan\alpha_2 d}$$

$$= \frac{Z_3 Z_2 - Z_2 Z_1 - \left(Z_2^2 - Z_3 Z_1\right)\left(\dfrac{\exp(j\alpha_2 d) - \exp(-j\alpha_2 d)}{\exp(j\alpha_2 d) + \exp(-j\alpha_2 d)}\right)}{Z_3 Z_2 + Z_2 Z_1 - \left(Z_2^2 + Z_3 Z_1\right)\left(\dfrac{\exp(j\alpha_2 d) - \exp(-j\alpha_2 d)}{\exp(j\alpha_2 d) + \exp(-j\alpha_2 d)}\right)}$$

$$R = \frac{(Z_3 + Z_2)(Z_2 - Z_1)\exp(-j\alpha_2 d) + (Z_3 - Z_2)(Z_2 + Z_1)\exp(j\alpha_2 d)}{(Z_3 + Z_2)(Z_2 + Z_1)\exp(-j\alpha_2 d) + (Z_3 - Z_2)(Z_2 - Z_1)\exp(j\alpha_2 d)} \tag{2.33}$$

For the special case in which the regions on both sides of the dielectric layer have same permittivity and permeability, i.e. $Z_1 = Z_3$, the reflection coefficient will be

$$R = \frac{\left(Z_2^2 - Z_3^2\right)\exp(-j\alpha_2 d) - \left(Z_2^2 - Z_3^2\right)\exp(j\alpha_2 d)}{\left(Z_2 + Z_3\right)^2 \exp(-j\alpha_2 d) + \left(Z_2 - Z_3\right)^2 \exp(j\alpha_2 d)}$$

$$R = \frac{Z_2^2 - Z_3^2}{Z_3^2 + Z_2^2 + 2jZ_3 Z_2 \cot\alpha_2 d} \tag{2.34}$$

If $d = 0$, i.e. when there is no dielectric layer, the expression (2.33) becomes $R = \dfrac{Z_3 - Z_1}{Z_3 + Z_1}$, the reflection coefficient at the boundary of semi-infinite regions, Medium 1 and Medium 3. The corresponding transmission coefficient is obtained below.

In Medium 3, the field-amplitude of the wave transmitted may be written as

$$E_{3y} = \mathrm{Y}\exp\left(-j\alpha_3 z + j\sigma_3 x\right) \tag{2.35}$$

where Y is a constant.

Since E_y is continuous at the boundary $z = 0$, $E_{3y} = E_{2y}$, equating (2.25) and (2.35) at $z = 0$, one gets

$$\mathrm{Y}\exp\left(j\sigma_3 x\right) = (C_1 + C_2)\exp\left(j\sigma_2 x\right)$$

$$\mathrm{Y} = C_1 + C_2 \quad (\text{since } \sigma_3 = \sigma_2) \tag{2.36}$$

Similarly, at $z = d$, for E_y to be continuous, equating (2.25) and (2.29)

$$C_3 + C_4 = C_1\exp\left(-j\alpha_2 d\right) + C_2\exp\left(j\alpha_2 d\right)$$

Since $R = \dfrac{C_4}{C_3}$

$$C_3\left(1 + R\right) = C_1\exp\left(-j\alpha_2 d\right) + C_2\exp\left(j\alpha_2 d\right) \tag{2.37}$$

Dividing (2.36) by (2.37), one gets

$$\frac{\mathrm{Y}}{C_3(1+R)} = \frac{C_1 + C_2}{C_1\exp\left(-j\alpha_2 d\right) + C_2\exp\left(j\alpha_2 d\right)}$$

$$\frac{\mathrm{Y}}{C_3} = (1+R)\frac{1 + C_2/C_1}{\exp\left(-j\alpha_2 d\right) + C_2/C_1\exp\left(j\alpha_2 d\right)}$$

Substituting for C_2/C_1 from (2.27) and for R from (2.33), one gets

$$T = \frac{\mathrm{Y}}{C_3} = \frac{1 + \dfrac{Z_3 - Z_2}{Z_3 + Z_2}}{\exp\left(-j\alpha_2 d\right) + \dfrac{Z_3 - Z_2}{Z_3 + Z_2}\exp\left(j\alpha_2 d\right)} \times$$

$$\times\left[1 + \frac{(Z_3 + Z_2)(Z_2 - Z_1)\exp\left(-j\alpha_2 d\right) + (Z_3 - Z_2)(Z_2 + Z_1)\exp\left(j\alpha_2 d\right)}{(Z_3 + Z_2)(Z_2 + Z_1)\exp\left(-j\alpha_2 d\right) + (Z_3 - Z_2)(Z_2 - Z_1)\exp\left(j\alpha_2 d\right)}\right]$$

$$= \frac{2Z_3}{(Z_3+Z_2)\exp(-j\alpha_2 d)+(Z_3-Z_2)\exp(j\alpha_2 d)} \times$$

$$\times \left[\frac{2Z_2\{(Z_3+Z_2)\exp(-j\alpha_2 d)+(Z_3-Z_2)\exp(j\alpha_2 d)\}}{(Z_3+Z_2)(Z_2+Z_1)\exp(-j\alpha_2 d)+(Z_3-Z_2)(Z_2-Z_1)\exp(j\alpha_2 d)} \right]$$

$$T = \frac{4Z_3 Z_2}{(Z_3+Z_2)(Z_2+Z_1)\exp(-j\alpha_2 d)+(Z_3-Z_2)(Z_2-Z_1)\exp(j\alpha_2 d)} \quad (2.38)$$

For $d = 0$, (2.38) becomes $T = \frac{2Z_3}{Z_3+Z_1}$, which is the transmission coefficient of the interface between Medium 3 and Medium 1.

2.2.3 Multiple reflections/transmissions at the boundaries

The reflection and transmission through a multilayered media can be derived in several ways. In this sub-section, the reflection/transmission of a plane EM wave is derived considering each boundary separately. A special case of normal incidence is considered. For a unit amplitude incident wave, the reflection from a layer is regarded as a superposition of several individual waves.

These waves can be described as follows: (i) waves that are reflected from (R_{21}) at the boundary between Medium 1 and Medium 2, as shown in Figure 2.6a; (ii) waves refracted through the front surface, waves reflected from the back surface (interface of Medium 2 and Medium 3) and waves refracted from Medium 2 to Medium 1.

The resultant wave is $T_{12}\ R_{32}\ T_{21}$ exp $(2jk_2d)$, as shown in Figure 2.6b, and (iii) waves penetrating the layer, undergoing two reflections at the back surface (Media 2 and 3) and one at the front surface (Media 2 and 1) and, then, transmitted to Medium 1. The resultant wave is $T_{12}\ R_{32}\ R_{12}\ R_{32}\ T_{21}$ exp $(4jk_2d)$ as shown in Figure 2.6c. Here, T_{12} denotes the transmission coefficient of boundary for a wave travelling from Medium 1 to Medium 2.

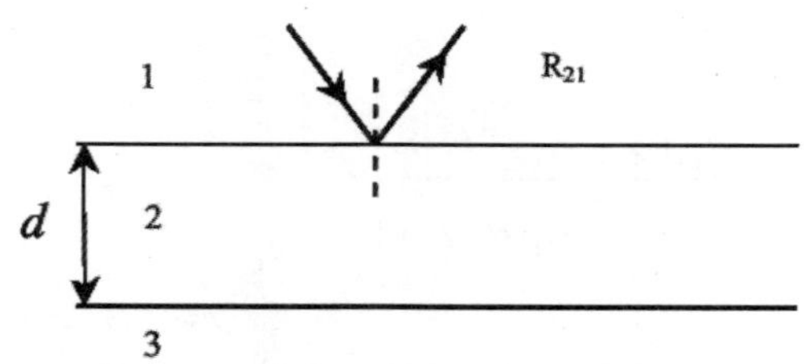

Figure 2.6a Wave reflection diagram (schematic) for the front surface of a dielectric layer

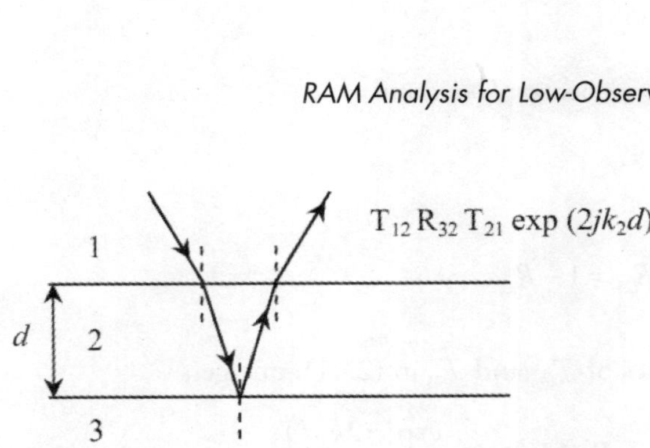

Figure 2.6b Wave reflection–refraction diagram (schematic) for the front surface of a finite-thickness dielectric layer

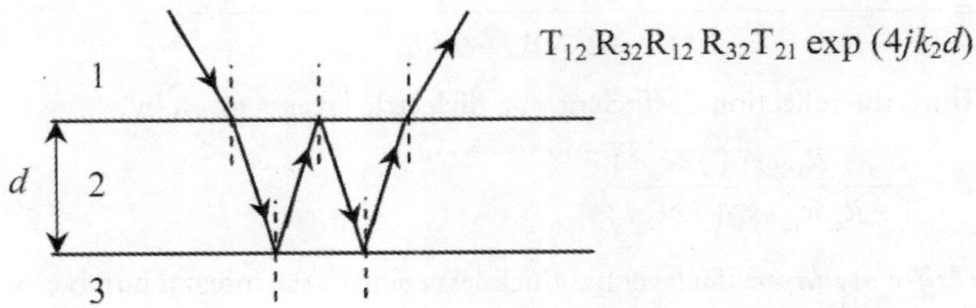

Figure 2.6c Multiple reflection–refraction diagram (schematic) for the front surface of a finite-thickness dielectric layer

The reflection coefficient R_{12} represents the wave reflection from Medium 2 to Medium 1. The factor $\exp(4jk_2d)$ accounts for the phase change and the attenuation of the wave if k_2 is complex, as it goes back and forth twice through the layer. Summing up each of the above waves, one has

$$R_{21} + T_{12}R_{32}T_{21}\exp\left(j2k_2d\right) + T_{12}R_{12}R_{32}^2T_{21}\exp\left(j4k_2d\right) + T_{12}R_{12}^2R_{32}^3T_{21}\exp\left(j6k_2d\right) + \ldots \quad (2.39)$$

Since the incident wave amplitude is unity, (2.39) will be equal to R, the reflection coefficient of the dielectric layer (Medium 2). Thus,

$$R = R_{21} + T_{12}T_{21}R_{32}\exp\left(j2k_2d\right)\sum_{n=0}^{\infty}\left[R_{12}R_{32}\exp\left(j2k_2d\right)\right]^n \quad (2.40)$$

Using $\sum_{n=0}^{\infty} a^n = \frac{1}{1-a}; a < 1$, one gets

$$R = R_{21} + T_{12}T_{21}R_{32}\exp\left(j2k_2d\right)\frac{1}{1 - R_{12}R_{32}\exp\left(j2k_2d\right)}$$

$$\text{or } R = R_{21} + T_{12}T_{21}R_{32}\frac{\exp\left(j2k_2d\right)}{1 - R_{12}R_{32}\exp\left(j2k_2d\right)} \quad (2.41)$$

And one has

$$T_{12} = \frac{2Z_1}{Z_2 + Z_1} = 1 + R_{21} = 1 - R_{12}$$

Substituting the values of T_{12} and T_{21} in (2.41), one gets

$$R = R_{21} + (1 + R_{21})(1 - R_{21}) R_{32} \frac{\exp(j2k_2 d)}{1 - R_{12} R_{32} \exp(j2k_2 d)}$$

$$= \frac{R_{21}\left(1 + R_{21} R_{32} \exp(j2k_2 d)\right) + \left(1 - R_{21}^2\right) R_{32} \exp(j2k_2 d)}{1 + R_{21} R_{32} \exp(j2k_2 d)}$$

Thus, the reflection coefficient of a dielectric layer is given by

$$R = \frac{R_{21} + R_{32} \exp(j2k_2 d)}{1 + R_{21} R_{32} \exp(j2k_2 d)} \tag{2.42}$$

Half-wave layer: If a layer has thickness equal to the integral number m half-wavelengths, i.e.

$$d = m\frac{\lambda_2}{2}$$

Then, $2k_2 d = 2 \times \frac{2\pi}{\lambda_2} \times m \frac{\lambda_2}{2} = 2\pi m$

According to (2.42), the reflection coefficient will be

$$R = \frac{R_{21} + R_{32} \exp(j2\pi m)}{1 + R_{21} R_{32} \exp(j2\pi m)}$$

Since $\exp(j2\pi m) = 1$,

$$R = \frac{R_{21} + R_{32}}{1 + R_{21} R_{32}} \tag{2.43}$$

Substituting $R_{21} = \frac{Z_2 - Z_1}{Z_2 + Z_1}$ and $R_{32} = \frac{Z_3 - Z_2}{Z_3 + Z_2}$ in the above equation one gets

$$R = \frac{\frac{Z_2 - Z_1}{Z_2 + Z_1} + \frac{Z_3 - Z_2}{Z_3 + Z_2}}{1 + \left(\frac{Z_2 - Z_1}{Z_2 + Z_1}\right)\left(\frac{Z_3 - Z_2}{Z_3 + Z_2}\right)} = \frac{Z_3 - Z_1}{Z_3 + Z_1} = R_{31}$$

This shows that the half-wave dielectric layer has no effect on the incident wave and the reflection coefficient is equal to the reflection coefficient at the boundary between Medium 1 and Medium 3. Furthermore, for the case of similar Medium 1 and Medium 3 (i.e. $Z_3 = Z_1$), R = 0. This will hold even for the case of oblique incidence except that the condition $2k_2d = 2\pi m$ is replaced by the condition $2k_2 d cos\theta_2 = 2\pi m$.

Quarter-wave transmitting layer: For such thickness, $2k_2d = 2\pi m$.

Let us consider the reflection coefficient expressed in (2.42) as

$$R = \frac{R_{21} + R_{32}\exp(j2k_2d)}{1 + R_{21}R_{32}\exp(j2k_2d)} = \frac{R_{21} + R_{32}\left(\dfrac{1 + j\tan(k_2d)}{1 - j\tan(k_2d)}\right)}{1 + R_{21}R_{32}\left(\dfrac{1 + j\tan(k_2d)}{1 - j\tan(k_2d)}\right)}$$

$$= \frac{(R_{21} + R_{32}) + j(R_{32} - R_{21})\tan(k_2d)}{(1 + R_{21}R_{32}) + j(R_{21}R_{32} - 1)\tan(k_2d)}$$

$$= \frac{(R_{21} + R_{32})\cos(k_2d) + j(R_{32} - R_{21})\sin(k_2d)}{(1 + R_{21}R_{32})\cos(k_2d) + j(R_{21}R_{32} - 1)\sin(k_2d)}$$

Taking the square of the modulus of the reflection coefficient, i.e. reflection loss, one gets

$$|R|^2 = \frac{(R_{21} + R_{32})^2\cos^2(k_2d) + (R_{32} - R_{21})^2\sin^2(k_2d)}{(1 + R_{21}R_{32})^2\cos^2(k_2d) + (R_{21}R_{32} - 1)^2\sin^2(k_2d)}$$

$$|R|^2 = \frac{(R_{21} + R_{32})^2 - 4R_{21}R_{2}\sin^2(k_2d)}{(1 + R_{21}R_{32})^2 - 4R_{21}R_{32}\sin^2(k_2d)} \tag{2.44}$$

For a quarter-wave layer,

$$d = m\frac{\lambda_2}{4}$$

which implies $2k_2d = \pi m$. The first minima will occur at $m = 1$, i.e. at $k_2d = \dfrac{\pi}{2}$.

Substituting $k_2d = \dfrac{\pi}{2}$ in (2.44), one gets

$$|R|^2 = \frac{(R_{21} + R_{32})^2 - 4R_{21}R_{32}\sin^2\dfrac{\pi}{2}}{(1 + R_{21}R_{32})^2 - 4R_{21}R_{32}\sin^2\dfrac{\pi}{2}} = \frac{(R_{21} - R_{32})^2}{(1 - R_{21}R_{32})^2}$$

$$\text{or } R_{\min} = \frac{R_{21} - R_{32}}{1 - R_{21}R_{32}} \tag{2.45}$$

If, $R_{21} = R_{32}$, the reflection from the dielectric layer will be completely absent. For this, the impedances can be related as follows:

$$R_{21} = R_{32}$$

$$\frac{Z_2 - Z_1}{Z_2 + Z_1} = \frac{Z_3 - Z_1}{Z_3 + Z_1}$$

$$Z_2 = \sqrt{Z_3 Z_1} \tag{2.46}$$

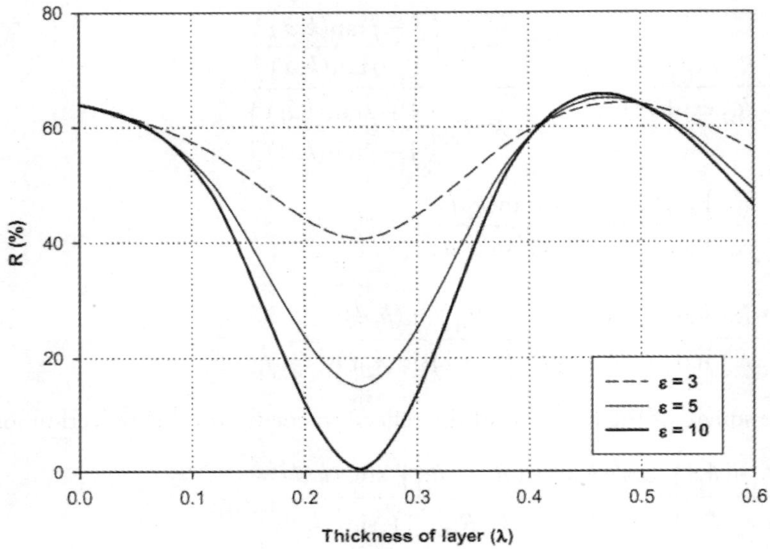

Figure 2.7 Percentage reflection of an EM wave due to dielectric layer between air and water

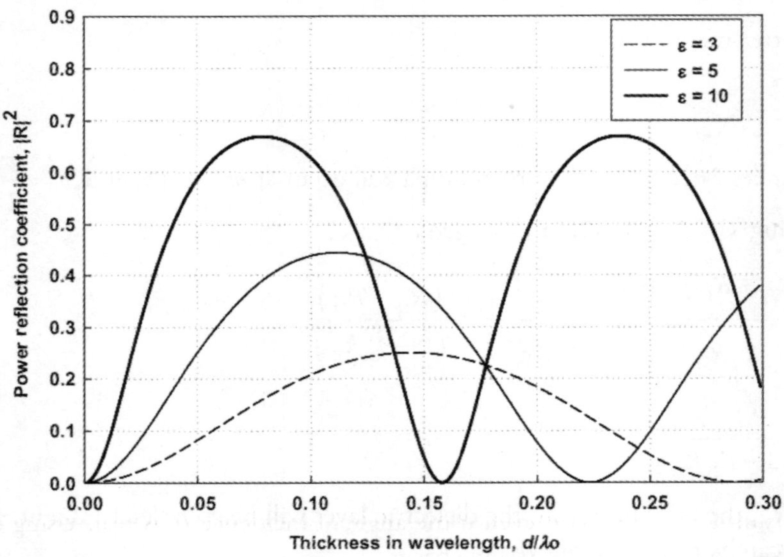

Figure 2.8 Power reflection coefficient of an EM wave from a dielectric layer

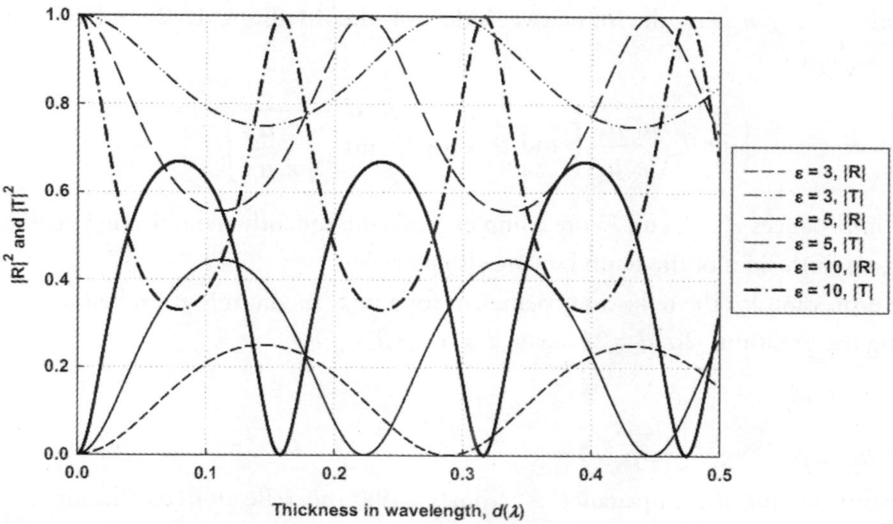

Figure 2.9 Power reflection and transmission coefficient of an EM wave from a dielectric layer

Thus, by inserting a quarter wave dielectric layer between two media, reflection at the boundary can be eliminated completely. The impedance of dielectric layer would be the geometric mean of the impedances of the two media. A similar effect will be observed if the thickness of the layer is equal to the odd multiples of quarter wavelengths, i.e. $\frac{3\lambda}{4}, \frac{5\lambda}{4}$, etc.

The percentage reflection for an EM wave travelling through air-dielectric-water media (ε_{air} = 1, $\varepsilon_{\text{water}}$ = 81) is presented in Figure 2.7. The layer thickness is varied from 0 to half-wavelength. For a lossless dielectric slab, the variation of $|R|^2$ with d/λ_0 for a different permittivity is shown in Figure 2.8.

Further, the dependence of R and T on the thickness of the dielectric slab is shown in Figure 2.9. Results are shown for different dielectrics. It may be observed that both the reflection and transmission coefficients are periodic with respect to the thickness of the dielectric layer.

2.2.4 Lossy dielectric layer

In the above sections, the transmission and reflection coefficients were derived for a lossless dielectric layer. The lossy nature of the medium in general is now taken into account by the complex wave numbers k_1, k_2 and k_3 in different media. It is known that the angles of refraction at the boundaries are related as

$$k_3 \sin\theta_3 = k_2 \sin\theta_2 = k_1 \sin\theta_1 \tag{2.47}$$

where θ_2 and θ_1 are complex, even if the angle of incidence θ_3 is real. Using Snell's law of refraction, for the incident angle θ_3, one has

$$n_3 \sin\theta_3 = n_2 \sin\theta_2 = n_1 \sin\theta_1$$

where n_1, n_2, n_3 are the refractive index of the Medium 1, 2 and 3, respectively and $n_i = \sqrt{\varepsilon_i \mu_i}$.

Thus, $\theta_2 = \sin^{-1}\left(\sin\theta_3 \sqrt{\frac{\varepsilon_3\mu_3}{\varepsilon_2\mu_2}}\right)$ and $\theta_1 = \sin^{-1}\left(\sin\theta_2 \sqrt{\frac{\varepsilon_2\mu_2}{\varepsilon_1\mu_1}}\right)$

The impedances Z_1, Z_2 and Z_3 are complex, and consequently the reflection coefficients R_{21} and R_{32} on both sides of the boundary are also complex.

The expression for the reflection coefficient for a lossy medium is given below.

Using the notations $2\alpha_2 d = 2k_2 \cos\theta_2 d = \alpha + j\beta$ (2.48)

$Z_i = \gamma_i + j\delta_i$, $i = 1,2,3$ (2.49)

and $R_{21} = \rho_{21} e^{j\phi_{21}}$, $R_{32} = \rho_{32} e^{j\phi_{32}}$ (2.50)

Substituting for the impedance Z_i from (2.49), the reflection coefficient at interface (Medium 1 and Medium 2) is given by

$$R_{21} = \frac{Z_2 - Z_1}{Z_2 + Z_1} = \frac{(\gamma_2 + j\delta_2) - (\gamma_1 + j\delta_1)}{(\gamma_2 + j\delta_2) + (\gamma_1 + j\delta_1)} = \frac{(\gamma_2 - \gamma_1) + j(\delta_2 - \delta_1)}{(\gamma_2 + \gamma_1) + j(\delta_2 + \delta_1)}$$

Taking the modulus, one has

$$|R_{21}| = \sqrt{\frac{(\gamma_2 - \gamma_1)^2 + (\delta_2 - \delta_1)^2}{(\gamma_2 + \gamma_1)^2 + (\delta_2 + \delta_1)^2}}$$

or $$|R_{21}|^2 = \rho_{21}^2 = \frac{(\gamma_2 - \gamma_1)^2 + (\delta_2 - \delta_1)^2}{(\gamma_2 + \gamma_1)^2 + (\delta_2 + \delta_1)^2} \tag{2.51}$$

Similarly, $$|R_{32}|^2 = \rho_{32}^2 = \frac{(\gamma_3 - \gamma_2)^2 + (\delta_3 - \delta_2)^2}{(\gamma_3 + \gamma_2)^2 + (\delta_3 + \delta_2)^2} \tag{2.52}$$

The phase of R_{21} is expressed as

$$\phi_{21} = \tan^{-1}\left(\frac{\delta_2 - \delta_1}{\gamma_2 - \gamma_1}\right) - \tan^{-1}\left(\frac{\delta_2 + \delta_1}{\gamma_2 + \gamma_1}\right) = \tan^{-1}\left(\frac{\frac{\delta_2 - \delta_1}{\gamma_2 - \gamma_1} - \frac{\delta_2 + \delta_1}{\gamma_2 + \gamma_1}}{1 + \frac{\delta_2 - \delta_1}{\gamma_2 - \gamma_1} \cdot \frac{\delta_2 + \delta_1}{\gamma_2 + \gamma_1}}\right)$$

Therefore, $$\tan\phi_{21} = \frac{\frac{\delta_2 - \delta_1}{\gamma_2 - \gamma_1} - \frac{\delta_2 + \delta_1}{\gamma_2 + \gamma_1}}{1 + \frac{\delta_2 - \delta_1}{\gamma_2 - \gamma_1} \cdot \frac{\delta_2 + \delta_1}{\gamma_2 + \gamma_1}} = \frac{(\delta_2 - \delta_1)(\gamma_2 + \gamma_1) - (\delta_2 + \delta_1)(\gamma_2 - \gamma_1)}{(\gamma_2 - \gamma_1)(\gamma_2 + \gamma_1) + (\delta_2 - \delta_1)(\delta_2 + \delta_1)}$$

$$\tan\phi_{21} = \frac{2(\delta_2\gamma_1 - \delta_1\gamma_2)}{\gamma_2^2 - \gamma_1^2 + \delta_2^2 - \delta_1^2} \tag{2.53}$$

Similarly, $$\tan\phi_{32} = \frac{2(\delta_3\gamma_2 - \delta_2\gamma_3)}{\gamma_3^2 - \gamma_2^2 + \delta_3^2 - \delta_2^2} \tag{2.54}$$

The reflection coefficient R in terms of its modulus and phase may be written as $R = \rho e^{j\phi}$. Using (2.42)

$$R = \frac{R_{21} + R_{32}\exp(j2\alpha_2 d)}{1 + R_{21}R_{32}\exp(j2\alpha_2 d)}$$

Using notations given in (2.48) and (2.49), the reflection coefficient is written as

$$R = \frac{\rho_{21}e^{j\phi_{21}} + \rho_{32}e^{j\phi_{32}}\exp[j(\alpha + j\beta)]}{1 + \rho_{21}e^{j\phi_{21}}\rho_{32}e^{j\phi_{32}}\exp[j(\alpha + j\beta)]} \tag{2.55}$$

$$= \frac{\rho_{21}e^{j\phi_{21}} + \rho_{32}e^{j\phi_{32}}e^{j\alpha}e^{-\beta}}{1 + \rho_{21}e^{j\phi_{21}}\rho_{32}e^{j\phi_{32}}e^{j\alpha}e^{-\beta}} = \frac{e^{j\phi_{21}}\left[\rho_{21} + \rho_{32}\left(e^{j(\phi_{32}+\alpha-\phi_{21})}\right)e^{-\beta}\right]}{1 + \rho_{21}\rho_{32}\left(e^{j(\phi_{32}+\alpha+\phi_{21})}\right)e^{-\beta}}$$

$$|R|^2 = \frac{\left[e^{j\phi_{21}}\left(\rho_{21} + \rho_{32}\left(e^{j(\phi_{32}+\alpha-\phi_{21})}\right)e^{-\beta}\right)\right]\left[e^{-j\phi_{21}}\left(\rho_{21} + \rho_{32}\left(e^{-j(\phi_{32}+\alpha-\phi_{21})}\right)e^{-\beta}\right)\right]}{\left[1 + \rho_{21}\rho_{32}\left(e^{j(\phi_{32}+\alpha+\phi_{21})}\right)e^{-\beta}\right]\left[1 + \rho_{21}\rho_{32}\left(e^{-j(\phi_{32}+\alpha+\phi_{21})}\right)e^{-\beta}\right]}$$

$$|R|^2 = \frac{\rho_{21}^2 + 2\rho_{21}\rho_{32}e^{-\beta}\cos(\phi_{32} - \phi_{21} + \alpha) + \rho_{32}^2 e^{-2\beta}}{1 + 2\rho_{21}\rho_{32}e^{-\beta}\cos(\phi_{32} + \phi_{21} + \alpha) + \rho_{21}^2\rho_{32}^2 e^{-2\beta}} \tag{2.56}$$

This expression gives the absolute reflection coefficient for a lossy dielectric layer. To obtain the phase associated with the reflection coefficient, let us consider (2.55) as

$$R = \frac{\rho_{21}e^{j\phi_{21}} + \rho_{32}e^{j\phi_{32}}\exp[j(\alpha + j\beta)]}{1 + \rho_{21}e^{j\phi_{23}}\rho_{32}e^{j\phi_{12}}\exp[j(\alpha + j\beta)]} = \frac{e^{j\phi_{21}}\left[\rho_{21} + \rho_{32}\left(e^{j(\phi_{32}+\alpha-\phi_{21})}\right)e^{-\beta}\right]}{1 + \rho_{21}\rho_{32}\left(e^{j(\phi_{32}+\alpha+\phi_{21})}\right)e^{-\beta}}$$

$$= e^{j\phi_{21}}\left[\frac{\rho_{21} + \rho_{32}e^{-\beta}\left(\cos(\phi_{32} + \alpha - \phi_{21}) + j\sin(\phi_{32} + \alpha - \phi_{21})\right)}{1 + \rho_{21}\rho_{32}e^{-\beta}\left(\cos(\phi_{32} + \alpha + \phi_{21}) + j\sin(\phi_{32} + \alpha + \phi_{21})\right)}\right]$$

$$\therefore \phi = \phi_{21} + \tan^{-1}\left[\frac{\rho_{32}e^{-\beta}\sin(\phi_{32} + \alpha - \phi_{21})}{\rho_{21} + \rho_{32}e^{-\beta}\cos(\phi_{32} + \alpha - \phi_{21})}\right] \tag{2.57}$$

$$- \tan^{-1}\left[\frac{\rho_{21}\rho_{32}e^{-\beta}\sin(\phi_{32} + \alpha + \phi_{21})}{1 + \rho_{21}\rho_{32}e^{-\beta}\cos(\phi_{32} + \alpha + \phi_{21})}\right]$$

The magnitude of the reflection coefficient expressed in (2.56) is determined for an EM wave reflected from a thin layer of water situated in air. Its dependence on the layer thickness is demonstrated in Figure 2.10. The wavelength in air is taken as $\lambda_1 = 10$ cm, while in dielectric layer, the wavelength is $\lambda_2 = 3$ cm and $k_2 = (5 + j0.45)$ cm^{-1}.

The results can be obtained directly using (2.33) instead of (2.56) as the lossy nature of dielectric layer is taken care by the complex quantities, more specifically in k_2. It may be noticed that the reflection coefficient of an absorbing layer shows oscillatory behaviour with the layer thickness. However, as the thickness increases, the amplitude of the oscillation is damped. For sufficiently large β, all the terms except the first one in (2.56) can be neglected and one obtains $R = \rho_{21}$. In other words, when the dielectric layer is very thick, the waves are absorbed completely and they do not reach the back surface of the layer at all.

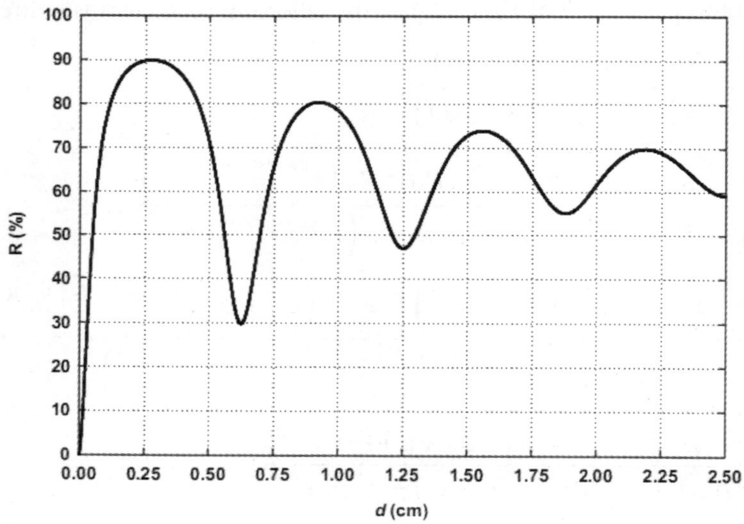

Figure 2.10 Reflection of an electromagnetic wave (λ= 10 cm) from a layer of water

2.2.5 Arbitrary number of dielectric layers

The analysis of reflection and transmission coefficients becomes complicated when the number of dielectric layers exceeds three. This is due to the multiple reflections and transmissions a wave may undergo while travelling through the multilayered media. In this section, a generalised expression for the input impedance $Z_{in}^{(n)}$ of a system of n layers is discussed. Using this input impedance, the reflection coefficient can be obtained easily. Let us consider a multilayered semi-infinite medium consisting of $(n + 1)$ layers, as shown in Figure 2.11.

In xz-plane, a plane EM wave impinges the first layer at an angle θ_1. The z-axis is assumed to be normal to the boundaries of the layers. Due to the multiple reflections at the boundaries, waves in opposite directions exist in each of the media, except Medium 1.

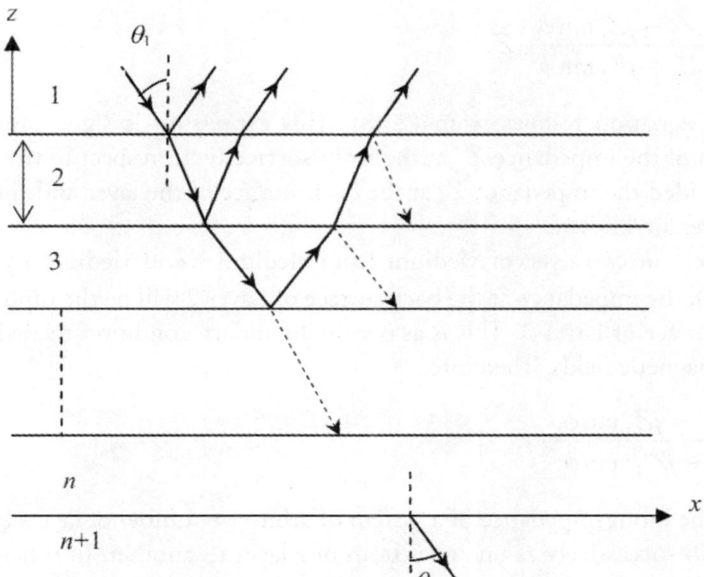

Figure 2.11 Schematic of multiply reflected–refracted ray diagram for a plane wave from a multi-layered dielectric media

The phase change in the wave within i^{th} layer is given by

$$\phi_i = \alpha_i d_i \tag{2.58}$$

where $\alpha_i = k_i \cos\theta_i$, $i = 1, 2,\ldots, n + 1$, is the z-component of the wave vector in the ith layer and $d_i = z_i - z_{i+1}$ is the layer thickness, and z_i is the coordinate boundary in between i^{th} and $(i + 1)^{th}$ layers.

The normal impedance of a layer is given by

$$Z_i = \frac{1}{\cos\theta_i}\sqrt{\frac{\mu_i}{\varepsilon_i}} \tag{2.59}$$

To begin with, let us analyse the reflection of an EM wave from the interface of Medium 1 and Medium 2, where Medium 1 extends to infinity in the positive z direction (Figure 2.12a). Then, the reflection coefficient is given by

$$R = \frac{Z_{in} - Z_1}{Z_{in} + Z_1},\ Z_{in} = Z_2 \tag{2.60a}$$

For boundary between Medium 2 and Medium 3, where Medium 2 stretches to infinity, the reflection coefficient is given by

$$R = \frac{Z_{in}^{(3)} - Z_2}{Z_{in}^{(3)} + Z_2}$$

where $Z_{in}^{(3)} = \dfrac{Z_2 - jZ_1 \tan\phi_1}{Z_1 - jZ_2 \tan\phi_1} \cdot Z_1$ (2.60b)

The above equation follows from (2.28). This expression is significant as it allows the determination of the impedance Z_{in} at the front surface (with respect to the incident wave) of the layer, provided the impedance Z_2 at the back surface of the layer and the phase change ϕ_1 within the layer are known.

Suppose there are two layers of Medium 2 and Medium 3, and Medium 1 stretches to infinity (Figure 2.12b), the impedance at the back surface of Layer 2 will be the input impedance $Z_{in}^{(3)}$ at the front surface of Layer 3. This is as per the boundary condition related to the tangential electric and magnetic fields. Therefore,

$$Z_{in}^{(2)} = \frac{Z_{in}^{(3)} - jZ_2 \tan\phi_2}{Z_2 - jZ_{in}^{(3)} \tan\phi_2} \cdot Z_2 \quad (2.61)$$

Similarly, the input impedance of a system of arbitrary number of layers can be determined by using (2.60) successively as one goes from one layer to another. In other words, the input impedance $Z_{in}^{(n)}$ at the front surface of the nth layer may be expressed as

$$Z_{in}^{(n)} = \frac{Z_{in}^{(n+1)} - jZ_n \tan\phi_n}{Z_{in} - jZ_{in}^{(n+1)} \tan\phi_n} \cdot Z_n \quad (2.62)$$

The corresponding reflection coefficient will be

$$R = \frac{Z_{in}^{(n)} - Z_{n-1}}{Z_{in}^{(n)} + Z_{n-1}} \quad (2.63)$$

It may be noted that the above expression has the same form as that of the reflection coefficient of a single layer.

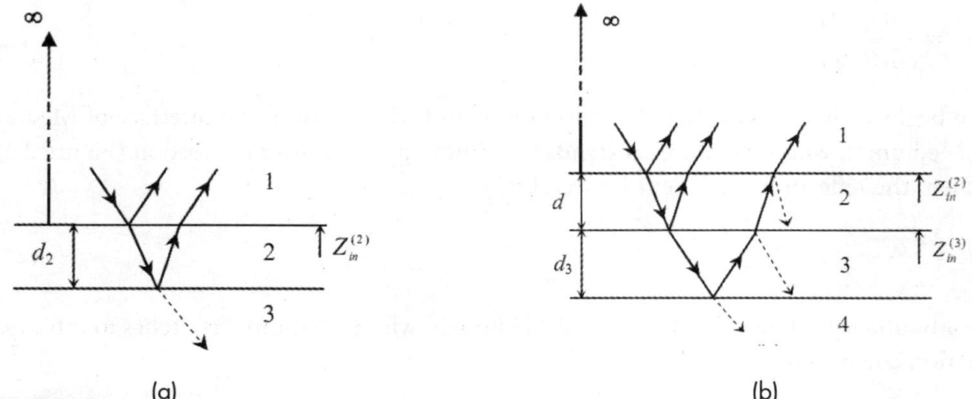

Figure 2.12 Schematic of a multilayered dielectric media

Figure 2.13 shows the reflection coefficient of a multilayered medium at 10 GHz. Two cases, (i) a three-layered medium consisting of air, plywood ($\varepsilon = 2.5 + j0.16$) and iron ($\varepsilon = 1 + j1.8 \times 10^7$, $\mu = 470$) and (ii) a four-layered medium consisting of air, lossy dielectric ($\varepsilon = 1.5 + j3$; thickness = 1 mm), plywood and iron are considered. The thickness of plywood is varied and the reflection coefficient is calculated for the normal incidence.

It is apparent that the magnitude of reflection coefficient is an oscillatory function with period nearly equal to the half-wavelength related to the plywood material. The minima and maxima are located approximately at $\lambda/4$, $\lambda/2$ and so on (Klement et al. 1988). For a four-layered medium, i.e. when a lossy dielectric layer is added, there is a change in the trend of oscillations in the reflection coefficient.

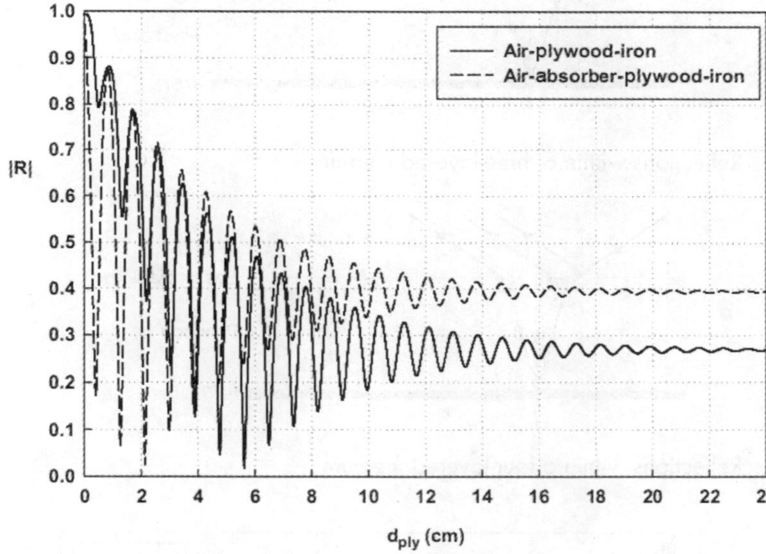

Figure 2.13 Reflection coefficient of a multilayered medium. Normal incidence; $f = 10$ GHz. Case 1: Three-layered medium: air, plywood ($\varepsilon_r = 3 + j0.2$) and iron ($\varepsilon_r = 1 + j2 \times 10^7$, $\mu_r = 480$). Case 2: Four-layered medium: free space, lossy dielectric ($\varepsilon_r = 1.6 + j2.6$; thickness = 1 mm), plywood, iron

In the case of the three-layered media, it may be observed that as the thickness of the plywood increases, the modulus of reflection coefficient decreases. However, the difference between the minima and the maxima of |R| increases with the plywood thickness until the thickness of approximately 6 cm ($\approx 2\lambda$) is reached. Specifically, an absolute minimum of |R| equal to 0.01 is observed.

This may be explained in terms of destructive superposition of the reflected waves (Figure 2.14a). The two reflected waves are (i) the wave reflected from the front surface of the plywood layer and (ii) the wave reflected from the surface of iron layer. These waves have different amplitudes and phases owing to different paths followed. Furthermore, as d_{ply} increases beyond 9 cm, $|R|$ goes on decreasing until a value of 0.275, which corresponds to the infinite thickness of the layer.

For a four-layered medium, i.e. when the lossy dielectric layer ($\varepsilon = 1.6 + j2.6$; thickness = 1 mm) is added to the plywood, the difference between the minima and maxima of oscillatory |R| is more pronounced as compared to the three-layered case. This is true only up to the plywood thickness of approximately 3 cm ($\approx\lambda$), which is due to the absorbing dielectric layer (Figure 2.14b). The absolute minimum is the first minimum and occurs when the thickness of the plywood is 1.4 mm (approximately). The reflection coefficient |R| approaches a value of 0.53 for a sufficiently large plywood thickness.

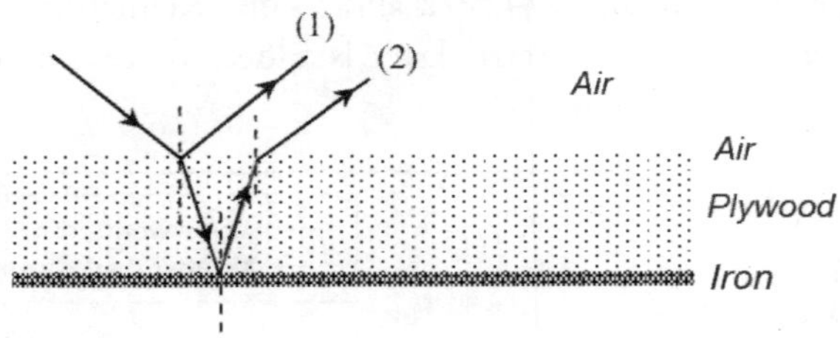

Figure 2.14a Reflections within a three-layered medium

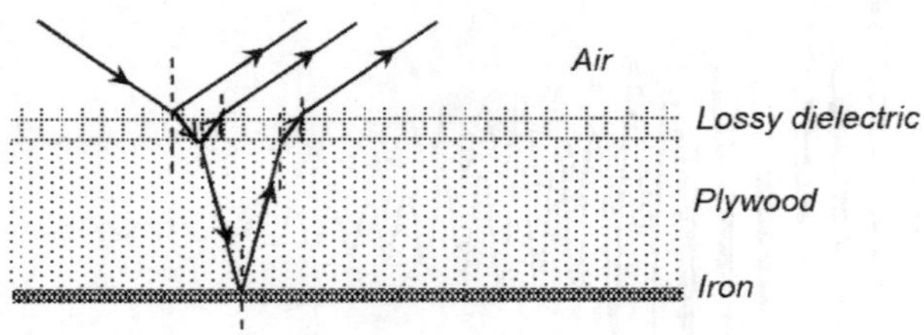

Figure 2.14b Reflections within a four-layered medium

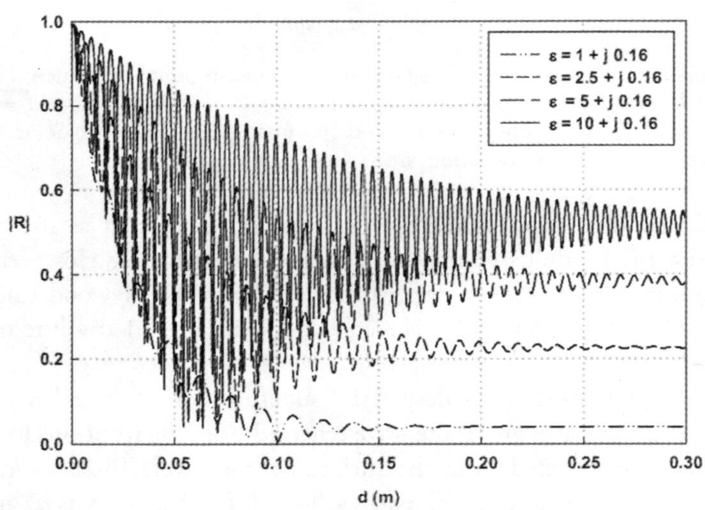

Figure 2.15 Reflection coefficient of a three-layered medium for different dielectric layers placed between the air and the iron; f = 10 GHz

A three-layered medium (air-dielectric-iron) is considered next. The reflection coefficient is determined for various dielectrics with different real parts of permittivity (Figure 2.15). It may be observed that as the real part of the permittivity of the Medium 2 increases, the difference between the maxima and minima of |R| for a three-layered medium also increases. The imaginary part of $j0.16$ signifies the lossy nature of the dielectric. Moreover, the |R| for a sufficiently large thickness of Medium 2 converges at a higher value with an increase in the real part of permittivity.

2.3 EM propagation in multilayered dielectric-metamaterial media

The EM radiation response of any material is determined in terms of constitutive parameters, *viz.* permittivity, ε, and permeability, μ. The materials are categorised on the basis of the real values of ε and μ; these are double-positive (DPS), epsilon-negative (ENG), mu-negative (MNG) and double-negative (DNG), as listed in Table 2.1. Materials having negative permittivity and permeability (Veselago 1968) are now well-known as metamaterials (or left-handed/backward-wave materials).

For a plane monochromatic EM wave propagating in a dielectric medium,

$$k \times E = \frac{\omega}{c} \mu H; \qquad k \times H = -\frac{\omega}{c} \varepsilon E \tag{2.64}$$

where (E, H) are the electric and magnetic field vectors, respectively, $k = \frac{2\pi}{\lambda}$, $\omega = 2\pi f$ and c is the velocity of propagation. For naturally occurring materials, $\varepsilon, \mu > 0$; E, H and k create a right-handed triplet (Figure 2.16a). However, if $\varepsilon, \mu < 0$, i.e. for metamaterials, these vectors create a left-handed triplet (Figure 2.16b). This is why naturally occurring materials are also called right-handed materials (RHM), while metamaterials are identified as left-handed materials (LHM).

Table 2.1 Classification of materials based on ε and μ

Type	ε	μ
DPS	+ve	+ve
ENG	–ve	+ve
MNG	+ve	–ve
DNG	–ve	–ve

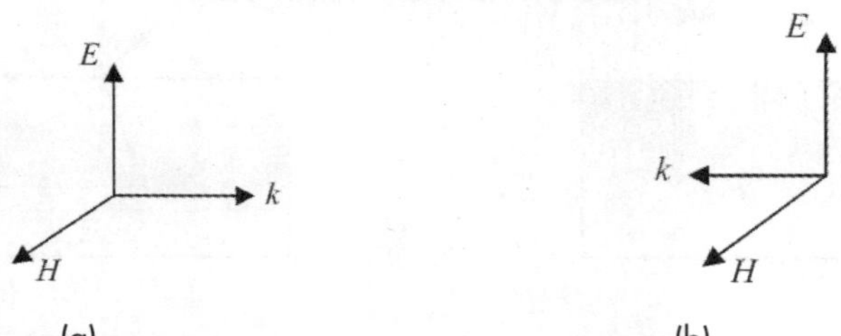

Figure 2.16 (E,H,k) triplet in (a) Right-handed material (b) Left-handed material

The refraction of an EM wave in metamaterials is different from the refraction classically encountered in a naturally occurring material and is governed (Pendry 2000) by

$$\frac{\sin\theta}{\sin\psi}=\frac{n_2}{n_1}=\frac{\sqrt{\varepsilon_2\mu_2}}{\sqrt{\varepsilon_1\mu_1}} \tag{2.65a}$$

where θ and ψ are the angles of incidence and refraction, respectively. If one of the media is left-handed, the refracted and incident rays will be in the same region with respect to the normal. This unique property exhibited by a metamaterial is known as negative refraction. Thus, for a metamaterial, the Snell's law can be written as (Shelby et al. 2001)

$$\frac{\sin\theta}{\sin\psi}=\frac{n_2}{n_1}=\frac{p_2}{p_1}\left|\frac{\sqrt{\varepsilon_2\mu_2}}{\sqrt{\varepsilon_1\mu_1}}\right| \tag{2.65b}$$

where p_1 and p_2 are the right-handedness of the first and second media, respectively. The value of p_1 and p_2 is 1 (for RHM) and –1 (for LHM). Thus, it may be inferred that the Snell's law gets reversed at the interface of a natural material and a metamaterial (Figure 2.17).

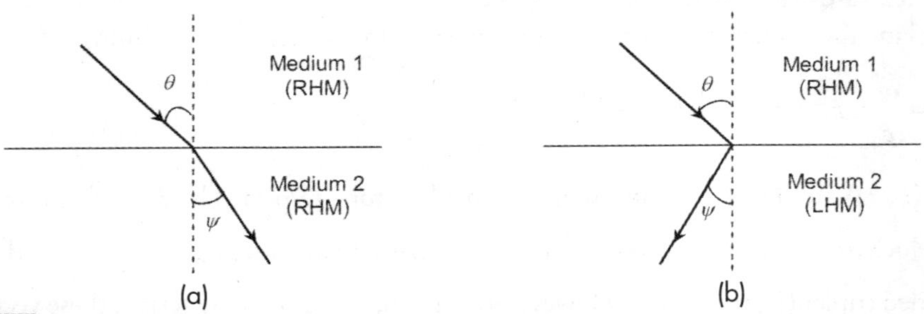

Figure 2.17 Wave propagation across the boundary between two media (a) Medium 1 and 2 are RHM (b) Medium 1 is RHM while Medium 2 is LHM

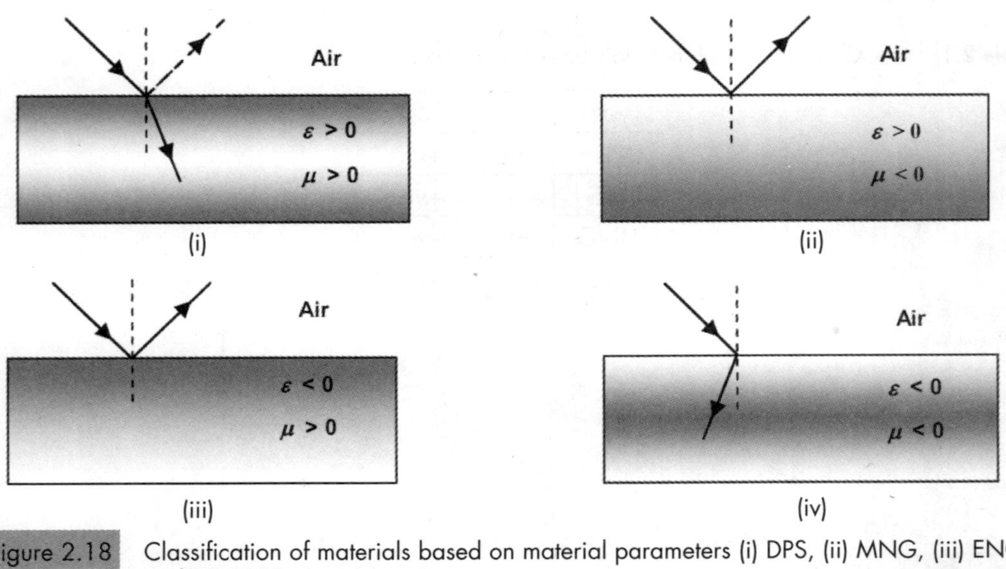

Figure 2.18 Classification of materials based on material parameters (i) DPS, (ii) MNG, (iii) ENG and (iv) DNG

Further, metamaterials can be classified as ENG, MNG and DNG materials. Natural materials are called DPS, with both ε and μ positive quantities. Figure 2.18 shows the nature of wave propagation in such media. It can be seen that for DPS materials, reflection and refraction occur conventionally. For DNG materials, since ε and μ are both negative, the refractive index remains a positive value ($n = \sqrt{\varepsilon\mu}$).

Although refraction (or transmission) does take place, it is in a manner opposite to that of DPS materials due to reversal of Snell's law. On the contrary, when either ε or μ are negative, the refractive index becomes negative and takes an imaginary value. This is the reason why these materials are electromagnetically opaque, and the wave is reflected back into the medium of incidence (Veselego 1967).

2.3.1 Reflection behaviour for dielectric-metamaterial layers

In this sub-section, the propagation of an EM wave in a multilayered dielectric-metamaterial structure is described. Figure 2.19a shows the schematic of wave propagation through metamaterial layers. It may be seen that the Snell's law of refraction is reversed and, thus, the wave is refracted onto the same region with respect to the normal at the boundary. When the media is a combination of dielectric and metamaterial layers, wave propagation will be different. A plane EM wave impinges on the $(n + 1)^{th}$ layer at an angle of θ_{n+1} in *xz*-plane. The *z*-axis is taken normal to the boundaries of the layers. Due to the multiple reflections at the boundaries, waves in opposite directions exist in each of the media, except Medium 1.

The phase change in the wave within i^{th} layer is expressed as $\phi_i = \alpha_i d_i$, where $\alpha_i = k_i \cos\theta_i$, $i = 1, 2, \ldots, n + 1$ is the *z*-component of the wave vector in the ith layer and $d_i = z_i - z_{i-1}$ is the layer thickness; z_i is the interface between i^{th} and $(i + 1)^{th}$ layers. The normal impedance of a layer is

for $Z_i = \dfrac{1}{\cos\theta_i}\sqrt{\dfrac{\mu_i}{\varepsilon_i}}$ perpendicular polarisation and $Z_i = \cos\theta_i\sqrt{\dfrac{\mu_i}{\varepsilon_i}}$ for parallel polarisation.

Figure 2.19b shows the four quadrants of the ε–μ diagram (Veselago 1968). The first quadrant covers all the isotropic dielectrics ($\varepsilon > 0$ and $\mu > 0$). The second quadrant includes the materials with $\varepsilon < 0$ and $\mu > 0$, e.g. plasmas, which are of dispersive nature. For such materials, the refractive index is negative, leading to the reflection of incident waves. The third and fourth quadrants include substances with negative permeability. Such substances include gyrotropic substances with tensor ε and μ, e.g. ferromagnetic metals and semiconductors.

The negative values of ε and μ can be realised when there is frequency dispersion (Veselago 1968). The negative values of permittivity and permeability may be realised in metamaterials by embedding thin metallic wires and square- and circular-shaped split ring resonators (SRR) in dielectrics via the Drude, Lorentz and resonance dispersion models. Numerous models have been proposed to depict the frequency dependence of a material. These models are based on the role of field vectors of an incident EM wave on the electric/magnetic dipole moment of the material. They provide the values of electric and magnetic susceptibility (hence ε, μ) of the material.

The Drude–Lorentz model is commonly used for metamaterials. The parameters f_{mp} and f_{ep} denote the magnetic and electric plasma frequencies, respectively, f_m is the damping frequency, f_{mo} is the magnetic resonant frequency, and γ_e and γ_m are the damping factors, related to electric and magnetic losses within the medium.

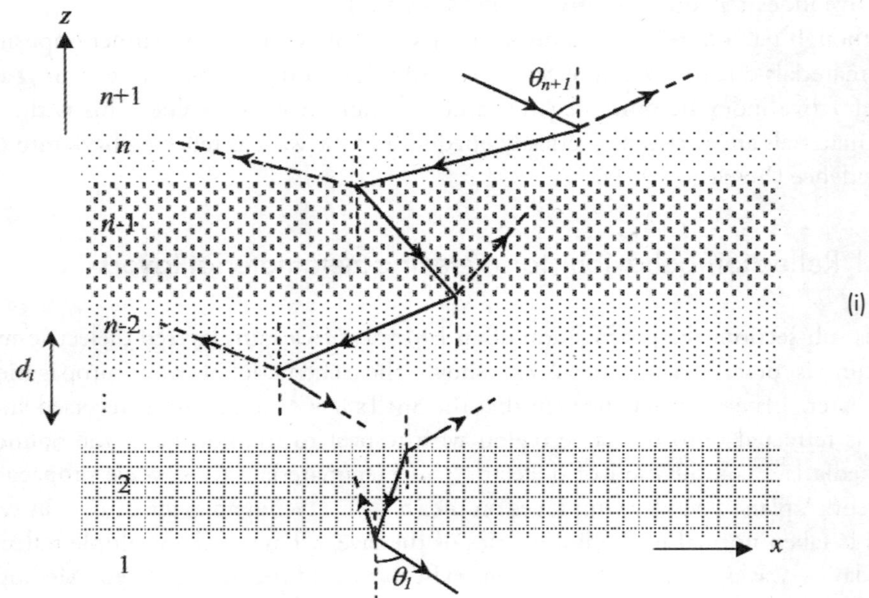

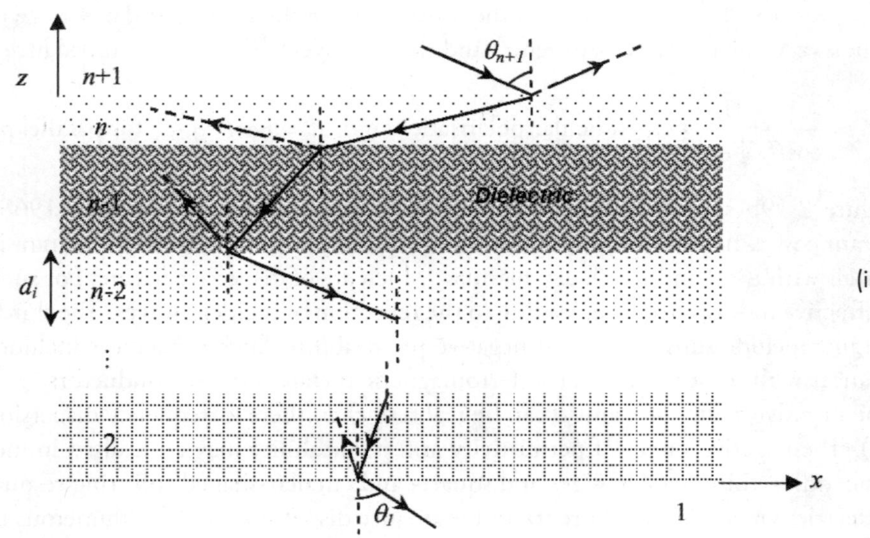

Figure 2.19a Reflection and transmission of a plane wave (i) Multilayered metamaterial media (ii) Multilayered dielectric-metamaterial media

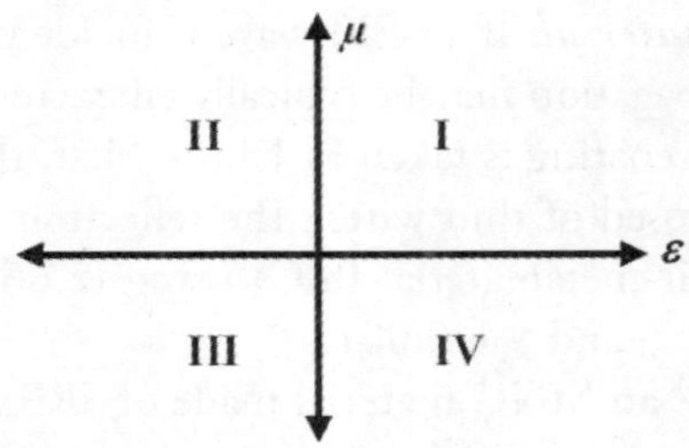

Figure 2.19b Four quadrants of ε–μ diagram

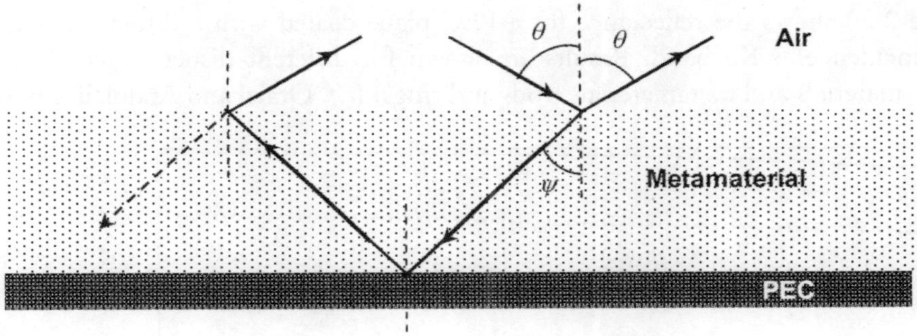

Figure 2.20 Wave propagation in a metamaterial-coated PEC

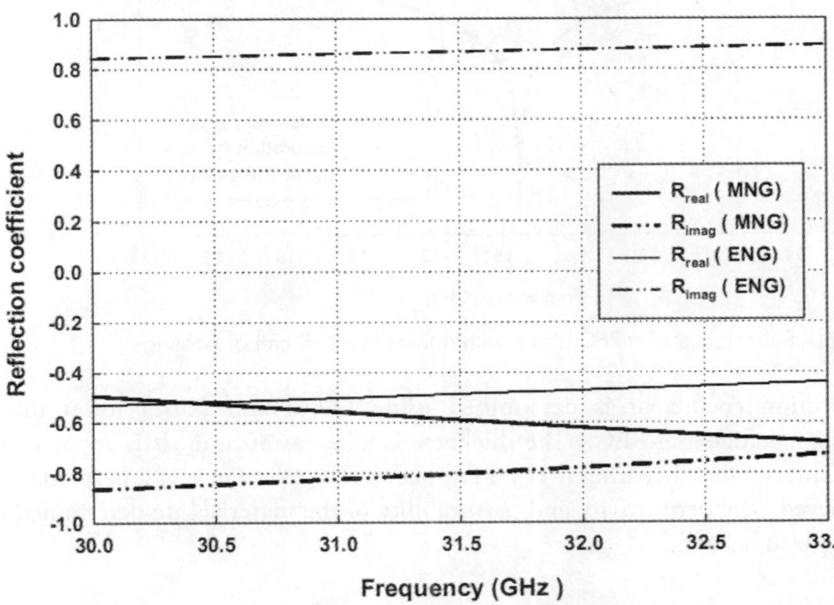

Figure 2.21 Reflection coefficient of a metal-backed metamaterial layer for a normal incidence; thickness of metamaterial layer = 1 mm

PEC coated with a metamaterial: If an EM wave is incident on a PEC backed thin layer of dispersive material, the propagation may be typically characterised as shown in Figure 2.20. The thickness of metamaterial coating is taken as 1 mm. First, the dispersive layer is assumed to be of a metamaterial composed of thin wires. The reflection coefficient is computed for a three-layered configuration (air-metamaterial-PEC) using (2.66) and (2.67). The dispersion parameters are taken as f_{ep} = 42.9 and γ_e = 0.001.

Next, the coating is taken as an MNG material made of SRRs with f_{ep} = 42.9, f_{mo} = 0.5 and γ_e = 0.001. The computed reflection coefficients of the above-mentioned cases (*c.f.* Oraizi and Abdolali 2010) are shown in Figure 2.21. It may be observed that the imaginary parts of the coefficients are negative of each other.

Figure 2.22 shows the reflectance for a PEC plane coated with a dispersive material, for normal incidence at Ku band. Results are shown for different dispersive coatings of both dielectric materials and metamaterials (rods and rings) (*c.f.* Oraizi and Abdolali 2010).

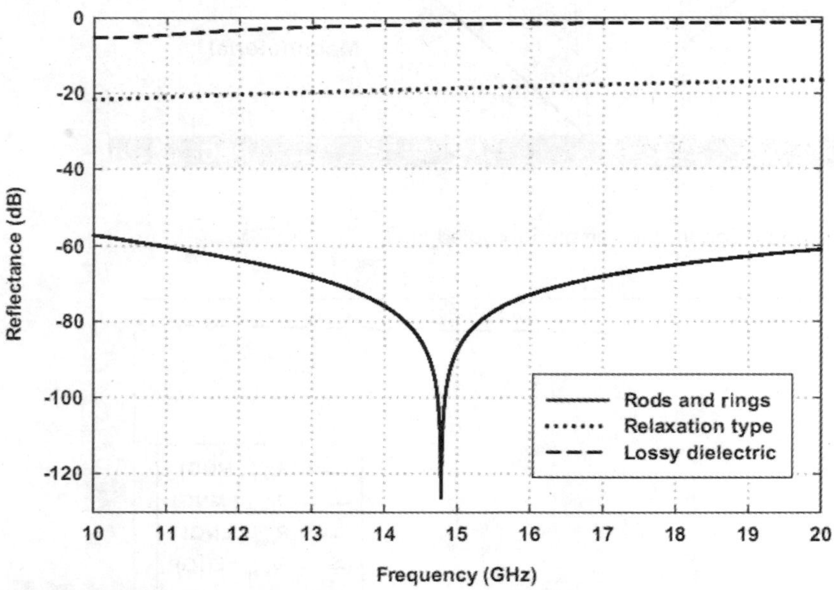

Figure 2.22 Reflectance of a PEC plane with a thick layer (3 cm) of coating

The reflection coefficient is determined using (2.66) and (2.67) for a three-layered medium (air-metamaterial-PEC). The thickness is taken as 30 mm. It is apparent that with appropriate metamaterial coating over a PEC plane, zero reflectance at a particular frequency can be achieved. The permittivity and permeability of the materials are determined using the expressions given below:

Lossy dielectric: $$\varepsilon = \frac{\varepsilon_r}{f^{\alpha}} - j\frac{\varepsilon_i}{f^{\beta}}; \quad \mu = \mu_r$$

Relaxation type: $\varepsilon = \varepsilon_r; \quad \mu = \dfrac{\mu_m\left(f_m^2 - jf_m f\right)}{f^2 + f_m^2}$

Rods and rings: $\varepsilon = 1 - \dfrac{f_{ep}^2}{f^2 - jf\gamma_e}; \quad \mu = 1 - \dfrac{f_{mp}^2 - f_{mo}^2}{f - f_{mo}^2 - jf\gamma_m}$

Two-layered planar metamaterial media: Here, a two-layered planar structure is considered. These two layers are first taken as DPS non-dispersive materials with thickness of 6 mm and 7 mm and $\varepsilon_1 = 5 - j0.2$, $\mu_1 = 1 - j0.1$, $\varepsilon_2 = 3 - j0.1$, $\mu_2 = 4 - j0.5$ (Figure 2.23). The real and imaginary part of reflection coefficient is calculated using (4) and (5) for a four-layered media configuration. As a second case, the DPS layers of same thicknesses are replaced by DNG with $\varepsilon_1 = -5 - j0.2$, $\mu_1 = -1 - j0.1$, $\varepsilon_2 = -3 - j0.1$, $\mu_2 = -4 - j0.5$ (Figure 2.24).

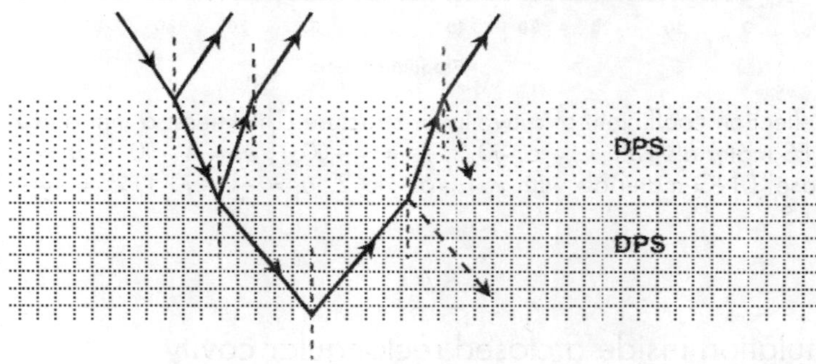

Figure 2.23 Wave propagation in a two-layered DPS structure

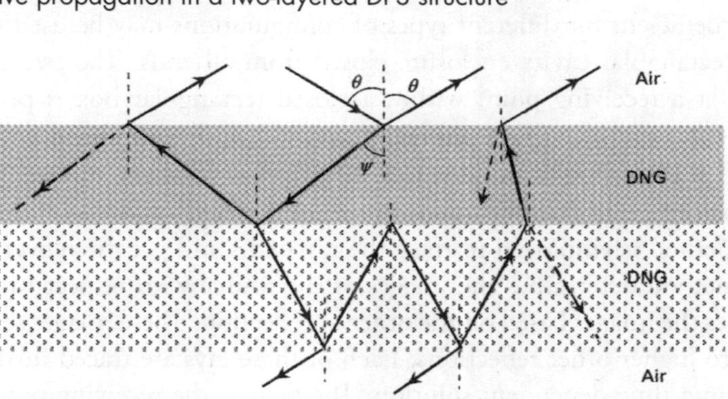

Figure 2.24 Wave propagation in two-layered DNG structure

Figure 2.25 shows the computed reflection coefficient over the frequency range (1 – 100 GHz) (*c.f.* Oraizi and Abdolali 2010). It may be observed that the real part of reflection coefficient remains the same when the DPS layers are replaced with DNG layers. The imaginary part of the reflection coefficient of the DNG structure is found to be a mirror image of DPS case.

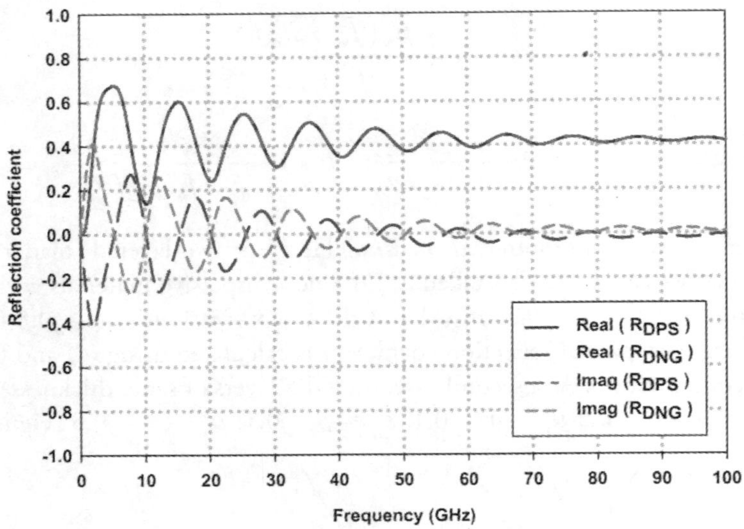

Figure 2.25 Reflection coefficient of a two-layer structure. Thickness: 6 mm and 7 mm. First case: DPS materials with ε_1 = 6 −j0.2, μ_1 = 1 − j0.1, ε_2 = 4 − j0.1, μ_2 = 5 − j0.5. Second case: DNG materials with ε_1 = −6 − j0.2, μ_1 = −1 − j0.1, ε_2 = −4 − j0.1, μ_2 = −5 − j0.5

2.3.2 RF simulation inside a closed rectangular cavity

The reflection coefficient for different types of configurations may be used to analyse the RF field within a rectangular cavity/enclosure closed from all ends. The estimation of building up of RF field at a receiving point within a closed rectangular box is presented here. The dimensions of the rectangular box are taken as 180 cm × 160 cm × 140 cm. The source (0,0,0) is placed at the centre of the sidewall, 20 cm away from the surface. The receiver point P(0,1.45,0) is placed exactly on the opposite wall of the box.

The image method is employed to trace the ray paths and get the reflection points within the box. This method determines the images of the source with respect to a reflecting plane and distinguishes the images based on the order of reflection. These images serve as a secondary source leading to higher order reflections. Each of these rays are traced further and separated in time to obtain a time-dependent solution. The field at the receiving point is estimated by summing up the individual contributions from the transmitter and all the images (Choudhury et al. 2013). The field build-up at the receiver point is determined against the propagation time and, hence, the number of bounces. The normalised RF field build-up in a closed metallic rectangular box (σ =10^4 S/m) is shown in Figure 2.26. Both parallel and perpendicular polarisations are considered.

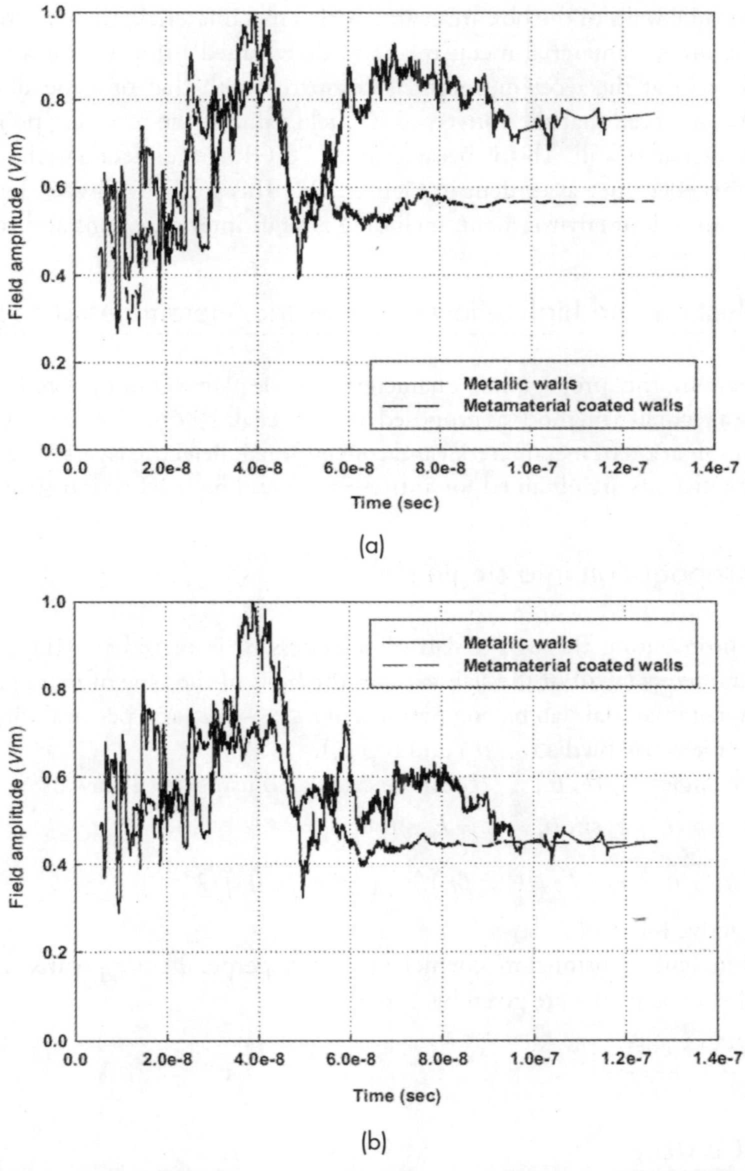

Figure 2.26 RF field build-up inside a rectangular box up to 20 bounces. Frequency = 15 GHz; metallic walls; $\sigma = 10^4$ S/m; metamaterial-coated walls; thickness = 30 mm (a) Perpendicular polarisation (b) Parallel polarisation

The operational frequency is taken to be 15 GHz. The medium inside the metallic box is assumed to be free space. A half-wave dipole antenna is taken as a transmitting element in the box. It may be observed from Figure 2.26 that the convergence of the RF field is not achieved. This is as per expectations due to the presence of metallic walls.

Next, the metallic walls of the box are coated with a metamaterial. The reflection coefficient of the wall, i.e. air-metamaterial-metal media is determined using (2.66) and (2.67). The normalised RF field at the receiving point is compared with that of a metallic box (Figure 2.26). It may be observed that the converged RF field value at the receiving point is lower for the metamaterial-coated wall. This is because at 14.78 GHz, the metamaterial configuration shows minimum reflectance as evident in Figure 2.22. This study has a wide scope in the RF field analysis in an indoor environment, including aircraft interior, cabins and cockpits.

2.4 Antireflection and high-reflection dielectric/metamaterial coatings

In this sub-section, the propagation characteristics of planar multilayered structures are discussed using a recursive method (as proposed by Cory et al. 1993). The multilayered structure is taken as a combination of metamaterial and conventional dielectric layers. The reflection and transmission coefficients are obtained for antireflection and high-reflection structures.

2.4.1 EM propagation in a single slab

An EM wave propagating through a slab of thickness d_2 is considered (Figure 2.27). The constitutive parameters (ε, μ) of the slab vary on the basis of the type of material chosen. It is assumed that a metamaterial slab having permittivity $\varepsilon_2 = -|\kappa|\varepsilon_0$ and permeability $\mu_2 = -\mu_0$ lies in between two dielectric media (ε_1, μ_0) and (ε_3, μ_0).

The interface angles θ_1, θ_2, θ_3, ... θ_N can be calculated using Snell's law as

$$k_1 \sin \theta_1 = k_2 \sin \theta_2 = k_3 \sin \theta_3 = \ldots = k_N \sin \theta_N \tag{2.66}$$

$$\text{where } k_i = \omega\sqrt{\varepsilon_i \mu_i} = \omega n_i / c = (2\pi f n_i)/c = \left(2\pi \times \sqrt{\varepsilon_i \mu_i}\right)/\lambda \tag{2.67}$$

n_i is the refractive index of ith layer.

The reflection and transmission coefficients for a perpendicular-polarised wave at the interface of a dielectric media are given by

$$r_{ij}^{\perp} = \frac{k_j \cos\theta_i - k_i \cos\theta_j}{k_j \cos\theta_i + k_i \cos\theta_j} \tag{2.68}$$

$$t_{ij}^{\perp} = \frac{2k_i \cos\theta_i}{k_j \cos\theta_i + k_i \cos\theta_j} \tag{2.69}$$

If a perpendicular-polarised wave propagates through a metamaterial slab ($\varepsilon_2 = -|\kappa|\varepsilon_0$) embedded in dielectric media (Figure 2.27), the reflection/transmission coefficient of the slab is expressed as

$$\rho_2^{\perp} = \frac{r_{21}^{\perp} e^{j2\phi_2} + r_{32}^{\perp}}{1 + r_{21}^{\perp} r_{32}^{\perp} e^{j2\phi_2}} \tag{2.70}$$

$$\tau_2^{\perp} = \frac{t_{21}^{\perp} t_{32}^{\perp} e^{j\phi_2}}{1 + r_{21}^{\perp} r_{32}^{\perp} e^{j2\phi_2}} \tag{2.71}$$

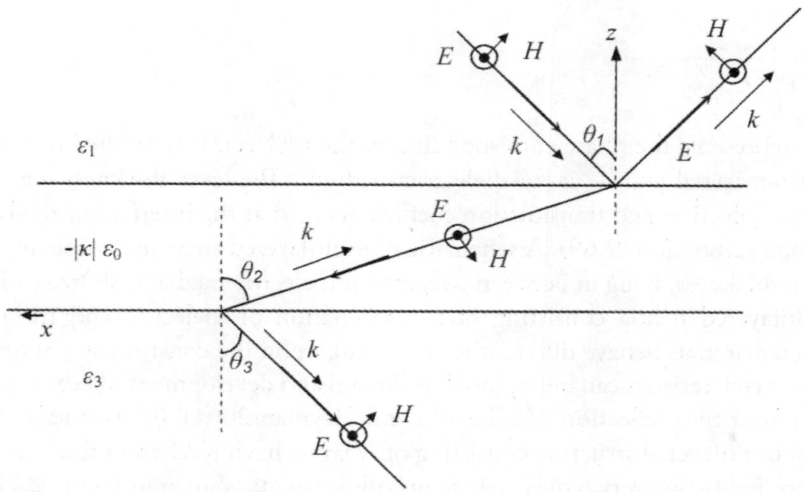

Figure 2.27 EM wave propagation in a metamaterial slab within semi-infinite dielectric media

The reflection and transmission coefficient $(r_{ij}^{\perp}, t_{ij}^{\perp})$ at the interface of the layers can be obtained using (2.68) and (2.69). The phase of EM wave within the slab is given by

$$\phi_i = \pm(\omega/c) n_i d_i \cos\theta_i \tag{2.72}$$

where '+' is for metamaterial slab and '–' denotes dielectric slab.

For a normal incidence, the propagation phase constant, ϕ_i, can be expressed as

$$\phi_i = k \, . n_i \, . \, d_i \tag{2.73}$$

If the optical length, $n_i l_i = \lambda_0/4$, the propagation phase constant is given by

$$\phi_i = \frac{2\pi}{\lambda} \cdot \frac{\lambda_o}{4} \tag{2.74}$$

2.4.2 EM propagation in a multilayered structure

Using iterative calculation of the reflections and transmissions of EM wave through the layers (Cory and Zach 2004), the total reflection and transmission coefficients of a multilayered medium for both perpendicular and parallel polarisations are given by

$$\rho_n^{\perp} = \frac{\rho_{n-1}^{\perp} e^{\pm 2j\phi_n} + r_{n+1,n}^{\perp}}{1 + \rho_{n-1}^{\perp} r_{n+1,n}^{\perp} e^{\pm 2j\phi_n}} \tag{2.75a}$$

$$\rho_n^{\parallel} = \frac{\rho_{n-1}^{\parallel} e^{\pm 2j\phi_n} + r_{n+1,n}^{\parallel}}{1 + \rho_{n-1}^{\parallel} r_{n+1,n}^{\parallel} e^{\pm 2j\phi_n}} \tag{2.75b}$$

$$\tau_n^{\perp} = \frac{\tau_{n-1}^{\perp} t_{n+1,n}^{\perp} e^{\pm j\phi_n}}{1 + \rho_{n-1}^{\perp} r_{n+1,n}^{\perp} e^{\pm j2\phi_n}} \tag{2.76a}$$

$$\tau_n^{||} = \frac{\tau_{n-1}^{||} t_{n+1,n}^{||} e^{\pm j\phi_n}}{1 + \rho_{n-1}^{||} r_{n+1,n}^{||} e^{\pm j2\phi_n}} \tag{2.76b}$$

where ϕ_n represents the phase corresponding to the nth layer. It is recalled that in '±' above, '+' is for metamaterial and '–' is for dielectric medium. The layer thickness may be same or different. The reflection and transmission coefficients (r, t) at the interface of the layers can be obtained using (2.68) and (2.69). A schematic of multilayered structure consisting of N layers having equal thickness, lying in between two similar dielectric media, is shown in Figure 2.28.

The multilayered media consisting of a combination of dielectric and metamaterial or different metamaterials behave differently, depending upon the constitutive parameters of the layers. These characteristics can be explored for design and development of coatings with either high-reflection or zero reflection (Ziolkowski and Heyman 2001), (Cory and Zach 2004). A schematic of multilayered structure consisting of N layers having different thicknesses d_1, d_2, d_3 … d_N are inserted between two dielectric semi-infinite media shown in Figure 2.29.

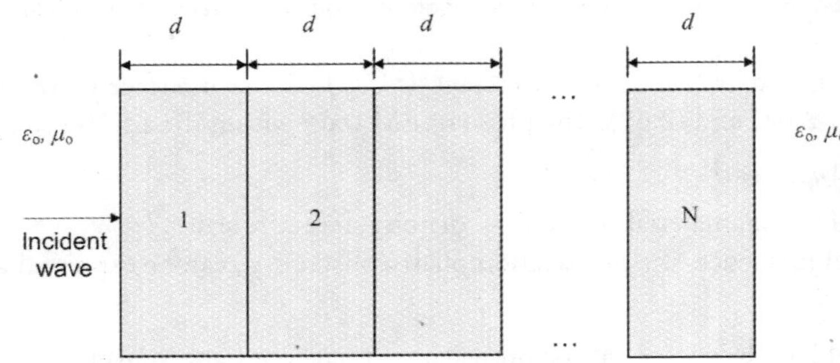

Figure 2.28 Dielectric slabs having same thickness, embedded in air

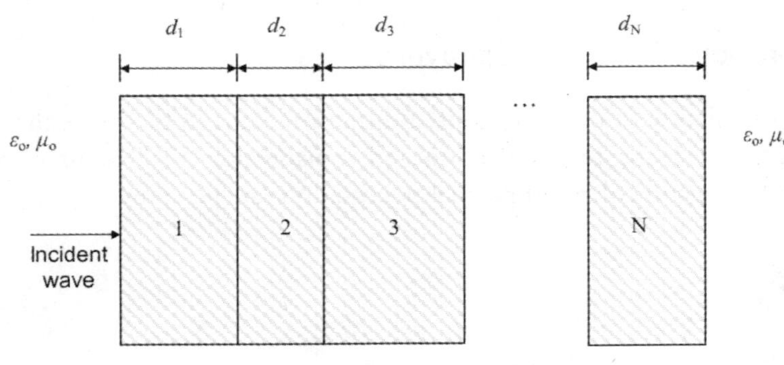

Figure 2.29 Dielectric slabs of different thicknesses, embedded in air

2.4.3 Antireflection coatings consisting of dielectrics

Antireflection coatings are coatings for which the reflection coefficient vanishes when applied to a structure. A dielectric slab of arbitrary thickness can be made into an 'anti-reflecting slab' by putting quarter-wave thick coatings on both sides. Particularly, for a half-wavelength slab, the medium should be same on either side of the slab (Oraizi and Abdolali 2010).

Figure 2.30(a) shows a semi-infinite dielectric medium, comprising air and glass. In Figure 2.30(b), a dielectric coating is inserted between these semi-infinite media. The reflection coefficient of these structures with different dielectric coatings is computed using (2.70), (2.75) and (2.76) in the visible wavelength region (400 nm to 700 nm). The cut-off wavelength λ_o is taken as 550 nm (*c.f.* Orfanidis 2002). Figure 2.31 shows the variation of reflection coefficient of magnesium fluoride (MgF_2) and cryolite (Na_3AlF_6) coatings having approximate refractive indices 1.4 and 1.2, respectively.

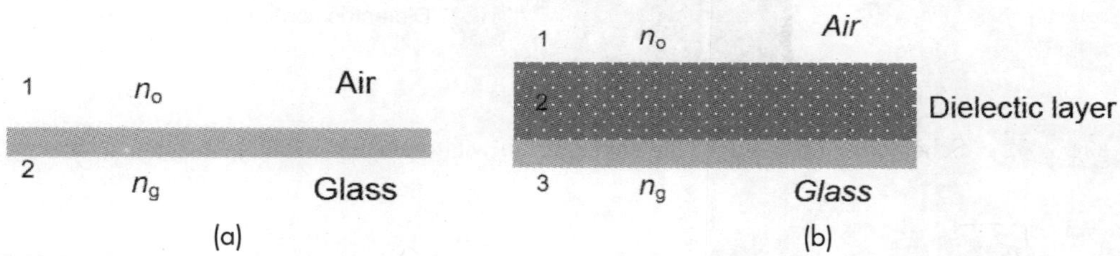

Figure 2.30 (a) Air–glass media, (b) Air–glass media with a dielectric layer inserted in between

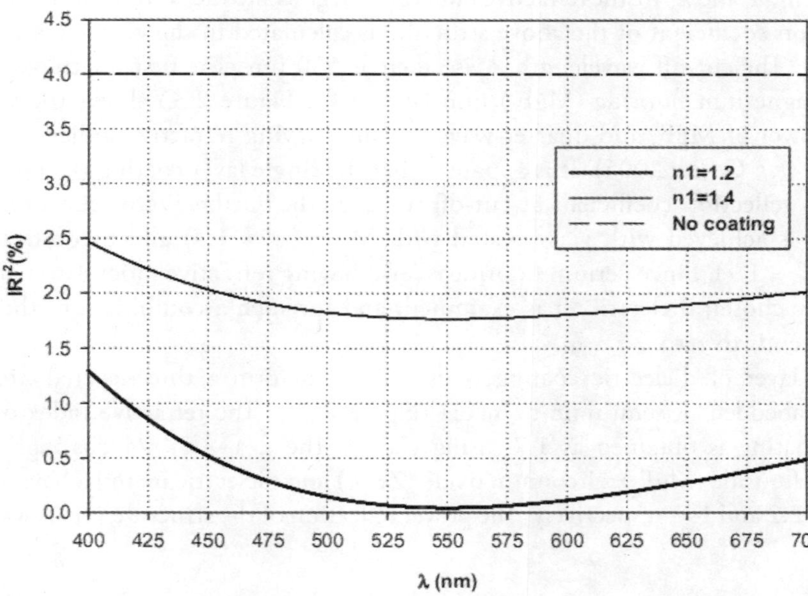

Figure 2.31 Reflection coefficient for an antireflection coating on glass

It may be noted that the reflection coefficient of the uncoated glass is non-zero and constant. However, the reflection coefficient of glass coated with cryolite (Na_3AlF_6) having refractive index value 1.2 becomes zero at cut-off wavelength (550 nm). Thus, cryolite acts as an antireflection coating at 550 nm. This is not true for the glass coated with magnesium fluoride (MgF_2) having refractive index of value 1.4. The reflection coefficient remains non-zero.

In Figure 2.32, two dielectric coating layers are embedded between these semi-infinite media. To design the structure as an antireflection coating, the appropriate value of the refractive index of the second dielectric coating is to be determined using

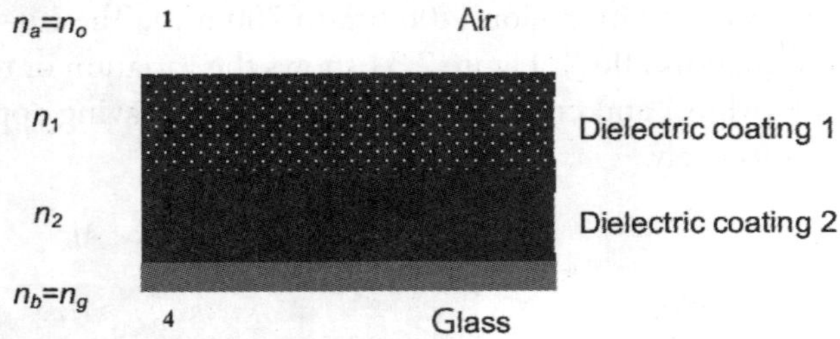

Figure 2.32 Schematic of four–layer dielectric media with dielectric coating

$$n_2 = \sqrt{\frac{n_1^2 n_b}{n_a}} \tag{2.77}$$

where n_a, n_b, n_1 and n_2 are the refractive indices of air, glass and dielectric coatings, respectively. The reflection coefficient of the above structure is calculated in the visible region (400 nm to 700 nm). The cut-off wavelength λ_o is taken as 550 nm. The first quarter-wave layer is taken as magnesium fluoride (MgF_2) film (n_1 = 1.4). Figure 2.33 shows the variation of reflected power of MgF_2 and quarter-wave coatings having refractive indices 1.4 and 1.7, respectively (*c.f.* Cotuk 2005). It is apparent that the single layer coating (n_1 = 1.2) on glass gives a zero reflection coefficient at cut-off wavelength. Further, zero reflection at cut-off wavelength is achieved with glass coated with MgF_2 (n_1 = 1.4) and theoretically chosen dielectric (n_2 = 1.7). Since cerium fluoride (CeF_3) having refractive index 1.6 is close to this theoretically chosen dielectric, it is examined; and for such a combination, the reflection coefficient tends to zero.

Another layer of dielectric coating is used next to form a three-layered $\lambda/4$–$\lambda/2$–$\lambda/4$ structure embedded in semi-infinite media (Figure 2.34). The refractive index of the third dielectric coating is obtained as 1.7, using (2.77). The $\lambda/4$–$\lambda/2$–$\lambda/4$ coatings consists of magnesium fluoride (MgF_2), zirconium oxide (ZrO_2) and dielectric material having refractive indices 1.4, 2.2 and 1.7, respectively. The power reflection of the structure versus wavelength is presented in Figure 2.35. The computed results (*c.f.* Orfanidis 2002) show that the reflection coefficient of the structure vanishes at the cut-off wavelength and, hence, can be used as an antireflection coating.

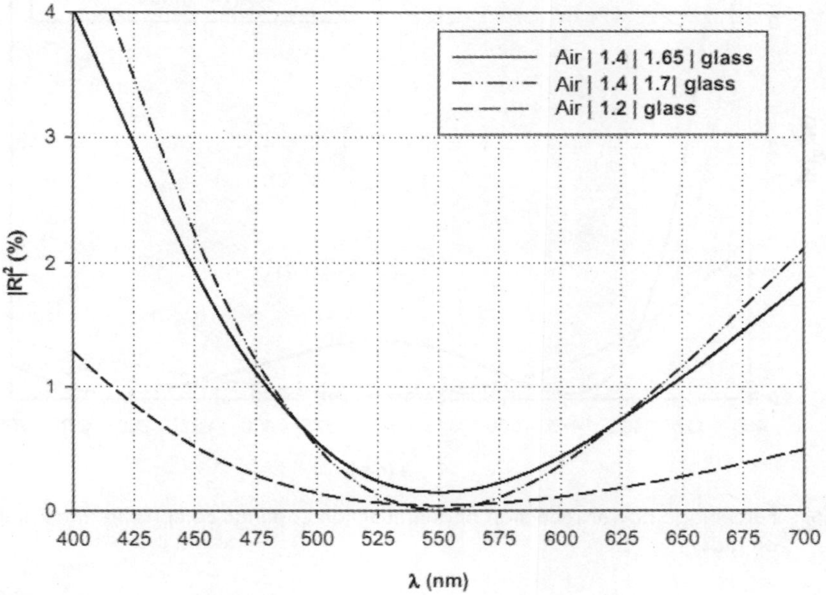

Figure 2.33 Percentage reflected power of antireflection coatings comprising two dielectric slabs on glass

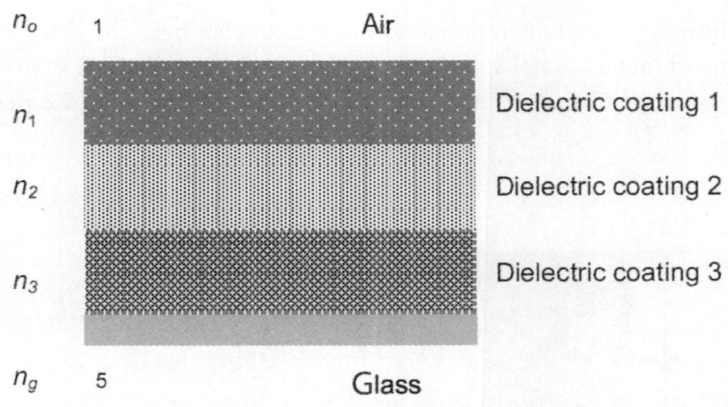

Figure 2.34 Schematic of four-layer dielectric media with $\lambda/4 - \lambda/2 - \lambda/4$ coatings

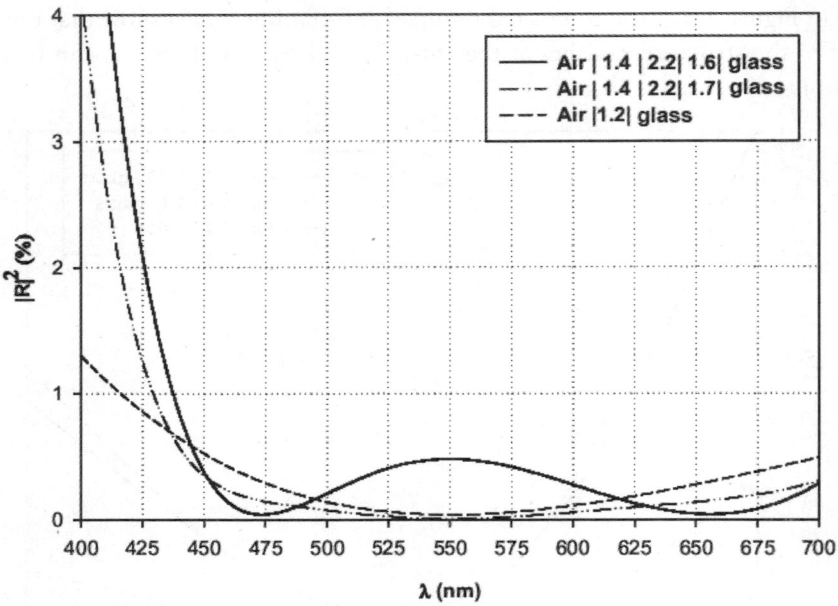

Figure 2.35 Percentage power reflection of antireflection coatings comprising three dielectric slabs on glass

2.4.4 Antireflection coatings consisting of metamaterials

This subsection introduces metamaterials as antireflection coatings. The design of antireflection coatings consisting of metamaterial and dielectric slabs embedded in air is shown in Figure 2.36.

The slabs are assumed to have same thickness (d = 0.1 m) but opposite permittivities $n_2 = |\kappa|\varepsilon_0$, $n_3 = -|\kappa|\varepsilon_0$ $(\varepsilon_{r_1} = 4, \mu_{r_1} = 1; \varepsilon_{r_2} = -4, \mu_{r_2} = -1)$.

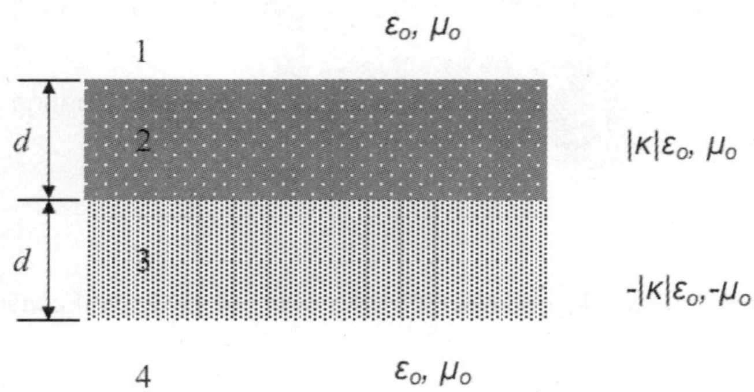

Figure 2.36 Dielectric-metamaterial slabs embedded in semi-infinite media

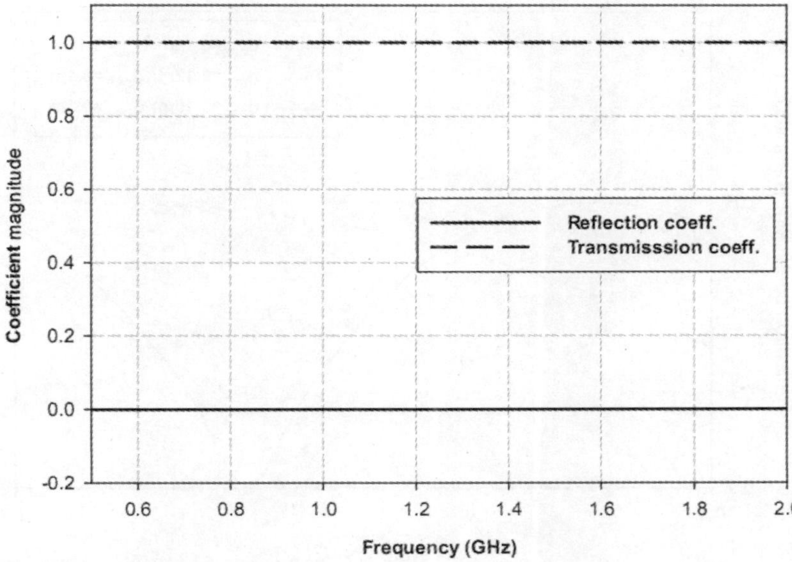

Figure 2.37 Reflection and transmission coefficients of a pair of dielectric-metamaterial slabs embedded in air versus frequency (normal incidence)

The variation of propagation coefficients at normal incidence for the above configuration shows that the reflection and transmission coefficient values are 0, and 1, respectively, over 0.5–2 GHz (Figure 2.37). This is a special case when the thickness of dielectric and metamaterial layers is same and the constitutive parameters (ε, μ) are equal and opposite in sign.

If the thickness of dielectric layers is not same, then the values of reflection and transmission coefficients will no more remain 0 and 1, respectively. This is demonstrated in Figures 2.38 and 2.39, respectively. The thickness of dielectric layer is taken as d = 0.1 m, 0.12 m and 0.15 m, keeping the thickness of metamaterial layer constant at d = 0.1 m. The constitutive parameters are taken same as in Figure 2.37.

If the layers are of different thicknesses, the reflection coefficient (Figure 2.38) becomes zero only for a particular frequency, e.g. with d = 0.15 m, reflection coefficient is zero at 1.5 GHz. If the frequency is increased further, the reflection coefficient will again attain zero value at another particular frequency. This repeats periodically. The same trend is valid for the transmission (Figure 2.39).

The variation in the coefficients of a two-layer dielectric-metamaterial media (Figure 2.36) for a perpendicularly polarised wave is shown next for different angles of incidence (Figure 2.40). The frequency is taken as 1 GHz. The thickness of dielectric and metamaterial (DNG) is kept same, i.e. d = 0.1 m. It may be observed that the reflection and transmission coefficient values vary between 0 and 1, respectively, from 0° to 90°. This holds for parallel polarisation as well.

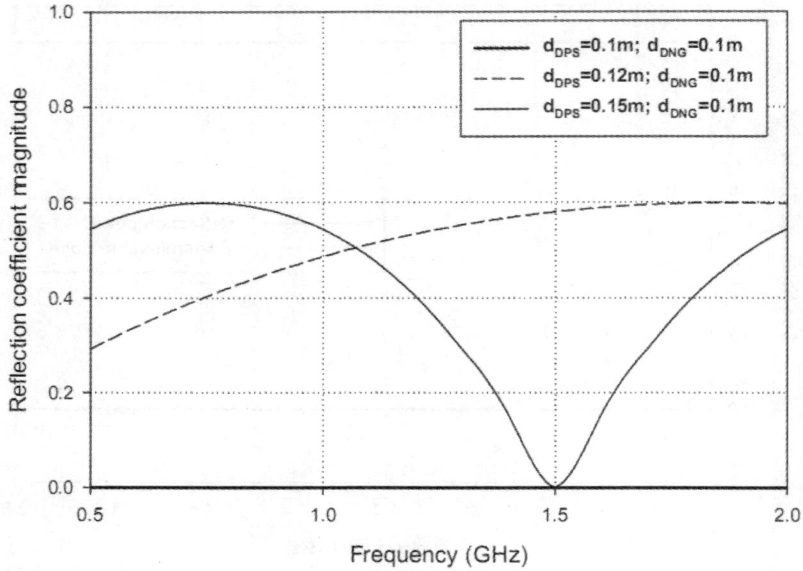

Figure 2.38 Reflection coefficient of two-layer dielectric (DPS)–metamaterial (DNG) slab for different thicknesses of dielectric layer (DPS) (normal incidence)

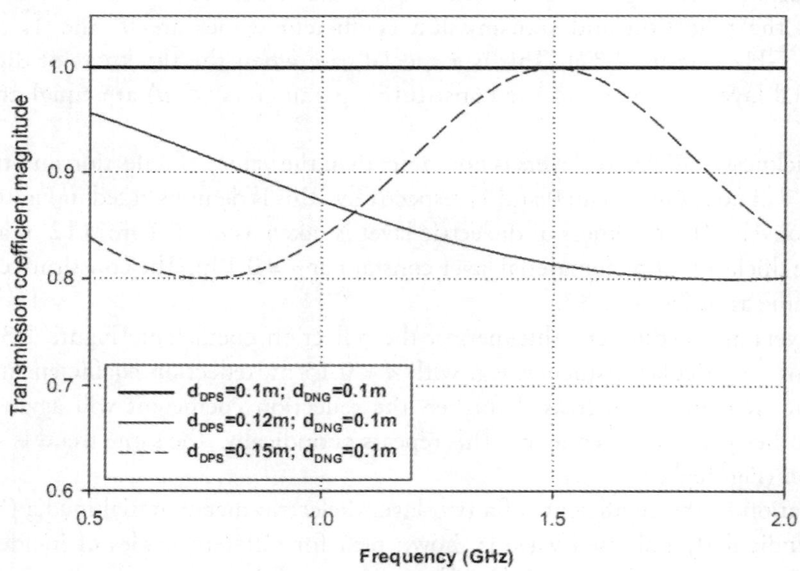

Figure 2.39 Transmission coefficient of two-layer dielectric (DPS)–metamaterial (DNG) slab for different thicknesses of dielectric layer (DPS) (normal incidence)

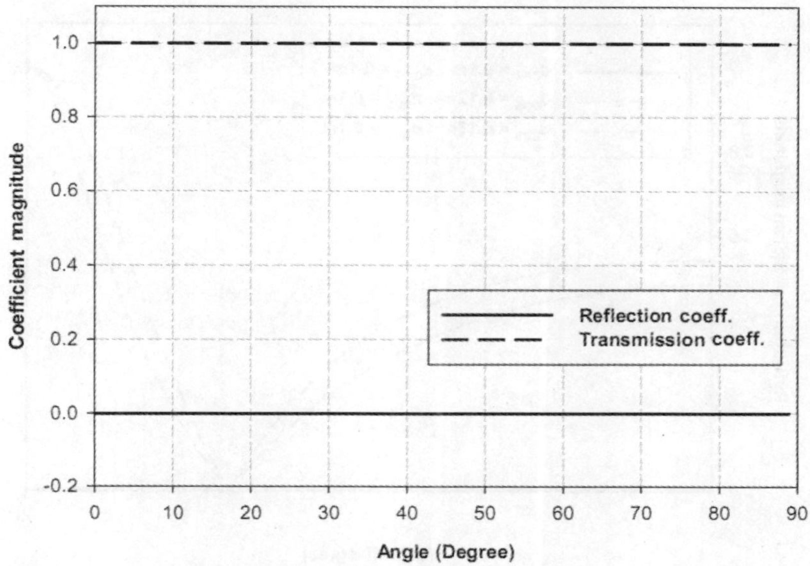

Figure 2.40 Reflection and transmission coefficients of a pair of dielectric and metamaterial slabs embedded in air versus angle of incidence (for perpendicular polarisation)

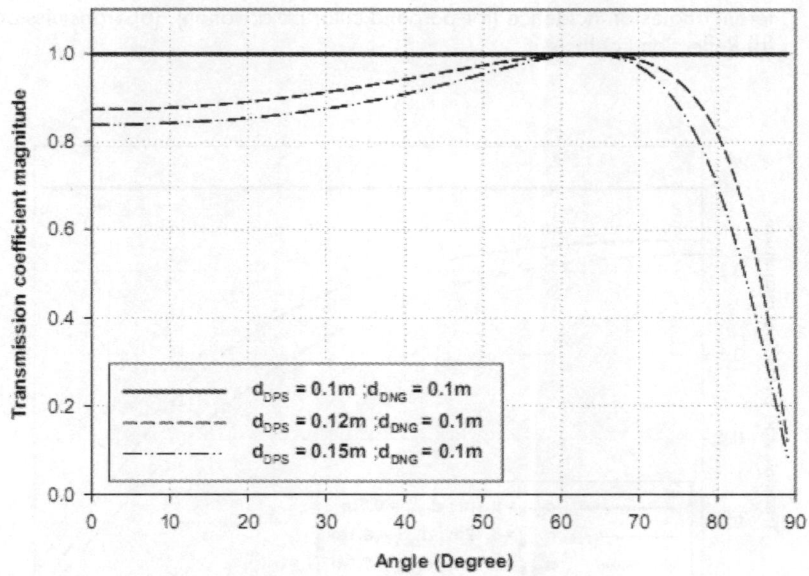

(a)

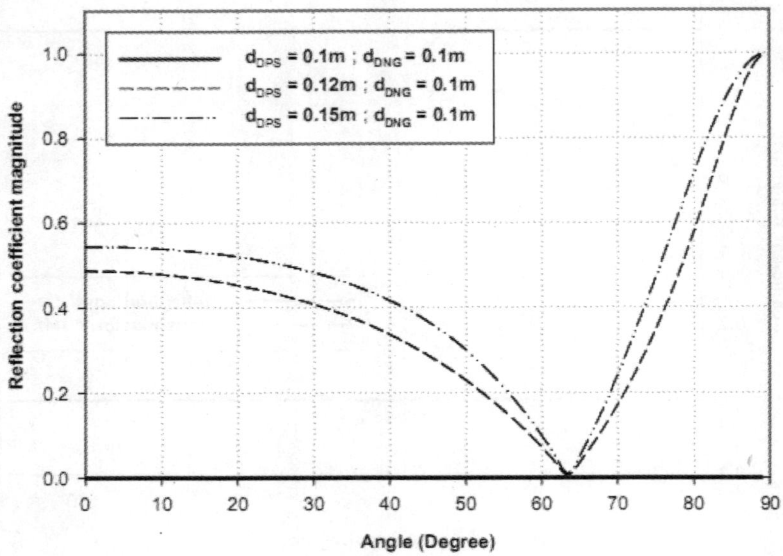

(b)

Figure 2.41 EM propagation through two-layer dielectric (DPS)–metamaterial (DNG) slab for different angles of incidence (for perpendicular polarisation). (a) Transmission coefficient (b) Reflection coefficient

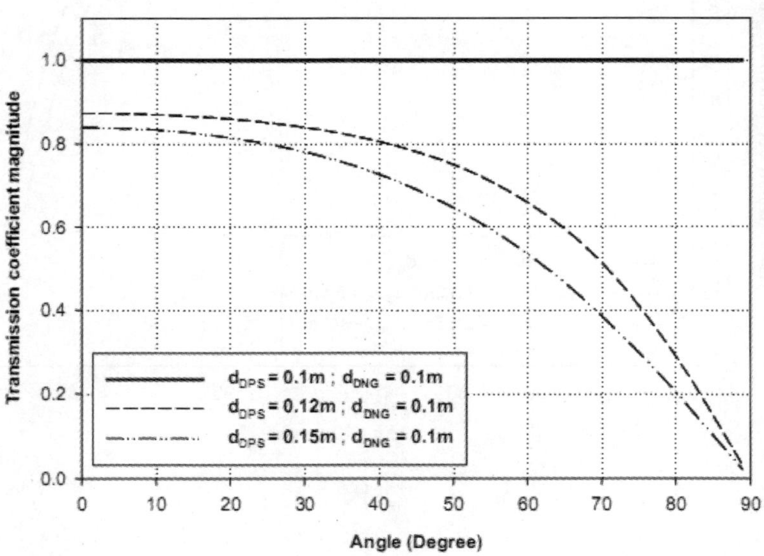

(a)

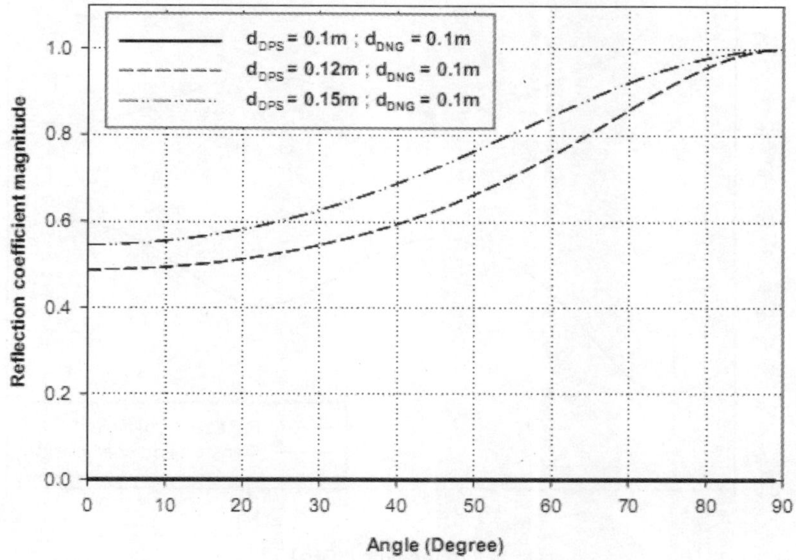

(b)

Figure 2.42 EM propagation through two-layer dielectric (DPS)–metamaterial (DNG) slab for different angles of incidence (for parallel polarisation). (a) Transmission coefficient (b) Reflection coefficient

If the thickness of dielectric layer is varied, the effect on the trend of coefficients with the angle of incidence will be different for perpendicular and parallel polarisations, as shown in Figures 2.41 and 2.42. The reflection and transmission coefficients are 0 and 1, respectively, only at 63°, for perpendicular polarisation. However, in case of parallel polarisation, the coefficients never attain these values at any angle of incidence (0° through 90°).

The constitutive parameters of dielectric and metamaterial layers were so far taken equal and opposite, i.e. (dielectric: $\varepsilon_{r_1} = 4, \mu_{r_1} = 1$; DNG: $\varepsilon_{r_2} = -4, \mu_{r_2} = -1$). This might not always be the case. Thus, the constitutive parameters of dielectric layer is changed to (ε_r = 2.4, μ_r = 1; d = 0.1 m) keeping metamaterial same (DNG: ε_r = –4, μ_r = –1; d = 0.1 m).

Figure 2.43 shows the coefficients for normal incidence for such configuration. It may be observed that both reflection and transmission have oscillatory behaviour. However, these coefficients are not mirror images of each other, owing to the material characteristics.

Figure 2.44 shows the variation of reflection and transmission coefficients with the incident angle for different thicknesses of dielectric layer ($\varepsilon_{r_1} = 2.4, \mu_{r_1} = 1$; d = 0.12 m) and metamaterial, ($\varepsilon_{r_2} = -4, \mu_{r_2} = -1$; d = 0.1 m). This establishes the dependence of the coefficients on the constitutive parameters of the medium.

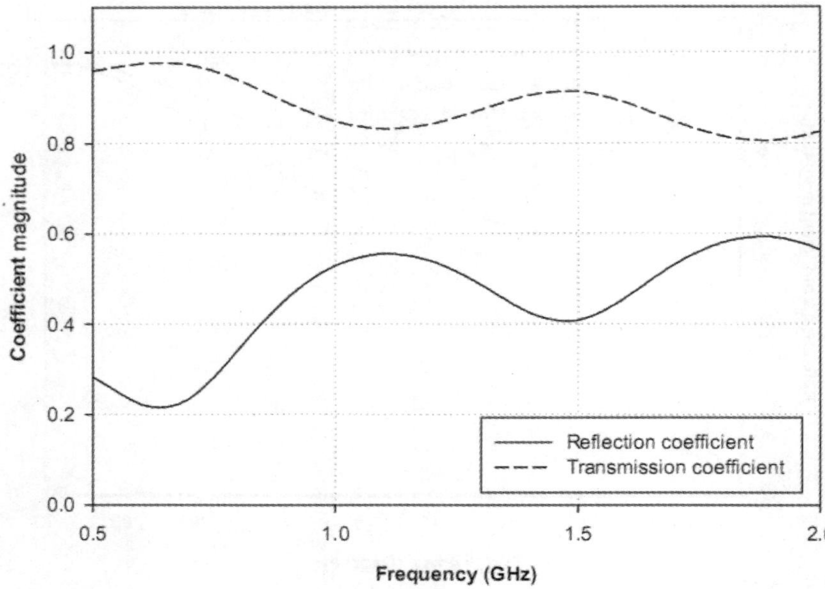

Figure 2.43 Coefficients of two-layer dielectric (DPS)–metamaterial (DNG) slab at normal incidence. DPS: $\varepsilon_r = 2.4$, $\mu_r = 1$; $d = 0.1$ m, DNG: $\varepsilon_r = -4$, $\mu_r = -1$; $d = 0.1$ m

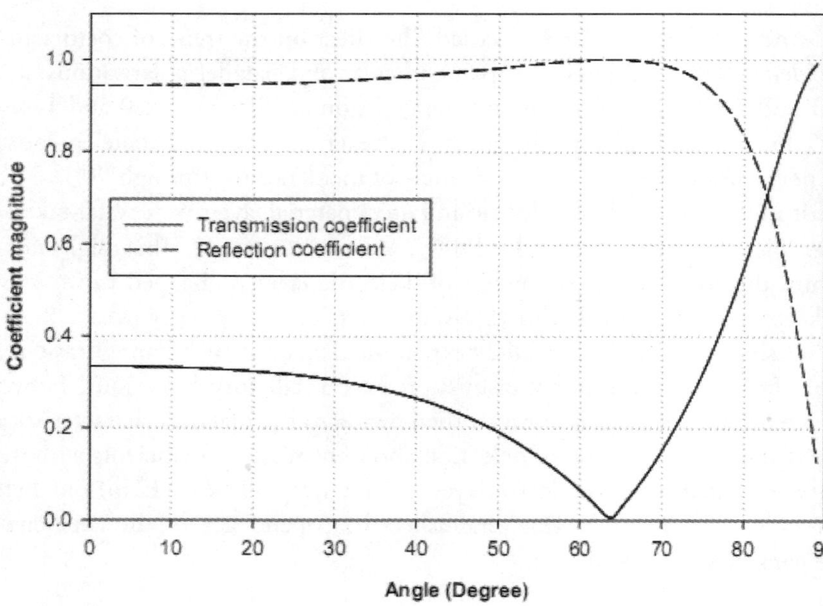

(a)

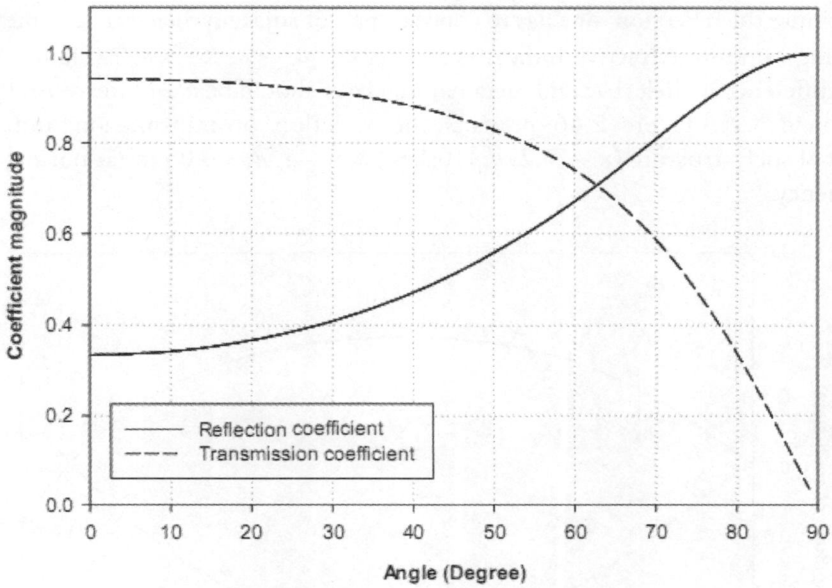

(b)

Figure 2.44 Coefficients of two-layer dielectric (DPS)–metamaterial (DNG) slab. DPS: $\varepsilon_r = 2.4$, $\mu_r = 1$; $d = 0.12$ m, DNG: $\varepsilon_r = -4$, $\mu_r = -1$; $d = 0.1$ m. (a) Perpendicular polarisation (b) Parallel polarisation

2.4.5 High-reflection coatings using dielectric and metamaterial

If a pair of slabs with opposite constitutive parameters but same thickness is considered, the overall reflection coefficient will either be zero (for identical initial and final media) or depend upon the angle of incidence and the refractive indices of the medium (for non-identical initial and final media) (Cory and Zach 2004). High-reflection coatings can be realised by choosing metamaterial and dielectric slabs appropriately.

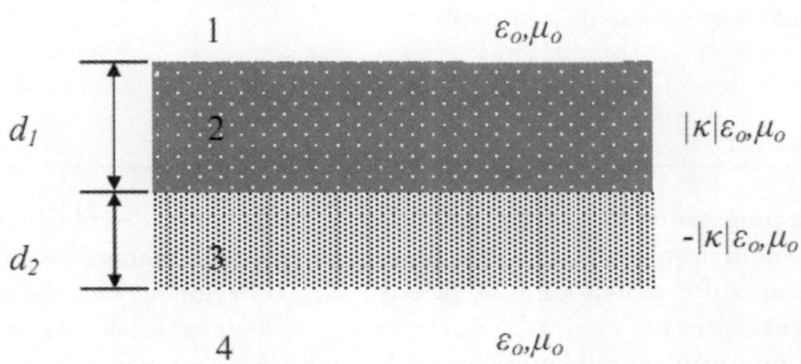

Figure 2.45 Pair of dielectric-metamaterial slab with opposite refractive indices

To maximise the reflection, one has to choose a pair of adjacent dielectric and metamaterial layers having opposite refractive indices (i.e. $\varepsilon_{r_4} < \varepsilon_{r_3}$; $\varepsilon_{r_3} > \varepsilon_{r_2}$; $\varepsilon_{r_2} < \varepsilon_{r_1}$) (Figure 2.45), and the phase difference in dielectric and metamaterial layer should be $\pi/2$ at the central frequency (normal incidence). Figure 2.46 presents the variation of transmission and reflection coefficient of such structure ($\varepsilon_{r1} = 0.2$, $d_1 = 0.1$ m; $\varepsilon_{r2} = -4$, $d_2 = 0.05$ m) at normal incidence with frequency.

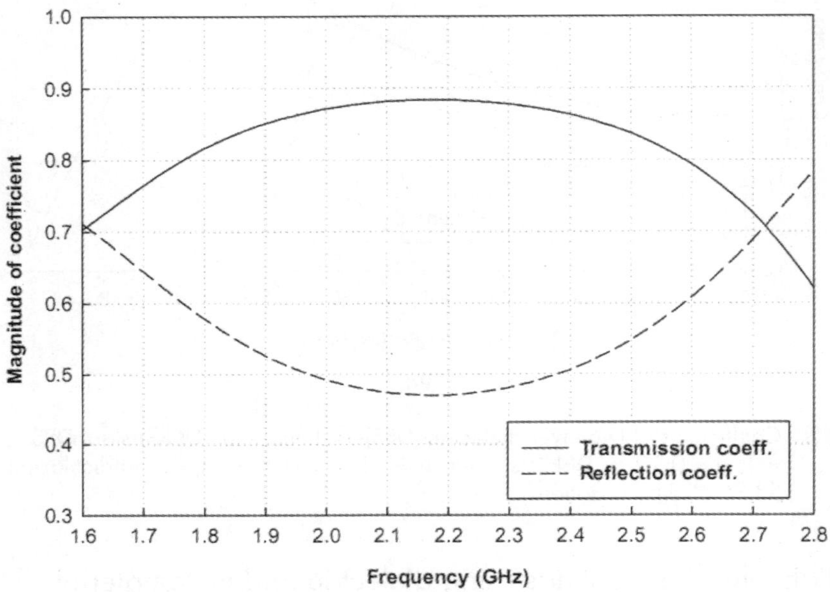

Figure 2.46 Reflection and transmission coefficients of a pair of dielectric and metamaterial slabs embedded in air

The central frequency is taken as 2.2 GHz. The maximum value of reflection coefficient is 0.88, while a minimum transmission coefficient of 0.47 is observed at the central frequency. Thus, one may achieve controlled transmission/reflection characteristics by using an appropriate combination of metamaterial–dielectric slabs.

2.5 Conclusion

The EM wave propagation in multilayered dielectric media is first described on the basis of the reflection and transmission coefficients for multilayered semi-infinite/finite medium. The results are discussed for various kinds of dielectric media. Beginning with the reflection and transmission coefficient for a normally incident wave at the interface of semi-infinite media, the EM wave propagation is discussed for lossless and absorbing media. It is observed that the reflection coefficient displays an oscillatory trend with variation in the thickness of the medium. For lossy media, the extent of damping depends on the number of layers and the constitutive parameters of the material. This concept is extended further to the estimation of transmission/reflection coefficients of multilayered dielectric-metamaterial media.

The EM wave propagation in a multilayered dielectric-metamaterial medium shows that a metamaterial layer offers significant reduction in reflectance in a particular frequency range. The simulations are discussed for different configurations like metal-backed metamaterial, DPS–DPS, DNG–DNG, etc. The real and imaginary components of reflection coefficient for ENG, MNG and DNG metamaterials establish that the real component of the reflection coefficient overlap for DPS and DNG layers. Whereas, the imaginary part of reflection coefficient forms a pair of mirror images. Further, the reflection coefficients of ENG and MNG layers over a PEC plane are shown to be complex conjugates of each other. This study has potential application in the analysis of RF field build-up within an enclosure such as a metallic enclosure coated with a metamaterial. The field strength at the receiver inside the box converges at lower amplitude on using a metamaterial coating over the walls. This is due to the minimum reflectance offered by the metamaterial coating at the specified frequency. This concept may be exploited for EM environment analysis in aircraft interiors, cabins and cockpits.

EM propagation through multilayered dielectric-metamaterial slab structures is described next in view of antireflection and high-reflection coatings. A dielectric slab of appropriate refractive index embedded between two semi-infinite media can act as an antireflection coating at a particular frequency (cut-off frequency). As the number of the dielectric layers is increased, the bandwidth over which minimum reflection can be achieved also increases.

Metamaterials has a potential application as antireflection/high-reflection structures. It is shown that a multilayered structure consisting of alternating pairs of dielectric-metamaterial slabs having same width but opposite permittivity acts as antireflection coatings. The overall reflection coefficient of the structure either vanishes or depends on the angle of incidence and the refractive index of the medium. These antireflection structures have strategic applications, e.g. in the design of antenna/radomes. In contrast, high-reflection coatings can be achieved by using a pair of dielectric-metamaterial layers with different refractive indices. The total reflection coefficient for such a structure is maximum in a particular frequency range. Such dielectric-metamaterial layers with proper design and constitutive parameters would serve as RAM for low-observable platforms.

References

Brekhovskikh, L. M., D. Lieberman, and R. T. Beyer. 1965. *Waves in Layered Media.* New York: Academic Press, 561.

Choudhury, B., H. Singh, J. P. Bommer, and R. M. Jha. 2013. 'RF field mapping inside large passenger aircraft cabin using refined ray-tracing algorithm.' *IEEE Antennas and Propagation Magazine* 55(1): 276–88.

Cory, H., S. Shiran, and M. Heilper. 1993. 'An iterative method for calculating the shielding effectiveness and light transmittance of multilayered media.' *IEEE Transactions on Electromagnetic Compatibility* 35: 451–56.

Cory, H. and C. Zach. 2004. 'Wave propagation in metamaterial multilayered structures.' *Microwave Optical Technological Letters* 40(6): 460–65.

Cotuk, U. 2005. *Scattering from Multi-Layered Metamaterials using Wave Matrices.* Monterey, CA: Master's Thesis Report, Naval Postgraduate School, 49.

Jordan, E. D. and K. C. Balmain. 1976. *Electromagnetic Waves and Radiating Systems.* Prentice-Hall of India, 753.

Klement, D., J. Preissner, and V. Stein. 1988. 'Special problems in applying the physical optics method for backscatter computations of complicated objects.' *IEEE Transactions on Antennas and Propagation* 36: 228–37.

Knott, E. F., J. F. Shaeffer, and M. T. Tuley. 1985. *Radar Cross Section.* Dedham, MA: Artech House Inc., 462.

Kong, J. A. 2002. 'Electromagnetic wave interaction with stratified negative isotropic media.' *Progress in Electromagnetic Research* 35: 1–52.

Oraizi, H. and A. Abdolali. 2008. 'Design and optimisation of planar multilayer antireflection metamaterial coatings at Ku band under circularly polarised oblique plane wave incidence.' *Progress in Electromagnetic Research C* 3: 1–18.

Oraizi, H. and A. Abdolali. 2010. 'Several theorems for reflection and transmission coefficients of plane wave incidence on planar multilayer metamaterial structures.' *IET Microwaves, Antennas and Propagation* 4: 1870–79.

Orfanidis, S. J. 2002. *Electromagnetic Waves and Antennas.* Piscataway, NJ: Rutgers University, 785.

Pendry, J. B. 2000. 'Negative refraction makes a perfect lens.' *Physical Review Letters* 85(18): 3966–69.

Ruck, G. T., D. E. Barrick, W. D. Stuart, and C. K. Krichbaum. 1970. *Radar Cross Section Handbook.* New York: Plenum Press, 2: 949.

Shelby, R. A., D. R. Smith, and S. Schultz. 2001. 'Experimental verification of a negative index of refraction.' *Science* 292(5514) 77–79.

Veselago, V. G. 1968. 'The electrodynamics of substances with simultaneously negative values of ε and μ.' *Soviet Physics Uspekhi* 10(4): 509–14.

Ziolkowski, R. W., and E. Heyman. 2001. 'Wave propagation in media having negative permittivity and permeability.' *Physical Review E* 64(5): 1–15.

3 Radar Cross Section of Phased Antenna Arrays

3.1 Introduction

Radar cross section (RCS) is the measure of detectability of a target which, in turn, depends on its external features and electromagnetic (EM) properties. It relates the EM energy reflected from the target at the receiver to the incident EM energy coming out from the source. Alternately, it is the power scattered by the target towards an illuminator. The RCS of a target depends on its physical shape, frequency, polarisation properties, orientation, constitutive materials and subcomponents such as antennas and other sensors.

Phased antenna array is an array of antenna elements such as aperture, slot, horn, microstrip patch, spiral, dipole, etc. used to achieve high directivity (Mallioux 1994). It also affects the RCS of a platform on which it is mounted. Thus, for the estimation followed by reduction of RCS of a phased antenna array, scattering for both in-band (i.e. the operational frequency of radar signal) and out-of-band threat frequencies should be considered.

Scattering analysis of phased antenna arrays includes two scattering modes: antenna (or radiation mode) and structural mode (Hansen 1989). The antenna mode occurs when an induced current gets reflected at the antenna feed point and then re-radiated. On the contrary, the structural mode is generated from the antenna surface induced currents. The two modes are difficult to identify and separate, when the array is mounted on a platform (Wang et al. 2010). Further, the radiation and scattering characteristics of an antenna array depend on the relative operational frequency range of an antenna array with respect to that of the radar. If the frequency of the radar signal falls within the operating band of the antenna array, then the impinging signal will match that of antenna elements, and then it will penetrate into the feed structure. This signal results in multiple reflections within the antenna system and hence contributes significantly to RCS of the antenna array (Jenn 1995). Alternately, if the frequency of the impinging signal falls outside the operating band of the antenna array, then there will be no match between the radar signal and that of the antenna elements, and reflections inside the feed will be insignificant.

Defense applications demand design of targets that exhibit the least possible RCS so as to be undetectable even to highly sensitive radars. For such a stealth or low RCS platform, not only the high performance of a phased antenna array (gain, sidelobe level, size and weight) but also achieving the RCS requirements is required. Conversely, it is essential for any active/passive technique to reduce RCS without degrading the performance of the antenna array.

This chapter is geared towards reducing the RCS of a phased antenna array mounted on an aerospace platform to ensure that their signatures do not become predominant (Lu et al. 2009). A typical example is a high-gain antenna on a low RCS platform. Nowadays integrated transmit-receive (T-R) modules or receive-only modules are collocated with antenna elements. Moreover, the self-calibrating and adaptive smart-skin conformal array are the preferred choice for onboard sensors/antennas. This technology has benefits of increased efficiency and reliability, besides low RCS, and thus promotes the use of conformal phased antenna arrays (Kuhn 2011).

The main objective is to estimate the in-band RCS of phased antenna arrays considering only the antenna mode scattering, which is dominant as compared to that of the structural mode of an in-band stand-alone antenna array. In order to achieve this objective, an efficient and accurate RCS model is required so that it can optimise the trade-off between antenna RCS and radiation performance.

3.2 Theoretical background

Antenna scattering has been the topic of interest since 1950s. However, until 1980s it was mainly focused on low-gain antennas. Then attention was shifted to the high-gain antennas because of their potential use on future low RCS platforms. Scattered field of an object is a combination of induced electric and magnetic currents, satisfying the boundary conditions. In fact, antennas with identical amplitude and phase radiation patterns can differ from each other on the basis of the way they scatter.

The radar cross section (RCS) of a given target—an aircraft, a ship, or any other aerospace body—can be attributed to its functioning as that of an antenna system. Therefore, RCS (also known as echo area) is the measure of power scattered in a given direction when a target is illuminated by an incident wave and this can be defined as (Knott et al. 1985)

$$\sigma(\theta_i) = \lim_{R\to\infty} 4\pi R^2 \frac{|W_s(\theta)|}{|W_i(\theta_i)|} = \lim_{R\to\infty} 4\pi R^2 \frac{|E_s(\theta)|^2}{|E_i(\theta_i)|^2}, \tag{3.1}$$

where R is the target-to-receiver distance, $W_s(\theta)$ is the scattered power in the direction of the receiver, $W_i(\theta_i)$ is the incident power, $E_s(\theta)$ is the scattered electric field in the direction of the receiver and $E_i(\theta_i)$ is the incident electric field. The term $\lim_{R\to\infty}$ in (3.1) signifies the limiting approximation for an incident wave to become a plane wave. It can also be inferred from the above fact that computation of RCS essentially finds the electric field scattered from the target, which further determines the current induced on it by the incident plane wave. Only monostatic RCS is considered here, i.e. $\theta = \theta_i$. The unit of RCS is necessarily the same as that of the area, is the square meter or m^2. However, the most commonly used unit is decibels relative to a square metre (dBsm). These units are inter-convertible as per the relation

$$\sigma \text{ in dBsm} = 10 \log\left(\sigma \text{ in m}^2\right). \tag{3.2}$$

Typical values of RCS range from 50 dBsm (10^5 m^2) for ships to –30 dBsm (10^{-3} m^2) for insects as mentioned in Table 3.1.

Table 3.1 Typical RCS Values

Object	**σ (m^2)**	**σ (dBsm)**
Ships and A/c carriers	10000-100000	40-50
Bomber aircraft	1000	30
Fighter aircraft	100	20
Tank	10	1
Human being	1	0
Guns	0.1	-10
Birds	0.01	-20
Insects	0.001	-30

RCS of a target (Gustafsson 2006) primarily depends on (i) frequency and polarisation of the incident wave, transmitter and receiver, (ii) angular orientation and shape of the scattering target (with respect to radar). These factors are to be considered while designing radar, which should detect a specific target, and also the targets, which are intended to face the specific radar. RCS of a target differs distinctly in the three frequency regions into which the incident wave is categorised. These regions are referred to as low, resonance and high-frequency regions, where low and high are defined considering the size of the target in terms of incident wavelength, rather than its physical size.

Low-frequency or Rayleigh region ($kL \ll 1$): In this region, the current induced on the body of a target remains approximately constant in amplitude and phase. At these frequencies, the shape of the body becomes insignificant.

Resonance or Mie region ($kL \approx 1$): Here, the current induced on the body of a target varies both in amplitude and phase. For these frequencies, whole target contributes to the scattering pattern.

High-frequency or optical region ($kL \gg 1$): The scattered field is angle-dependent here because a significant variation in current occurs across the body of the target. At these frequencies, distinct points on the target result in peak scattering levels. Here, k ($= 2\pi/\lambda$) is the wave number, λ represents the wavelength and L is the characteristic length of the target.

The RCS estimation of the phased array can fall into any of the above three regions, depending on the wavelength of the radar relative to the operational frequency band of the antenna. Further, the scattering characteristics of an antenna can be associated with three frequency regions interspersed with the transition bands, as indicated in Figure 3.1.

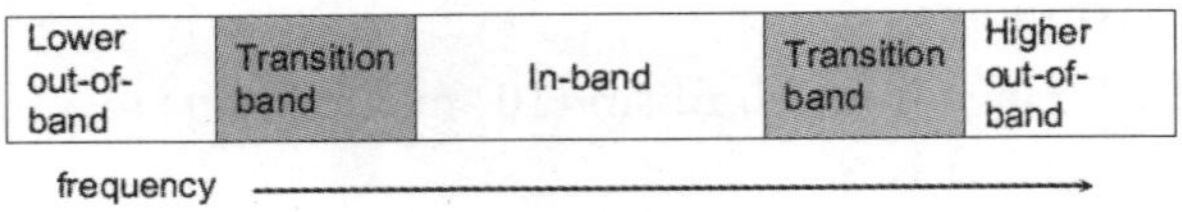

Figure 3.1 Frequency regions for RCS estimation

3.2.1 Antenna scattering

An antenna mounted on a platform has a two-fold impact on the RCS of that platform. It significantly disturbs the continuity of the surface (e.g. nuts and bolts are required to mount an antenna array on the surface of an aircraft, which increases the number of edges). Further, a well-designed antenna has the potential to absorb almost all the energy incident on its operating band. These two factors collectively give rise to wide-angle scattering. Moreover, the radar signal can enter the feed of the antenna system and get reflected at each impedance mismatch. Even if these mismatches were of small magnitudes within the operating band, they may result in multiple scattering sources for a large array. These individual contributions may add constructively. This scattering depends on the feed network and its components, and hence complicates the overall RCS prediction.

The three frequency regions for antenna scattering (Jenn 1995) are described as follows:

Lower out-of-band region: In this region, antenna elements are closely placed and there is no reflection. Moreover, radiating elements are not resonant; hence, negligible amount of energy enters the array feed. If $\lambda >> l$, where l is the element length, scattering from the element itself becomes negligible. The incident wave travels through the element without being disturbed and is reflected by the surfaces behind the aperture such as the ground plane.

In-band region: The antenna elements are well matched in this region; thus, most of the incident energy enters the feed. For a well-designed antenna, the RCS in the operating band should be low. However, there can be few small-level internal reflections, which might add coherently, giving rise to a significant RCS.

Higher out-of-band region: Scattering becomes complicated and difficult to predict in this region. Here, the antenna elements are not matched at most of the frequencies, and the reflections within the feed are, in general, not significant. However, if the probing frequency is a harmonic of the antenna frequency, then antenna elements might be resonant. Furthermore, electrical dimensions of antenna elements can be large and may be placed far apart. This might result in a significant RCS value due to diffraction/scattering according to Bragg's condition $2d\sin\theta = n\lambda$, where d is inter-element spacing. The basic equation of antenna scattering (Lo and Lee 1993), for a linearly polarised antenna, is expressed as

$$\vec{E}^{s}(Z_L) = \vec{E}^{s}(Z_a^{*}) + \left[\frac{j\eta_0}{4\lambda R_a} \vec{h}\left(\vec{h}.\vec{E}^{i}\right) \frac{e^{-jk_o R}}{R} \right] \Gamma_o. \tag{3.3}$$

Here, the antenna port is assumed to be terminated with load Z_L. $\vec{E}^{s}$ is the scattered field, $\vec{E}^{i}$ is the incident field, $Z_a = R_a + jX_a$ is the radiation impedance with $R_a = R_r + R_d$; R_a: antenna resistance, R_r: radiation resistance and R_d: ohmic resistance, Z_L is the load impedance and $\eta_o = 120\pi \approx 377\ \Omega$ is the free space impedance, $\vec{h} = h\hat{x}$ is the effective height of a x-polarised antenna element, R is the target-to-receiver distance and Γ_o is the modified reflection coefficient, expressed as (Wang et al. 2010)

$$\Gamma_o = \frac{Z_L - Z_a^{*}}{Z_L + Z_a^{*}}, \tag{3.4}$$

where * indicates the complex conjugate.

When the load impedance Z_L becomes equal to the complex conjugate of the radiation impedance Z_a (i.e. $Z_L = Z_a^{*}$), the antenna is said to be conjugate matched. Under such a condition, the modified reflection coefficient Γ_o becomes zero. It may be observed from (3.3) that the in-band antenna RCS consists of two antenna scattering modes-structural mode (first term) and antenna or radiation mode (second term). Structural mode is generated by the currents induced on the antenna and platform when the terminating load is equal to the complex conjugate of the antenna impedance and is primarily due to edge effects, *viz.* ground plane edge diffraction and mutual coupling changes near the edges of the array. In contrast, the radiation properties determine the antenna mode, which is proportional to the gain of the antenna in a given direction and also to the modified reflection coefficient Γ_o.

For a well-designed antenna array, both antenna and structural mode scattering need to be small within its operating band. However, for an array with an inter-element spacing of 0.5λ or less, the effect of mutual coupling becomes significant. Due to this, variation of Z_a for any element in the antenna array depends on the angle of arrival of the incident wave. This makes it practically impossible to force Γ_o to zero at every angle simultaneously. Such a mismatch between load and radiation impedance results in antenna mode reflections. Further, for a stand-alone phased array, e.g. an array of dipoles over a finite ground plane, structural mode scattering dominates and hence cannot be ignored.

Usually, the antenna systems are chosen to have a higher gain accompanied with low scattering in order to ensure that they serve the requirements of both the high performance as well as the low RCS. Further, the antenna impedance is conjugate matched with its load impedance in practice, so as to enable maximum power transfer. Although terminating the antennas of a large array by a matched receiver helps in reducing the antenna scattering mode, yet it fails to reduce RCS. This is because, for such a case, the scattered power from the antenna exceeds the absorbed power. Both scattered and absorbed power become equal only for minimum-scattering antennas (Kahn and Kurss 1965). In any case, the scattered power cannot be less than the absorbed power. For example, in a dipole antenna, conjugate matching results in half the power being scattered and the other half being absorbed. This is due to the dominance of its structural scattering mode. This fact shows that no conjugate-matched antenna can absorb more than what it can scatter.

Low RCS of an antenna demands for a slight mismatch between antenna terminals to generate sufficient antenna modes leading to cancellation of its structural mode. While doing so, the gain of the antenna array reduces, which might affect its radiation performance. An antenna can absorb more than what it scatters, only if its gain in the backward direction exceeds that in the forward direction (Green 1966). Thus, all high-performance phased arrays should necessarily satisfy this condition within their operating bands.

3.2.2 Formulation for antenna RCS

The RCS of a target is directly related to its scattered electric field. When a free-space antenna is installed on a platform, the surface that would contribute to the structural mode is not necessarily illuminated. Thus, for a phased antenna array, the structural mode scattering becomes negligible, and the RCS can be approximated to comprise of only the antenna radiation mode, which can be expressed by modifying (3.3) as

$$\vec{E}^s(Z_L) = \left[\frac{j\eta_0}{4\lambda R_a}\vec{h}\left(\vec{h}.\vec{E}^i\right)\frac{e^{-jk_oR}}{R}\right]\Gamma_o, \text{ assuming } \vec{E}^s(Z_a^*) = 0.$$

This equation shows that for a single element (m, n) in a planar array (Figure 3.2), the monostatic radiation mode (Jenn and Flokas 1996) may be expressed as

$$\vec{E}^s_{mn}(\theta,\phi) = \left\{\frac{j\eta_o}{4\lambda R_a}\vec{h}\left[\vec{h}.\vec{E}^i(\theta,\phi)\right]\frac{e^{-jk_oR}}{R}\right\}\Gamma_{mn}(\theta,\phi) \tag{3.5}$$

where Γ_{mn} represents the total reflected signal towards the antenna aperture (m,n) and (θ, ϕ) indicates the direction of incident wave. The equation (3.5) is based on the assumption that the radiation impedance is real and with zero ohmic loss, i.e. $X_a = 0$ and $R_d = 0$ (Schindler et al. 1965). Mathematically, it can be expressed as

$$Z_a = R_a + jX_a = \left(R_r + R_d\right) + jX_a = R_r.$$

For an incident TM_z-polarised wave, $\vec{E}^i$ consists of only θ component. Thus, for a unit magnitude plane wave

$$\vec{h}.\vec{E}^i(\theta,\phi) \approx \vec{h}.\vec{E}^i(\theta) = \left(\hat{x}.\hat{\theta}\right)\left|\vec{h}\right|e^{-j\vec{k}.\vec{d}_{mn}}, \tag{3.6}$$

where $\vec{k} = k\left(\hat{x}\sin\theta\cos\phi + \hat{y}\sin\theta\sin\phi + \hat{z}\cos\theta\right)$ and $\vec{d}_{mn}$ is the position vector to element (m,n). Substituting (3.6) in (3.5), one gets,

$$\vec{E}^{s}_{mn}(\theta,\phi)=\left[\frac{j\eta_o}{4\lambda R_a}\vec{h}\left\{\left(\hat{x}.\hat{\theta}\right)\left|\vec{h}\right|e^{-j\vec{k}.\vec{d}_{mn}}\right\}\frac{e^{-jk_oR}}{R}\right]\Gamma_{mn}(\theta,\phi) \tag{3.7}$$

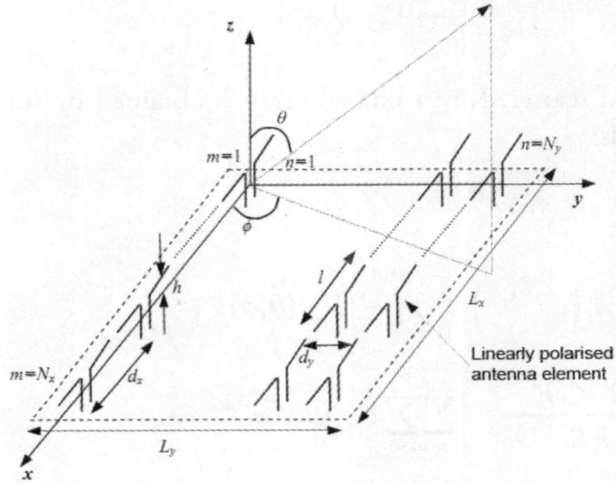

Figure 3.2 Planar array geometry

The relationship between the effective height and maximum effective area of a single element linearly polarised along the *x*-direction (Hansen 1989) is given by

$$\vec{h}=\left|\vec{h}\right|\hat{x}=2\sqrt{\frac{A_{em}R_r}{\eta}}\hat{x}, \tag{3.8}$$

where, A_{em} is the maximum effective area of a single element.
Substituting (3.8) into (3.7), one obtains

$$\vec{E}^{s}_{mn}(\theta,\phi)=\left[\frac{j\eta_0}{4\lambda R_a}\left(2\sqrt{\frac{A_{em}R_r}{\eta}}\hat{x}\right)\left\{\left(\hat{x}.\hat{\theta}\right)\left(2\sqrt{\frac{A_{em}R_r}{\eta}}\right)e^{-j\vec{k}.\vec{d}_{mn}}\right\}\frac{e^{-jk_oR}}{R}\right]\Gamma_{mn}(\theta,\phi)$$

$$=\left[\frac{j\eta_0}{4\lambda R_a}\left(4\frac{A_{em}R_r}{\eta}\hat{x}\right)\left\{\left(\hat{x}.\hat{\theta}\right)e^{-j\vec{k}.\vec{d}_{mn}}\right\}\frac{e^{-jk_oR}}{R}\right]\Gamma_{mn}(\theta,\phi)$$

Assuming $\eta_0 = \eta$ and $R_a = R_r$,

$$\vec{E}^{s}_{mn}(\theta,\phi)=\frac{j}{\lambda}\hat{x}.\hat{\theta}\,A_{em}\left(e^{-j\vec{k}.\vec{d}_{mn}}\right)\left(\frac{e^{-jk_0R}}{R}\right)\Gamma_{mn}(\theta,\phi)\hat{x}. \tag{3.9}$$

The effective elemental area, A_e presented by a planar array towards the incident wave (Chu 1991) is given by

$$A_e=\hat{x}\hat{\theta}\,A_{em}\approx d_xd_y\cos\theta, \tag{3.10}$$

where $A_{em} \approx d_x d_y$; d_x and d_y are the inter-element spacing along the *x*- and *y*-axis, respectively. Substituting (3.10) into (3.9) results in

$$\vec{E}^s_{mn}(\theta,\phi) = \frac{j}{\lambda} A_e \left(e^{-j\vec{k}.\vec{d}_{mn}}\right) \left(\frac{e^{-jk_0 R}}{R}\right) \Gamma_{mn}(\theta,\phi)\hat{x}$$

Thus, the total field scattered by a phased array is obtained by summing over individual element scattered fields:

$$\vec{E}^s(\theta,\phi) = \sum_{m=1}^{N_x}\sum_{n=1}^{N_y} \vec{E}^s_{mn}(\theta,\phi) \tag{3.11}$$

$$= \sum_{m=1}^{N_x}\sum_{n=1}^{N_y} \frac{j}{\lambda} A_e \left(e^{-j\vec{k}.\vec{d}_{mn}}\right) \left(\frac{e^{-jk_0 R}}{R}\right) \Gamma_{mn}(\theta,\phi)\hat{x}$$

$$= \left(\frac{j\left(N_x N_y A_e\right) e^{-jk_0 R}}{\lambda R}\right) \sum_{m=1}^{N_x}\sum_{n=1}^{N_y} \Gamma_{mn}(\theta,\phi) e^{-j\vec{k}.\vec{d}_{mn}}\,\hat{x}$$

$$= \left(\frac{jAe^{-jk_0 R}}{\lambda R}\right) \sum_{m=1}^{N_x}\sum_{n=1}^{N_y} \Gamma_{mn}(\theta,\phi) e^{-j\vec{k}.\vec{d}_{mn}}\,\hat{x}, \tag{3.12}$$

where $A \approx N_x N_y A_e = N_x N_y d_x d_y \cos\theta$ represents the projected area of the antenna array, while $A_p = N_x N_y d_x d_y$ represents its physical area.

For a closed body, with the origin of the coordinate system located within it,

$\hat{k}.\hat{d}_{mn} < 0 \Rightarrow e^{-j\vec{k}.\vec{d}_{mn}} = e^{j\vec{k}.\vec{d}_{mn}}$,

This modifies (3.12) as

$$\vec{E}^s_\theta(\theta,\phi) = \left(\frac{jAe^{-jk_0 R}}{\lambda R}\right) \sum_{m=1}^{N_x}\sum_{n=1}^{N_y} \Gamma_{mn}(\theta,\phi) e^{j\vec{k}.\vec{d}_{mn}}\,\hat{x}, \tag{3.13}$$

Substitution of (3.13) into the general RCS (3.1) yields

$$\sigma(\theta,\phi) = \lim_{R\to\infty} 4\pi R^2 \frac{\left|\left(\frac{jAe^{-jk_0 R}}{\lambda R}\right) \sum_{m=1}^{N_x}\sum_{n=1}^{N_y} \Gamma_{mn}(\theta,\phi) e^{j\vec{k}.\vec{d}_{mn}}\right|^2}{\left|E_i(\theta,\phi)\right|^2}. \tag{3.14}$$

For a unit magnitude plane incident wave, the denominator of (3.14) is unity. Thus,

$$\sigma(\theta,\phi) = \lim_{R\to\infty} 4\pi R^2 \left|\left(\frac{jAe^{-jk_0 R}}{\lambda R}\right) \sum_{m=1}^{N_x}\sum_{n=1}^{N_y} \Gamma_{mn}(\theta,\phi) e^{j\vec{k}.\vec{d}_{mn}}\right|^2.$$

Considering only the magnitude term, one gets

$$\sigma(\theta,\phi)=\lim_{R\to\infty}4\pi R^2\left(\frac{A^2e^{-2jk_0R}}{\lambda^2R^2}\right)\left|\sum_{m=1}^{N_x}\sum_{n=1}^{N_y}\Gamma_{mn}(\theta,\phi)e^{j\vec{k}.\vec{d}_{mn}}\right|^2$$

$$=\frac{4\pi A^2}{\lambda^2}\left|\sum_{m=1}^{N_x}\sum_{n=1}^{N_y}\Gamma_{mn}(\theta,\phi)e^{j\vec{k}.\vec{d}_{mn}}\right|^2\lim_{R\to\infty}\left(e^{-2jk_0R}\right). \tag{3.15}$$

Neglecting the phase term, (3.15) can be expressed as

$$\sigma(\theta,\phi)=\frac{4\pi A^2}{\lambda^2}\left|\sum_{m=1}^{N_x}\sum_{n=1}^{N_y}\Gamma_{mn}(\theta,\phi)e^{j\vec{k}.\vec{d}_{mn}}\right|^2=\frac{4\pi A_p^2}{\lambda^2}\left|\sum_{m=1}^{N_x}\sum_{n=1}^{N_y}\Gamma_{mn}(\theta,\phi)e^{j\vec{k}.\vec{d}_{mn}}\right|^2\cos^2\theta. \tag{3.16}$$

The equation (3.16) represents the RCS of a phased array with linearly polarised elements radiating along the *x*-axis when a θ-polarised wave is incident on it. However, in more generalised terms, the total RCS for a *N*-element linear array may be expressed as

$$\sigma(\theta,\phi)=\underbrace{\frac{4\pi A_p^2}{\lambda^2}}_{1^{\text{st}}\text{ factor}}\underbrace{\left|\sum_{n=1}^{N}\Gamma_n(\theta,\phi)e^{j\vec{k}.\vec{d}_n}\right|^2}_{2^{\text{nd}}\text{ factor}}\underbrace{\left|F_{\text{norm}}(\theta,\phi)\right|^2}_{3^{\text{rd}}\text{ factor}} \tag{3.17}$$

and for an $N_x \times N_y$ planar array, it is expressed as

$$\sigma(\theta,\phi)=\underbrace{\frac{4\pi A_p^2}{\lambda^2}}_{1_{\text{st}}\text{ factor}}\underbrace{\left|\sum_{m=1}^{N_x}\sum_{n=1}^{N_y}\Gamma_{mn}(\theta,\phi)e^{j\vec{k}.\vec{d}_{mn}}\right|^2}_{2_{\text{nd}}\text{ factor}}\underbrace{\left|F_{\text{norm}}(\theta,\phi)\right|^2}_{3^{\text{rd}}\text{ factor}} \tag{3.18}$$

where σ is the RCS of the phased antenna array, $\vec{d}_n=\hat{x}(n-1)d$ is the position vector to element *n* and F_{norm} is the normalised element scattering pattern designated as form factor. The equations (3.17) and (3.18) represent the RCS obtained due to scattering of all the signals, which enter the array and return to the aperture, after being reflected by the feed mismatches (Jenn 1995).

For the contribution of every antenna element to RCS, the first and the third factors of equation (θ, ϕ) remain the same, but the second factor (designated as array factor of the couplers) varies. It may be noted that the first factor in both equations is essentially the same as that in the conventional RCS definition. From (3.16) through (3.18), it can be inferred that the third factor is taken to be $\cos\theta$ for an array with *x*-polarised elements when the θ-polarised wave is incident on it, i.e. $F_{\text{norm}}(\theta,\phi)=\cos\theta$. This shows that the form factor depends on the direction of polarisation and thus should be chosen appropriately. The array factor in (3.17) and (3.18) depends on the type of element considered as well as on the type of the array (linear or planar). Array factors for a linear array of N_x elements along the *x*-axis and for other components (phase shifters/couplers) in the feed network are given as

$$\text{for radiation elements}: AF_r = \rho_r \frac{1}{N_x^2}\left[\frac{\sin N_x \alpha}{\sin \alpha}\right], \tag{3.19}$$

$$\text{for phase shifters:} AF_p = \frac{\rho_p \tau_r^2}{N_x^2}\left[\frac{\sin N_x \alpha}{\sin \alpha}\right], \tag{3.20}$$

for the side arms of the first-level of couplers:

$$AF_c = \frac{\tau_r^2 \tau_p^2 \rho_c}{N_x^2}\left[\frac{\sin N_x \zeta_x}{\sin \zeta_x}\right], \tag{3.21}$$

where $\alpha = -k_o d_y \sin\theta \cos\phi$ is the inter-element delay for the incident wave, using the first antenna element as a phase reference and $\zeta = \alpha + \alpha_s; \alpha_s = -k_o d_x \sin\theta_s \cos\phi_s$, is the inter-element phase to scan the antenna beam along the *x*-direction.

Likewise, for a planar array ($N_x \times N_y$), all the scattering terms for a linear array of *y*-dimension must be multiplied by an array factor. The multiplication factor for scattering sources ahead of phase shifters is given by

$$AF_y = \frac{1}{N_y^2}\left[\frac{\sin N_y \beta}{\sin \beta}\right], \tag{3.22a}$$

where the one for the scattering sources behind the phase shifters is

$$AF_y = \frac{1}{N_y^2}\left[\frac{\sin N_y \zeta_y}{\sin \zeta_y}\right], \tag{3.22b}$$

where $\beta = -k_o d_y \sin\theta \sin\phi$ is the inter-element space delay for the incident wave, using the first antenna element as a phase reference and $\zeta_y = \beta + \beta_s; \beta_s = -k_o d_y \sin\theta_s \sin\phi_s$ is the inter-element phase to scan the antenna beam along the *y*-direction.

Thus, the new set of array factors for planar ($N_x \times N_y$) antenna array elements/components is suggested as

$$\text{for radiation elements: } AF_r = \frac{\rho_r}{N_x N_y}\left[\frac{\sin N_x \alpha}{\sin \alpha}\right]\left[\frac{\sin N_y \beta}{\sin \beta}\right], \tag{3.23}$$

$$\text{for phase shifters: } AF_p = \frac{\rho_p \tau_r^2}{N_x N_y}\left[\frac{\sin N_x \alpha}{\sin \alpha}\right]\left[\frac{\sin N_y \beta}{\sin \beta}\right], \tag{3.24}$$

for the side arms of first-level couplers:

$$AF_c = \frac{\tau_r^2 \tau_p^2 \rho_c}{N_x N_y}\left[\frac{\sin N_x \zeta_x}{\sin \zeta_x}\right]\left[\frac{\sin N_y \zeta_y}{\sin \zeta_y}\right]. \tag{3.25}$$

The scattered field Γ in (3.17) and (3.18) may be obtained by either the network matrix method such as scattering parameters or by tracing signals through the feed network at different stages. The network matrix method is computationally intense as it includes all the interactions between the feed devices that lead to a matrix equation. The matrix dimension increases with the number of antenna array elements, making the corresponding solution computationally intensive. This approach of tracing signals through the feed has more advantages as it is simple and computationally efficient.

When an incident wave enters into the feed network (Figure 3.3) of an in-band phased array, tracing signals through the feed depends on the internal junctions and devices at which reflections occur. As evident from Figure 3.3, the first scattering source encountered by a plane wave incident at an angle θ is the radiating antenna element with reflection coefficient ρ_r. For a well-designed in-band operating phased antenna array, ρ_r corresponding to a normal incidence wave should be small. If not, a fraction of the incident signal, as determined by the reflection coefficient, ρ_r will be reradiated. The remaining signal, τ_r is transmitted further into the feed network towards the phase shifter.

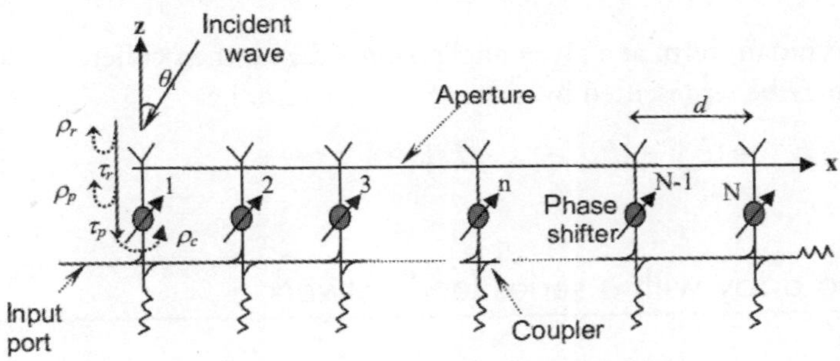

Figure 3.3 Series feed for a phased array of N antenna elements

If the phase shifter and the transmission line are not matched, then a reflected signal, ρ_p returns to the antenna aperture. The antenna being a reciprocal device re-reflects a fraction of this signal, giving rise to second-order reflections. However, the second-order reflections are neglected for the estimation of RCS.

The signal that enters the phase shifter suffers a phase shift, represented by the transmission coefficient, τ_p. This signal enters the coupler through the transmission line. Scattering at the coupler depends on the feed configuration. This is discussed for series feed networks in the subsequent sections. The sources of scattering in case of feed networks are aperture, ρ_r, phase shifter inputs, ρ_p, and inputs of couplers, ρ_c. Other typical sources of scattering include devices that are not perfectly matched due to their physical limitations, surface roughness, errors built within the antenna, edge effects, etc.

Scattering at various mismatches continues as the signal propagates deeper into the feed network. The vector sum of all these individual scattered fields that return to the aperture and reradiate is equal to the total RCS.

Assumptions made for an approximate model: The feed network of a phased array has a complex structure with every junction being a significant contributor to reflections. Tracing the propagating signal inside such a feed network becomes tedious, especially when multiple reflections are taken into account. Several approximations are considered to explain how to simplify the estimation process (Jenn and Flokas 1996). These are as follows:

1. All identical devices have same reflection and transmission coefficients. Since elements are non-ideal, $\rho_r = 0$. Furthermore, the phases of these coefficients except phase shifters are taken as zero. The transmission coefficient of the phase shifter is given by

$$\tau_{P_{mn}} = \left|\tau_{P}\right| e^{j\chi_{mn}}, \tag{3.26}$$

where $\chi_{mn} = (m-1)\alpha_s + (n-1)\beta_s$; $\alpha_s = -k_o d_x \sin\theta_s \cos\phi_s$ and $\beta_s = -k_o d_y \sin\theta_s \sin\phi_s$ are the inter-element phase along the *x*- and *y*-directions, respectively, and (θ_s, ϕ_s) represents the direction of the beam scan.

2. All the feed network components are matched, i.e. $\rho \ll 1$. The higher-order reflections $(\rho^2, \rho^3, \ldots)$ are not considered.
3. $|\rho|^2 + |\tau|^2 = 1$
4. Edge effects and the mutual coupling effects are ignored.

If only one dominant term at a given angle is considered, then coherent scattered fields and hence the RCS may be represented by a non-coherent sum, i.e.

$$|E_1 + E_2 + \ldots + E_n|^2 \approx |E_1|^2 + |E_2|^2 + \ldots + |E_n|^2$$

3.3 A phased array with a series feed network

In-band scattering of a phased antenna array significantly depends on the type of feed employed. Two major techniques of feeding the phased array are network feed and space feed (Volakis 2007), respectively. The space feed consists of a horn antenna or a small array used for illuminating an array aperture. This array acts like a lens, and it transforms the incident spherical wave front into a planar one by adjusting the amplitude and phase of the incident wave properly. Space feed can either be two-dimensional (2-D) or three-dimensional (3-D) in configuration. Bootlace and Rotman lenses are a few examples of 2-D space-feeds. Three-dimensional space feeds are mainly used for ground-based applications. Network feeds are again of two types; *viz.* series feed and parallel feed. Parallel feed networks consist of couplers that combine the signals from adjacent antenna elements in an array. In a series feed, the power to each antenna element is tapped off from the feed line sequentially with the help of the couplers (Figure 3.3).

Coupling coefficients of these couplers are adjusted accordingly to provide a desired amplitude distribution (uniform, cosine squared on a pedestal, etc.). Further, their values depend upon the desired sidelobe level (e.g. 13 dB for uniform distribution). Following the couplers, there are phase shifters arranged either along the main line or branch lines (Rudge et al. 1983). These phase shifters facilitate beam scanning as they vary the phase of signals fed to the antenna elements. It is observed that the main feed line of the series feed network is terminated by a matched load. This causes some power to reflect back from the load. Reflection coefficients (ρ_r, ρ_p, ρ_c) correspond to radiating elements, the phase shifters and the couplers of feed network, respectively. Similarly, τ_r and τ_p represent the transmission coefficients of the radiating elements and phase shifters, respectively.

3.3.1 RCS formulation with isotropic array elements

In case of the series-fed phased antenna array (Lee 2008), the fraction of the signal entering the element array and re-radiating to the aperture is

$$\Gamma_n(\theta,\phi) = \rho_r e^{j(n-1)\alpha} + \tau_r^2 \rho_p e^{j(n-1)\alpha} + \tau_r^2 \tau_p^2 \rho_c e^{j2(n-1)\alpha_s} e^{j(n-1)\alpha} + \tau_r^2 \tau_p^2 e^{j(n-1)\alpha_s} E_n^s, \tag{3.27}$$

where E_n^s represents the signal re-radiated from the series feed at the nth antenna element. The first, second and third terms of (3.27) correspond to the contribution of radiating elements, phase shifters and couplers to the scattered field, respectively. The RCS equation corresponding to these terms can be obtained readily on the basis of array theory.

RCS due to radiating elements: The general RCS equation for a series-fed linear phased array, expressed in (3.17), is

$$\sigma(\theta,\phi) = \frac{4\pi A^2}{\lambda^2} \left| \sum_{n=1}^{N} \Gamma_n(\theta,\phi) e^{jk.d_n} \right|^2 \left| F_{\text{norm}}(\theta,\phi) \right|^2,$$

where $F_{\text{norm}}(\theta,\phi) = \cos\theta$. Thus

$$\sigma(\theta,\phi) \approx \frac{4\pi A^2 \cos^2\theta}{\lambda^2} \left| \sum_{n=1}^{N} \Gamma_n(\theta,\phi) e^{j\vec{k}.\vec{d}_n} \right|^2 \tag{3.28}$$

Here, $\vec{k} = k\left(\hat{x}\sin\theta\cos\phi + \hat{y}\sin\theta\sin\phi + \hat{z}\cos\theta\right)$ is the wave vector and $\vec{d}_n = \hat{x}(n-1)d$ is the position vector to element n. Thus

$$\vec{k}.\vec{d}_n = k\left(\hat{x}\sin\theta\cos\phi + \hat{y}\sin\theta\sin\phi + \hat{z}\cos\theta\right).\hat{x}(n-1)d$$

$$= (n-1)kd\sin\theta\cos\phi \qquad \because\ \hat{x}.\hat{x} = 1 \text{ and } \hat{x}.\hat{y} = \hat{x}.\hat{z} = 0$$

$$\vec{k}.\vec{d}_n = (n-1)\alpha \tag{3.29a}$$

The scattered field due to radiating elements only, as given by the first term of (3.27) is expressed as

$$\Gamma_r(\theta,\phi) = \rho_r e^{j(n-1)\alpha}. \tag{3.29b}$$

Substituting (3.29a) and (3.29b) in Γ factor of (3.28) yields

$$\sum_{n=1}^{N} \Gamma_n(\theta,\phi) e^{j\vec{k}.\vec{d}_n} = \sum_{n=1}^{N} \rho_r e^{j(n-1)\alpha} e^{j(n-1)\alpha} = \rho_r \sum_{n=1}^{N} e^{j2(n-1)\alpha}. \tag{3.30}$$

Changing the index of summation,

$$\sum_{n=1}^{N}\Gamma_n(\theta,\phi)e^{j\vec{k}.\vec{d}_n}=\rho_r\sum_{m=0}^{N-1}\left(e^{j2\alpha}\right)^m \tag{3.31}$$

$$=\rho_r\frac{1-\left(e^{j2\alpha}\right)^N}{1-e^{j2\alpha}}=\rho_r\left(\frac{e^{j\alpha N}e^{-j\alpha N}-e^{j\alpha N}e^{j\alpha N}}{e^{j\alpha}e^{-j\alpha}-e^{j\alpha}e^{j\alpha}}\right)$$

$$=\rho_r\left(\frac{e^{j\alpha N}}{e^{j\alpha}}\right)\left(\frac{e^{-j\alpha N}-e^{j\alpha N}}{e^{-j\alpha}-e^{j\alpha}}\right)$$

$$=\rho_r e^{j(N-1)\alpha}\frac{\sin(N\alpha)}{\sin(\alpha)}. \tag{3.32}$$

Normalisation of (3.32) by the number of antenna elements, N results in

$$\sum_{n=1}^{N}\Gamma_n(\theta,\phi)e^{j\vec{k}.\vec{d}_n}=\rho_r\frac{\sin(N\alpha)}{N\sin(\alpha)}e^{j(N-1)\alpha}. \tag{3.33}$$

Substituting (3.33) in (3.28), and considering the effect of only radiating elements, one gets

$$\sigma_r\approx\frac{4\pi A^2\cos^2\theta}{\lambda^2}\left|\rho_r\frac{\sin(N\alpha)}{N\sin(\alpha)}e^{j(N-1)\alpha}\right|^2,$$

$$\sigma_r\approx\frac{4\pi A^2\cos^2\theta}{\lambda^2}\rho_r^2\left[\frac{\sin(N\alpha)}{N\sin(\alpha)}\right]^2. \tag{3.34}$$

Following the same steps as mentioned above, the RCS equation considering the effect of phase shifters can be expressed as

$$\sigma_p\approx\frac{4\pi A^2\cos^2\theta}{\lambda^2}\rho_p^2\tau_r^4\left[\frac{\sin(N\alpha)}{N\sin(\alpha)}\right]^2. \tag{3.35}$$

However, the contribution to the scattered field by the couplers, i.e. the third term of (3.27) has an additional exponential factor, $e^{j2(n-1)\alpha_s}$. Thus,

$$\sum_{n=1}^{N}\Gamma_n(\theta,\phi)e^{j\vec{k}.\vec{d}_n}=\sum_{n=1}^{N}\tau_r^2\tau_p^2\rho_c e^{j2(n-1)\alpha_s}e^{j(n-1)\alpha}e^{j(n-1)\alpha}=\tau_r^2\tau_p^2\rho_c\sum_{n=1}^{N}e^{j2(n-1)\alpha_s}e^{j2(n-1)\alpha}$$

$$=\tau_r^2\tau_p^2\rho_c\sum_{n=1}^{N}e^{j2(n-1)(\alpha_s+\alpha)}=\tau_r^2\tau_p^2\rho_c\sum_{n=1}^{N}e^{j2(n-1)\zeta}. \tag{3.36}$$

The exponential factor within the summation in (3.36) is similar to that in (3.30) except α being replaced by ζ. This shows that a similar derivation, as in (3.31) through (3.34), is applicable, and one gets

$$\sigma_c \approx \frac{4\pi A^2 \cos^2\theta}{\lambda^2} \rho_c^2 \tau_r^4 \tau_p^4 \left[\frac{\sin(N\zeta)}{N\sin(\zeta)}\right]^2. \qquad (3.37)$$

The equation (3.37) gives the contribution of the couplers to the scattered field of a series-fed N-element linear phased array.

The fourth term of (3.27) consists of E_n^s that indicates the cumulative effect produced due to the signals received from (i) the elements towards input ($m < n$), (ii) elements towards termination ($m > n$) and (iii) reflection at the load of coupler n. As these signals travel along the main feed line, the couplers affect their path. Hence, it is necessary to know the signal distribution at the coupler ports.

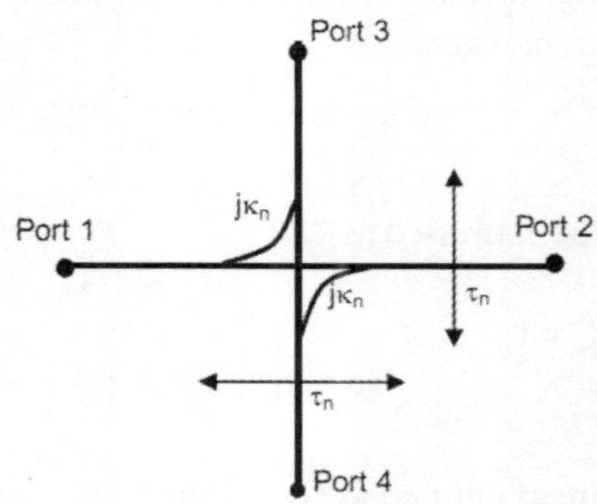

Figure 3.4a Coupling and transmission paths for a four-port coupler

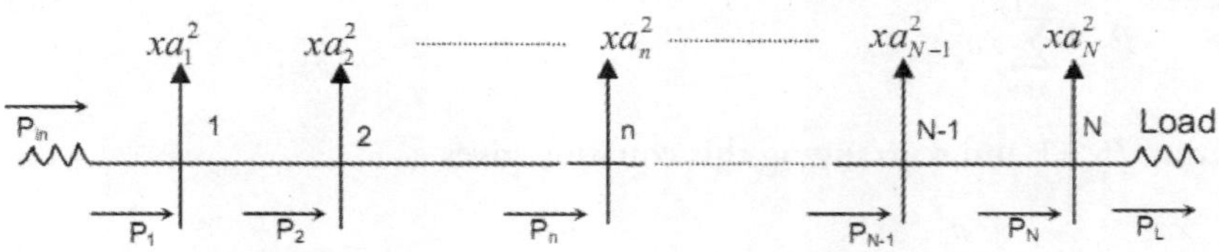

Figure 3.4b A schematic of lossless power feed line

Figure 3.4a shows the signal distribution of an nth four-port coupler with coupling coefficient, κ_n and transmission coefficient, τ_n used in the series feed networks. The scattering matrix of such a coupler is (Lee 1994)

$$K_n = \begin{bmatrix} 0 & \tau_n & j\kappa_n & 0 \\ \tau_n & 0 & 0 & -j\kappa_n \\ j\kappa_n & 0 & 0 & \tau_n \\ 0 & -j\kappa_n & \tau_n & 0 \end{bmatrix}. \qquad (3.38)$$

The coupling coefficients, κ_n, are determined as per the desired amplitude distribution, a_n. For example, a uniform distribution with a 13 dB sidelobe level requires a_n to be 1 for all couplers. However, for having an amplitude distribution of cosine squared on a pedestal with a 32 dB sidelobe level, the amplitude coefficients (Mallioux 1994) are given by

$$a_n = a_o + (1 - a_o)\sin^2\left(\frac{n-1}{N-1}\pi\right), \tag{3.39}$$

where $\alpha_o = 0.2$ is the height of the pedestal on which the cosine squared distribution takes place. For a lossless feed of a phased array (Figure 3.4b), the input power is given by

$$P_{\text{in}} = \sum_{i=1}^{N} x a_i^2 + P_L. \tag{3.40}$$

Here x is the fraction of the power, which goes into each coupler. P_{in} and P_L represent the total input power and the power into the load, respectively, where P_L is taken to be 0.05. After rearranging the equation, one gets

$$P_{\text{in}} - P_L = x\sum_{i=1}^{N} a_i^2. \tag{3.41}$$

Further, this equation can be expressed as

$$x = \frac{P_{\text{in}} - P_L}{\sum_{i=1}^{N} a_i^2} = \frac{1 - P_L}{\sum_{i=1}^{N} a_i^2}, \text{ if } P_{in} = 1.$$

The coupling coefficients are as defined as (Lee 2008)

$$\kappa_n^2 = \frac{x a_n^2}{P_n} = \frac{x a_n^2}{P_{in} - \sum_{q=1}^{n-1} x a_q^2}. \tag{3.42}$$

Substituting $P_{\text{in}} = 1$ and rearranging this equation gives

$$\kappa_n^2 = \frac{x a_n^2}{1 - x\sum_{q=1}^{n-1} a_q^2} = \frac{a_n^2}{\frac{1}{x} - \sum_{q=1}^{n-1} a_q^2}. \tag{3.43}$$

Substituting x into the above equation yields

$$\kappa_n^2 = \frac{a_n^2}{\dfrac{1}{\frac{1-P_L}{\sum_{i=1}^{N} a_i^2}} - \sum_{q=1}^{n-1} a_q^2} = \frac{a_n^2}{\dfrac{\sum_{i=1}^{N} a_i^2}{1-P_L} - \sum_{q=1}^{n-1} a_q^2}. \tag{3.44}$$

Assuming lossless and matched couplers, with perfect isolation between the coupled and through ports, the transmission coefficients are expressed as

$$\tau_n^2 = 1 - \kappa_n^2. \tag{3.45}$$

The considered approximate model includes only the reflection from coupler loads, neglecting the reflections originated from the junctions between adjacent couplers. To facilitate the determination of E_n^s, the received signal is decomposed into forward and backward travelling waves (Figure 3.5).

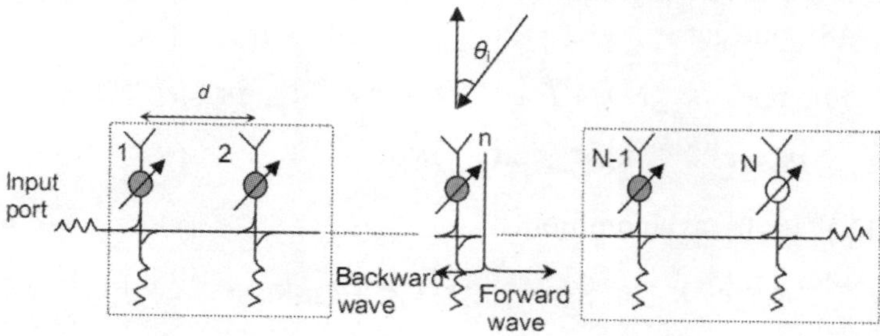

Figure 3.5 Travelling waves in a series-fed phased array

The signal received by an nth element acts as an excitation for (i) $(N - n)$ elements towards the load, constituting forward travelling wave, (ii) $(n - 1)$ elements towards the input, constituting backward travelling wave, (iii) other scattering components such as a self-reflected wave (due to the reflection from the load terminating the nth element) and an input load reflected wave (due to the reflection from the input load). Thus, for each antenna element in the array, four beams are re-radiated. The beam-widths and the peak levels of the lobes depend on the element location in the array.

The total scattered field due to series feed is obtained by the superposition of forward, backward, self-reflected and input load-reflected waves from all the elements (Zhang et al. 2010). Thus,

$$\left|E^s\right|^2 = \left|E_f + E_b + E_{\text{self}} + E_{\text{in}}\right|^2. \tag{3.46}$$

If only one term dominates at a given angle, it can be approximated to

$$\left|E^s\right|^2 \approx \left|E_f\right|^2 + \left|E_b\right|^2 + \left|E_{\text{self}}\right|^2 + \left|E_{\text{in}}\right|^2. \tag{3.47}$$

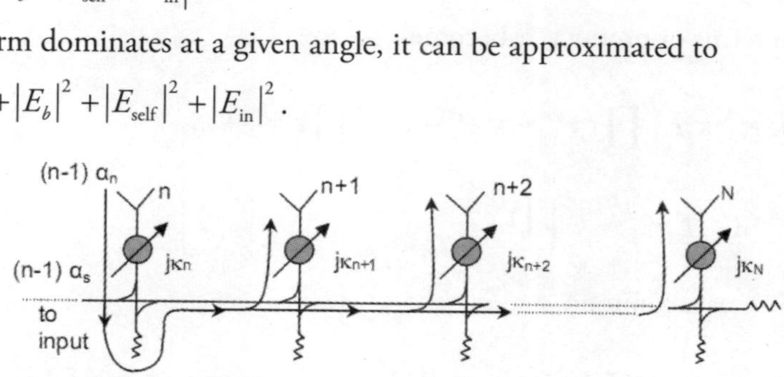

Figure 3.6 Forward travelling wave towards N^{th} antenna

Scattered field due to forward travelling wave: For an incident signal at the nth element, a forward wave travels towards the Nth element, through $(N - n)$ elements (Figure 3.6). The scattered field due to this forward travelling wave for element n is expressed as

$$E_{f_n} \approx \Gamma_l e^{j(n-1)\zeta} j\kappa_n \tau_n \begin{Bmatrix} e^{j\psi} j\kappa_{n+1} e^{jn\zeta} + e^{j2\psi} \tau_{n+1} j\kappa_{n+2} e^{j(n+1)\zeta} + \dots \\ + e^{j(N-n)\psi} \tau_{n+1} \tau_{n+2} \dots \tau_{N-1} j\kappa_N e^{j(N-1)\zeta} \end{Bmatrix}, \tag{3.48}$$

where $\psi = kL$ is the electrical length between the couplers
Expanding (3.48), one gets

$$E_{f_n} \approx \Gamma_l e^{j(n-1)\zeta} j\kappa_n \tau_n e^{j\psi} j\kappa_{n+1} e^{jn\zeta} + \Gamma_l e^{j(n-1)\zeta} j\kappa_n \tau_n e^{j2\psi} \tau_{n+1} j\kappa_{n+2} e^{j(n+1)\zeta} + \dots$$
$$+ \Gamma_l e^{j(n-1)\zeta} j\kappa_n \tau_n e^{j(N-n)\psi} \tau_{n+1} \tau_{n+2} \dots \tau_{N-1} j\kappa_N e^{j(N-1)\zeta}.$$

Keeping $\left(j^2 \Gamma_l\right) = -\Gamma_l$ as a common,

$$E_{f_n} \approx -\Gamma_l \begin{Bmatrix} e^{j(n-1)\zeta} \kappa_n \tau_n e^{j\psi} \kappa_{n+1} e^{jn\zeta} + e^{j(n-1)\zeta} \kappa_n \tau_n e^{j2\psi} \tau_{n+1} \kappa_{n+2} e^{j(n+1)\zeta} + \dots \\ + e^{j(n-1)\zeta} \kappa_n \tau_n e^{j(N-n)\psi} \tau_{n+1} \tau_{n+2} \dots \tau_{N-1} \kappa_N e^{j(N-1)\zeta} \end{Bmatrix}. \tag{3.49}$$

For $n = 1$ (contribution to the scattered field due to backward travelling wave by the first element of the phased array), (3.49) can be expressed as

$$E_{f_1} \approx -\Gamma_l \begin{Bmatrix} e^{j0\zeta} \kappa_1 \tau_1 e^{j\psi} \kappa_2 e^{j\zeta} + e^{j0\zeta} \kappa_1 \tau_1 e^{j2\psi} \tau_2 \kappa_3 e^{j2\zeta} + \dots \\ + e^{j0\zeta} \kappa_1 \tau_1 e^{j(N-1)\psi} \tau_2 \tau_3 \dots \tau_{N-1} \kappa_N e^{j(N-1)\zeta} \end{Bmatrix}.$$

Rearranging and regrouping the above equation, it results in

$$E_{f_1} \approx -\Gamma_l \begin{Bmatrix} \kappa_1 e^{j0\zeta} \kappa_2 e^{j\zeta} \tau_1 e^{j\psi} + \kappa_1 e^{j0\zeta} \kappa_3 e^{j2\zeta} \tau_1 \tau_2 e^{j2\psi} + \dots \\ + \kappa_1 e^{j0\zeta} \kappa_N e^{j(N-1)\zeta} \tau_1 \tau_2 \tau_3 \dots \tau_{N-1} e^{j(N-1)\psi} \end{Bmatrix}$$
$$\approx -\Gamma_l \begin{Bmatrix} \kappa_1 e^{j0\zeta} \kappa_2 e^{j\zeta} \left(\tau_1 e^{j\psi}\right) + \kappa_1 e^{j0\zeta} \kappa_3 e^{j2\zeta} \left(\tau_1 e^{j\psi} \tau_2 e^{j\psi}\right) + \dots \\ + \kappa_1 e^{j0\zeta} \kappa_N e^{j(N-1)\zeta} \left(\tau_1 e^{j\psi} \tau_2 e^{j\psi} \tau_3 e^{j\psi} \dots \tau_{N-1} e^{j\psi}\right) \end{Bmatrix}.$$

Expressing in terms of product, it becomes

$$E_{f_1} \approx -\Gamma_l \begin{Bmatrix} \kappa_1 e^{j0\zeta} \kappa_2 e^{j\zeta} \prod_{i=1}^{1} \tau_i e^{j\psi} + \kappa_1 e^{j0\zeta} \kappa_3 e^{j2\zeta} \prod_{i=1}^{2} \tau_i e^{j\psi} + \dots \\ + \kappa_1 e^{j0\zeta} \kappa_N e^{j(N-1)\zeta} \prod_{i=1}^{N-1} \tau_i e^{j\psi} \end{Bmatrix}.$$

Keeping $\kappa_1 e^{j0\zeta}$ as a common,

$$E_{f_1} \approx -\Gamma_l \left[\kappa_1 e^{j0\zeta} \left\{ \kappa_2 e^{j\zeta} \prod_{i=1}^{1} \tau_i e^{j\psi} + \kappa_3 e^{j2\zeta} \prod_{i=1}^{2} \tau_i e^{j\psi} + \dots + \kappa_N e^{j(N-1)\zeta} \prod_{i=1}^{N-1} \tau_i e^{j\psi} \right\} \right].$$

In the summation form, it becomes

$$E_{f_1} \approx -\Gamma_l \left[\kappa_1 e^{j0\zeta} \left\{ \sum_{m=2}^{N} \kappa_m e^{j(m-1)\zeta} \left(\prod_{i=1}^{m-1} \tau_i e^{j\psi} \right) \right\} \right] \tag{3.50}$$

Similarly, for element 2

$$E_{f_2} \approx -\Gamma_l \left\{ \begin{array}{l} e^{j\zeta}\kappa_2\tau_2 e^{j\psi}\kappa_3 e^{j2\zeta} + e^{j\zeta}\kappa_2\tau_2 e^{j2\psi}\tau_3\kappa_4 e^{j3\zeta} + \dots \\ + e^{j\zeta}\kappa_2\tau_2 e^{j(N-2)\psi}\tau_3\tau_4\dots\tau_{N-1}\kappa_N e^{j(N-1)\zeta} \end{array} \right\}$$

$$\approx -\Gamma_l \left\{ \begin{array}{l} \kappa_2 e^{j\zeta}\kappa_3 e^{j2\zeta}\left(\tau_2 e^{j\psi}\right) + \kappa_2 e^{j\zeta}\kappa_4 e^{j3\zeta}\left(\tau_2 e^{j\psi}\tau_3 e^{j\psi}\right) + \dots \\ + \kappa_2 e^{j\zeta}\kappa_N e^{j(N-1)\zeta}\left(\tau_2 e^{j\psi}\tau_3 e^{j\psi}\tau_4 e^{j\psi}\dots\tau_{N-1}e^{j\psi}\right) \end{array} \right\}$$

$$\approx -\Gamma_l \left\{ \kappa_2 e^{j\zeta}\kappa_3 e^{j2\zeta}\prod_{i=2}^{2}\tau_i e^{j\psi} + \kappa_2 e^{j\zeta}\kappa_4 e^{j3\zeta}\prod_{i=2}^{3}\tau_i e^{j\psi} + \dots + \kappa_2 e^{j\zeta}\kappa_N e^{j(N-1)\zeta}\prod_{i=2}^{N-1}\tau_i e^{j\psi} \right\}$$

$$\approx -\Gamma_l \left[\kappa_2 e^{j\zeta} \left\{ \kappa_3 e^{j2\zeta}\prod_{i=2}^{2}\tau_i e^{j\psi} + \kappa_4 e^{j3\zeta}\prod_{i=2}^{3}\tau_i e^{j\psi} + \dots + \kappa_N e^{j(N-1)\zeta}\prod_{i=2}^{N-1}\tau_i e^{j\psi} \right\} \right]$$

$$E_{f_2} \approx -\Gamma_l \left[\kappa_2 e^{j\zeta} \left\{ \sum_{m=3}^{N} \kappa_m e^{j(m-1)\zeta} \left(\prod_{i=2}^{m-1} \tau_i e^{j\psi} \right) \right\} \right]. \tag{3.51}$$

Likewise, for element 3,

$$E_{f_3} \approx -\Gamma_l \left[\kappa_3 e^{j2\zeta} \left\{ \sum_{m=4}^{N} \kappa_m e^{j(m-1)\zeta} \left(\prod_{i=3}^{m-1} \tau_i e^{j\psi} \right) \right\} \right] \tag{3.52}$$

$$\vdots$$

$$E_{f_{N-1}} \approx -\Gamma_l \left[\kappa_{N-1} e^{j(N-2)\zeta} \left\{ \sum_{m=N}^{N} \kappa_m e^{j(m-1)\zeta} \left(\prod_{i=N-1}^{m-1} \tau_i e^{j\psi} \right) \right\} \right]. \tag{3.53}$$

Thus, the total scattered field due to forward travelling wave can be obtained by summing individual contributions of all elements in a linear array. Mathematically, it can be expressed as

$$E_f \approx E_{f_1} + E_{f_2} + E_{f_3} + \dots + E_{f_{N-1}}. \tag{3.54}$$

Substituting (3.50) through (3.53) in (3.54), and removing $-\Gamma_l$ as common yields, we get

$$E_f \approx -\Gamma_l \left[\begin{array}{l} \kappa_1 e^{j0\zeta} \left\{ \sum_{m=2}^{N} \kappa_m e^{j(m-1)\zeta} \left(\prod_{i=1}^{m-1} \tau_i e^{j\psi} \right) \right\} + \kappa_2 e^{j\zeta} \left\{ \sum_{m=3}^{N} \kappa_m e^{j(m-1)\zeta} \left(\prod_{i=2}^{m-1} \tau_i e^{j\psi} \right) \right\} \\ + \kappa_3 e^{j2\zeta} \left\{ \sum_{m=4}^{N} \kappa_m e^{j(m-1)\zeta} \left(\prod_{i=3}^{m-1} \tau_i e^{j\psi} \right) \right\} + \dots \\ \qquad + \kappa_{N-1} e^{j(N-2)\zeta} \left\{ \sum_{m=N}^{N} \kappa_m e^{j(m-1)\zeta} \left(\prod_{i=N-1}^{m-1} \tau_i e^{j\psi} \right) \right\} \end{array} \right]. \tag{3.55}$$

Expressing (3.55) in terms of summation, one gets

$$E_f \approx -\Gamma_l \sum_{n=1}^{N-1} \kappa_n e^{j(n-1)\zeta} \left\{ \sum_{m=n+1}^{N} \kappa_m e^{j(m-1)\zeta} \left(\prod_{i=n}^{m-1} \tau_i e^{j\psi} \right) \right\}. \tag{3.56}$$

Scattered field due to the backward travelling wave: Similar to the forward wave, a backward wave travels towards the input (element 1), as shown in Figure 3.7. The field scattered by this backward travelling wave for n^{th} antenna element is given by

$$E_{b_n} \approx \Gamma_l e^{j(n-1)\zeta} \left\{ \begin{array}{l} j\kappa_n j\kappa_{n-1}\tau_{n-1} e^{j\psi} e^{j(n-2)\zeta} + j\kappa_n(\tau_{n-1}e^{j\psi}) j\kappa_{n-2} e^{j\psi} \tau_{n-2} e^{j(n-3)\zeta} + \ldots \\ + j\kappa_n e^{j\psi} \left(\prod_{i=2}^{n-1} \tau_i e^{j\psi} \right) j\kappa_1 \tau_1 e^{j0\zeta} \end{array} \right\}. \tag{3.57}$$

Expanding and taking Γ_l as common, one gets

$$E_{b_n} \approx \left\{ \begin{array}{l} \Gamma_l e^{j(n-1)\zeta} j\kappa_n j\kappa_{n-1}\tau_{n-1} e^{j\psi} e^{j(n-2)\zeta} \\ +\Gamma_l e^{j(n-1)\zeta} j\kappa_n(\tau_{n-1}e^{j\psi}) j\kappa_{n-2} e^{j\psi}\tau_{n-2} e^{j(n-3)\zeta} + \ldots \\ +\Gamma_l e^{j(n-1)\zeta} j\kappa_n e^{j\psi} \left(\tau_2 e^{j\psi} \tau_3 e^{j\psi} \ldots \tau_{n-1} e^{j\psi} \right) j\kappa_1 \tau_1 e^{j0\zeta} \end{array} \right\} \tag{3.58}$$

$$\approx \Gamma_l \left\{ \begin{array}{l} e^{j(n-1)\zeta} j\kappa_n j\kappa_{n-1}\tau_{n-1} e^{j\psi} e^{j(n-2)\zeta} \\ +e^{j(n-1)\zeta} j\kappa_n(\tau_{n-1}e^{j\psi}) j\kappa_{n-2} e^{j\psi}\tau_{n-2} e^{j(n-3)\zeta} + \ldots \\ +e^{j(n-1)\zeta} j\kappa_n e^{j\psi} \left(\tau_2 e^{j\psi} \tau_3 e^{j\psi} \ldots \tau_{n-1} e^{j\psi} \right) j\kappa_1 \tau_1 e^{j0\zeta} \end{array} \right\}. \tag{3.59}$$

For first element, i.e. $n = 1$, equation (3.59) can be expressed as

$$E_{b_1} \approx \Gamma_l \{0\} = 0. \tag{3.60}$$

For second element,

$$\begin{aligned} E_{b_2} &\approx \Gamma_l \left\{ e^{j\zeta} j\kappa_2 j\kappa_1 \tau_1 e^{j\psi} e^{j0\zeta} \right\} \\ &\approx \Gamma_l \left[j\kappa_2 e^{j\zeta} \left\{ \sum_{m=1}^{1} j\kappa_m e^{j(m-1)\zeta} \left(\prod_{i=m}^{1} \tau_i e^{j\psi} \right) \right\} \right] \end{aligned} \tag{3.61}$$

Likewise, for $(N-1)^{\text{th}}$ element,

$$E_{b_{N-1}} \approx \Gamma_l \left\{ \begin{array}{l} e^{j(N-2)\zeta} j\kappa_{N-1} j\kappa_{N-2}\tau_{N-2} e^{j\psi} e^{j(N-3)\zeta} \\ +e^{j(N-2)\zeta} j\kappa_{N-1}(\tau_{N-2}e^{j\psi}) j\kappa_{N-3} e^{j\psi}\tau_{N-3} e^{j(N-4)\zeta} \\ +\ldots + e^{j(N-2)\zeta} j\kappa_{N-1} e^{j\psi} \left(\tau_2 e^{j\psi} \tau_3 e^{j\psi} \ldots \tau_{N-3} e^{j\psi} \tau_{N-2} e^{j\psi} \right) j\kappa_1 \tau_1 e^{j0\zeta} \end{array} \right\}.$$

Rearrangement and regrouping the above equation, one gets

$$E_{b_{N-1}} \approx \Gamma_l \begin{Bmatrix} j\kappa_{N-1}e^{j(N-2)\zeta}\, j\kappa_{N-2}e^{j(N-3)\zeta}\left(\tau_{N-2}e^{j\psi}\right) \\ +j\kappa_{N-1}e^{j(N-2)\zeta}\, j\kappa_{N-3}e^{j(N-4)\zeta}\left(\tau_{N-2}e^{j\psi}\tau_{N-3}e^{j\psi}\right) \\ +\ldots+j\kappa_{N-1}e^{j(N-2)\zeta}\, j\kappa_{1}e^{j0\zeta}\left(\tau_{1}e^{j\psi}\tau_{2}e^{j\psi}\tau_{3}e^{j\psi}\ldots\tau_{N-3}e^{j\psi}\tau_{N-2}e^{j\psi}\right) \end{Bmatrix}.$$

Expressing transmission and associated factors in terms of product, one gets

$$E_{b_{N-1}} \approx \Gamma_l \begin{Bmatrix} j\kappa_{N-1}e^{j(N-2)\zeta}\, j\kappa_{N-2}e^{j(N-3)\zeta}\prod_{i=N-2}^{N-2}\tau_i e^{j\psi} + \\ j\kappa_{N-1}e^{j(N-2)\zeta}\, j\kappa_{N-3}e^{j(N-4)\zeta}\prod_{i=N-3}^{N-2}\tau_i e^{j\psi} \\ +\ldots+j\kappa_{N-1}e^{j(N-2)\zeta}\, j\kappa_{1}e^{j0\zeta}\prod_{i=1}^{N-2}\tau_i e^{j\psi} \end{Bmatrix}$$

Considering $j\kappa_{N-1}e^{j(N-2)\zeta}$ as common and removing it, one gets

$$E_{b_{N-1}} \approx \Gamma_l \left[j\kappa_{N-1}e^{j(N-2)\zeta} \begin{Bmatrix} j\kappa_{N-2}e^{j(N-3)\zeta}\prod_{i=N-2}^{N-2}\tau_i e^{j\psi} + \\ j\kappa_{N-3}e^{j(N-4)\zeta}\prod_{i=N-3}^{N-2}\tau_i e^{j\psi} +\ldots+ j\kappa_{1}e^{j0\zeta}\prod_{i=1}^{N-2}\tau_i e^{j\psi} \end{Bmatrix} \right]$$

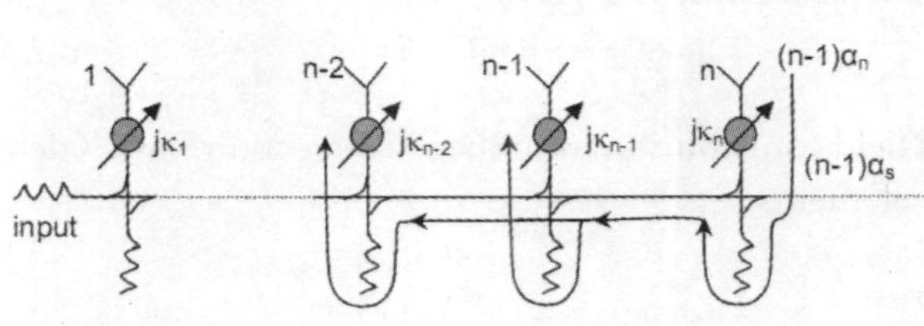

Figure 3.7 Backward travelling wave

Expressing the same in terms of summation, one gets

$$E_{b_{N-1}} \approx \Gamma_l \left[j\kappa_{N-1}e^{j(N-2)\zeta} \left\{ \sum_{m=1}^{N-2} j\kappa_m e^{j(m-1)\zeta} \left(\prod_{i=m}^{N-2} \tau_i e^{j\psi} \right) \right\} \right]. \tag{3.62}$$

Similarly, for the Nth element it may be written by replacing (N–1) in equation (3.62) by N and it is given as

$$E_{b_N} \approx \Gamma_l \left[j\kappa_N e^{j(N-1)\zeta} \left\{ \sum_{m=1}^{N-1} j\kappa_m e^{j(m-1)\zeta} \left(\prod_{i=m}^{N-1} \tau_i e^{j\psi} \right) \right\} \right] \tag{3.63}$$

Thus, the total scattered field due to the backward travelling wave can be obtained by summing the individual contributions of all elements present in the linear array.

Mathematically, it can be expressed as

$$E_b = E_{b_1} + E_{b_2} + \ldots + E_{b_{N-1}} + E_{b_N} \tag{3.64}$$

Substituting (3.60) through (3.63) into (3.64), one gets

$$E_b = \Gamma_l \begin{bmatrix} 0 + j\kappa_2 e^{j\zeta} \left\{ \sum_{m=1}^{1} j\kappa_m e^{j(m-1)\zeta} \left(\prod_{i=m}^{1} \tau_i e^{j\psi} \right) \right\} + \ldots \\ + j\kappa_{N-1} e^{j(N-2)\zeta} \left\{ \sum_{m=1}^{N-2} j\kappa_m e^{j(m-1)\zeta} \left(\prod_{i=m}^{N-2} \tau_i e^{j\psi} \right) \right\} \\ + j\kappa_N e^{j(N-1)\zeta} \left\{ \sum_{m=1}^{N-1} j\kappa_m e^{j(m-1)\zeta} \left(\prod_{i=m}^{N-1} \tau_i e^{j\psi} \right) \right\} \end{bmatrix}.$$

Expressing it in terms of summation, one gets

$$E_b = \Gamma_l \left[\sum_{n=1}^{N} j\kappa_n e^{j(n-1)\zeta} \left\{ \sum_{m=1}^{n-1} j\kappa_m e^{j(m-1)\zeta} \left(\prod_{i=m}^{n-1} \tau_i e^{j\psi} \right) \right\} \right]. \tag{3.65}$$

Scattered field due to the self-reflected wave: Some portion of the incident signal gets reflected due to the load at the nth coupler. The scattered field contribution from the self-reflected wave at n^{th} antenna element is given by

$$E_{self_n} \approx \Gamma_l \tau_n^2 e^{j2(n-1)\zeta}. \tag{3.66}$$

The total scattered field contribution from the self-reflection wave is determined by summing (3.56) over the array elements:

$$E_{\text{self}} \approx \Gamma_l \left\{ \sum_{n=1}^{N} \tau_n^2 e^{j2(n-1)\zeta} \right\}. \tag{3.67}$$

Scattered field from reflections at the input load: A part of the received incident signal also gets reflected from the input load. The field scattered by the reflections at the input load for the nth element is given by

$$E_{\text{in}_n} \approx j\kappa_n e^{j(n-1)(\zeta+\psi)} \Gamma_l \left[\prod_{m=1}^{n-1} \tau_m \right]. \tag{3.68}$$

In this case, the antenna array acts as a receiver antenna that collects the signal. The scattering pattern of this RCS component is the gain pattern squared. Thus, the scattered field due to the input load reflected wave from all elements (Lee 1994) is

$$E_{\text{in}} \approx -\Gamma_l \tau_r^2 \tau_p^2 \left\{ \sum_{n=1}^{N} \kappa_n e^{j(n-1)(\zeta+\psi)} \prod_{m=1}^{n-1} \tau_m \right\}^2. \tag{3.69}$$

Substituting (3.57), (3.65), (3.67) and (3.69) into (3.28) one gets the total scattered field due to all four individual contributions. Thus, the RCS due to the antenna scattered field can be expressed as

$$\sigma_s \approx \frac{4\pi A^2 \cos^2\theta}{\lambda^2} \tau_r^4 \tau_p^4 \left| \frac{E^s}{N} \right|^2 . \tag{3.70}$$

3.3.2 RCS pattern analysis

Phased arrays with low gain in their out-band region are considered in this section. This assumption serves to neglect wide-angle scattering and thus nullifies the effect of structural mode. The assumption of low gain in the out-band region eliminates the antenna mode scattering in case of a perfect match between antenna elements. This is because the antenna mode scattering loses its dominance over perfect matching even if its modified reflection coefficient remains large. However, mismatches within the feed network during fabrication are almost unavoidable and thus the antenna mode remains a dominant RCS contributor in case of a phased antenna array operating within its frequency band.

In the case of linear arrays, all elements are assumed to be aligned along the x-axis with equal inter-element spacing, d. The z-axis is taken as the broadside of the array. The radar frequency and operating frequency of the antenna array are taken to be the same. The RCS of a phased array consisting of N identical elements with series feed is computed using the formulation described above. All reflection coefficients are assumed to be the same. The effective scattering height of antenna elements, in the plane transverse to the array axis, is taken as 0.5λ.

The RCS trend is analysed by varying the number of antenna elements, spacing between these elements, beam scan angle, electrical length between couplers and their coefficients.

Number of antenna elements: The broadside RCS of a linear series-fed phased array for $d = 0.4\lambda$, $\theta_s = 0$, $l = 0.5\lambda$ and $\psi = \pi/4$ is determined. Antenna excitation is assumed to be unit amplitude uniform distribution with a sidelobe level (SLL) of 13 dB. Coupling coefficients are chosen appropriately. RCS behaviour is analysed by varying the number of antenna elements.

Figure 3.8 shows the broadside RCS of a 16-element linear phased array. The highest spike at $\theta = 0°$ represents the scattering from antenna aperture, phase shifters and coupler inputs (called specular lobe) while the spike at about -20° can be attributed to the reflection from the input load.

Figures 3.9 represents the broadside RCS of a linear phased antenna array with 64 antenna elements. It may be observed that the location of lobes is similar in both Figures 3.8 and 3.9. However, the lobes are not sharp and well pronounced in the $N = 16$ array as compared to the $N = 64$ array. The sidelobe level also increases with the increase in physical (and hence effective) area of the phased antenna array, and the increase in the number of antenna elements ($Ap = Ndl$). The numbers of sidelobes in the RCS pattern is also directly proportional to the number of antenna elements.

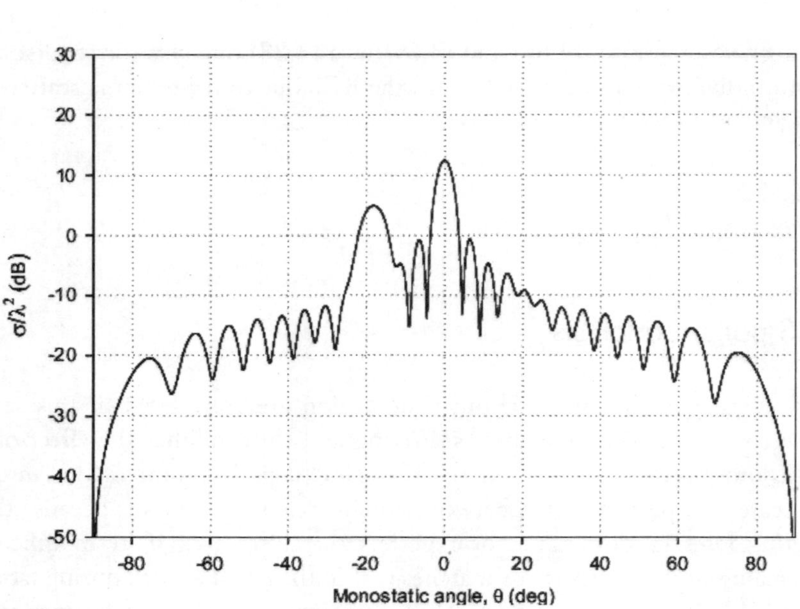

Figure 3.8 RCS of linear array with series feed, N=16, d=0.4, λ θ_s=0°, ψ=π/4, and l =0.5λ with uniform amplitude distribution (unit amplitude)

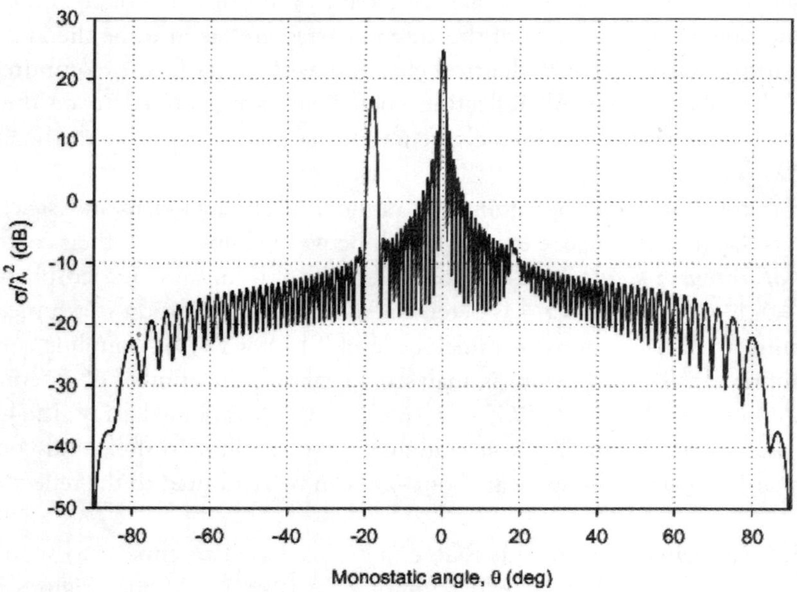

Figure 3.9 RCS of a linear array of N = 64, θ_s=0°, ψ=π/4, d=0.4λ and l =0.5λ with unit amplitude uniform distribution and series feed network

Inter-element spacing: RCS behaviour is analysed by varying the spacing between antenna elements along the array axis while keeping the other parameters fixed for a series-fed phased array, *viz.* uniform distribution (unit amplitude, 13 dB SLL), N =50, θ_s = 0°, l = 0.5λ and $\psi = \pi/4$. Figures 3.10 and 3.11 show the broadside RCS of the linear phased antenna array with inter-element spacing of 0.4λ and 0.5λ, respectively.

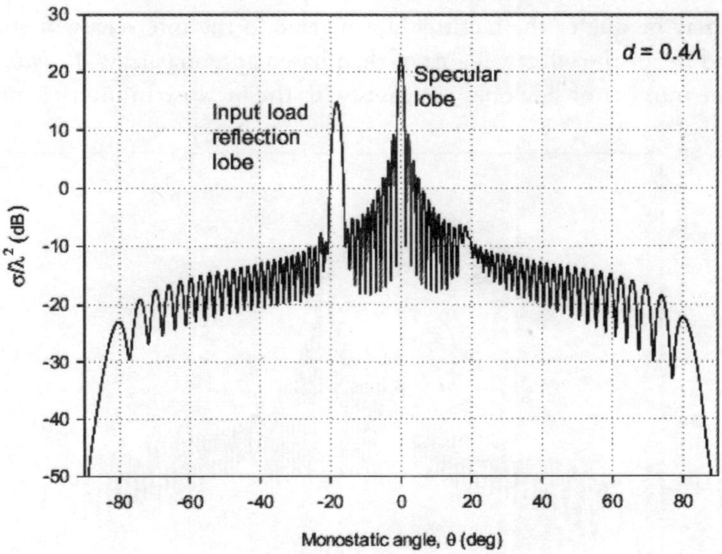

Figure 3.10 RCS of a series-fed linear array of N = 50, θ_s = 0°, $\psi = \pi/4$, d = 0.4λ and l = 0.5λ with uniform amplitude distribution (unit amplitude)

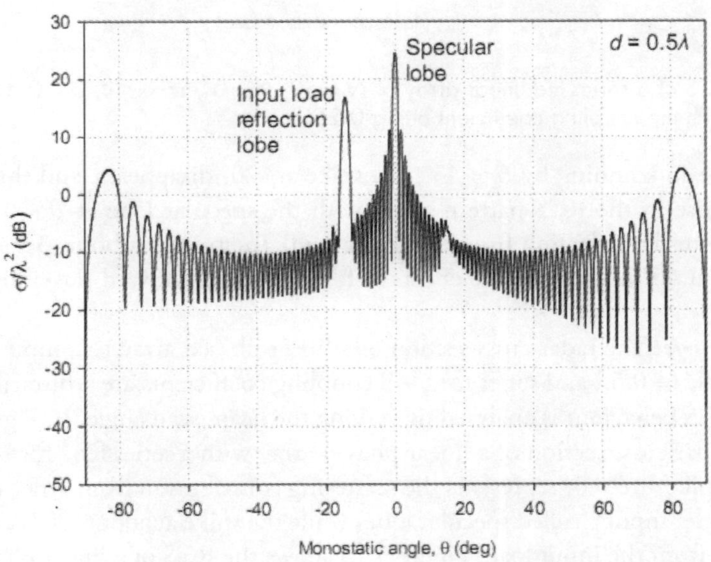

Figure 3.11 RCS of a series-fed linear array of N = 50, θ_s = 0°, $\psi = \pi/4$, d = 0.5λ and l = 0.5λ with uniform amplitude distribution (unit amplitude)

It may be observed that the location of lobes (both specular and input load reflection lobes) remains the same as shown in Figures 3.8 and 3.9. However, the level of both these lobes increases by about 2 dB, when the spacing between the antenna elements increases by 0.1λ. In other words, when the antenna elements are far apart, the lobes in the RCS pattern become narrower. This may be due to the fact that an increase in the inter-element spacing increases the physical (and hence the effective) area of the phased antenna array, $A = Ndl$. It may also be observed that the number of sidelobes increases with the increase in inter-element spacing d.

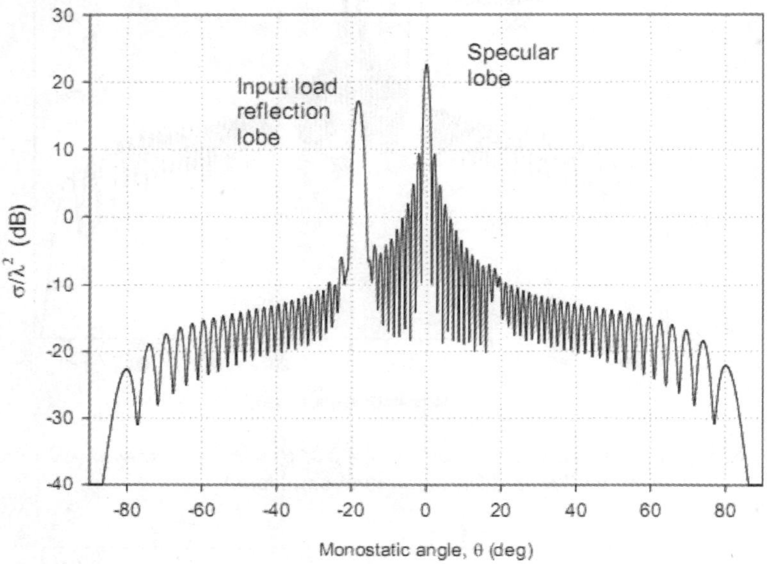

Figure 3.12 RCS of a series-fed linear array of $N = 50$, $\theta_s = 0°$, $\psi = \pi/4$, $d = 0.4\lambda$ and $l = 0.5\lambda$ with the coupling coefficient being 0.25

Due to the beam scanning by $\theta_s = 45°$, the spike at -20° disappears, and three spikes (20°, -35° and 45°) arise in the RCS pattern along with the specular lobe at $\theta = 0°$. The spike at about 20° is due to the reflection from the input load. The spike at about -35° is due to Bragg diffraction, and at 45° it is due to the effects of forward and backward travelling waves on the feed line.

Beam scan angle: The radar cross section of a linear phased array is computed for $N = 50$, $d = 0.4\lambda$, $\psi = \pi/4$, $l = 0.5\lambda$ and series feed. All coupling coefficients are arbitrarily chosen to be 0.25, and the RCS behaviour is analysed by varying the beam scan angle, θ_s. Figure 3.12 shows the broadside radar cross section of a linear phased array with a series feed for $\theta_s = 0°$.

The highest spike at $\theta = 0°$ represents the scattering contribution from array aperture, phase shifters and coupler inputs (called specular lobe) while the spike at about -20° can be attributed to the reflection from the input load. Figure 3.13 shows the RCS of a linear phased array with a series feed network, for $\theta_s = 45°$. The highest spike at $\theta = 0°$ is due to the specular scattering from an aperture, phase shifters and coupler inputs, same as shown in Figure 3.12.

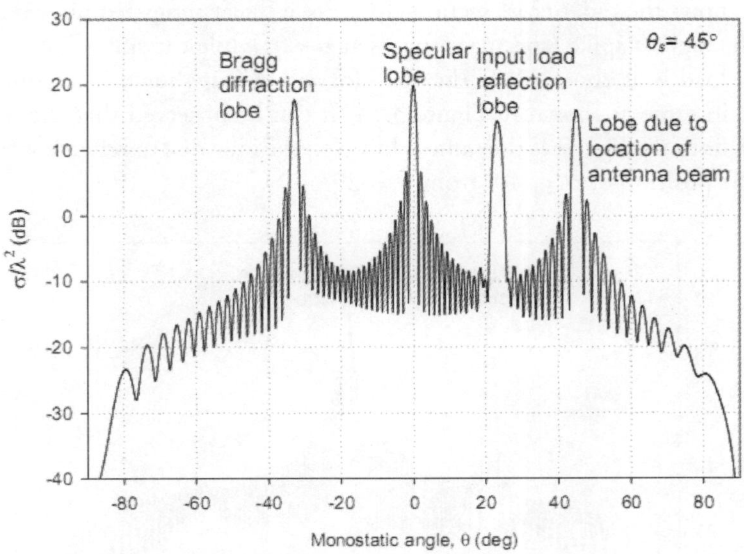

Figure 3.13 RCS of a series-fed linear array of $N = 50$, $\theta_s = 45°$, $\psi = \pi/4$, $d = 0.4\lambda$, $l = 0.5\lambda$ with coupling coefficients being 0.25

Electrical length between the couplers: RCS behaviour is analysed by varying the electrical length between the couplers. Here, the broadside RCS of the 64-element linear phased antenna array with a series feed is computed for $\theta_s = 0°$, $d = 0.4\lambda$ and $l = 0.5\lambda$. The amplitude distribution is taken to be a uniform distribution and coupling coefficients are calculated appropriately.

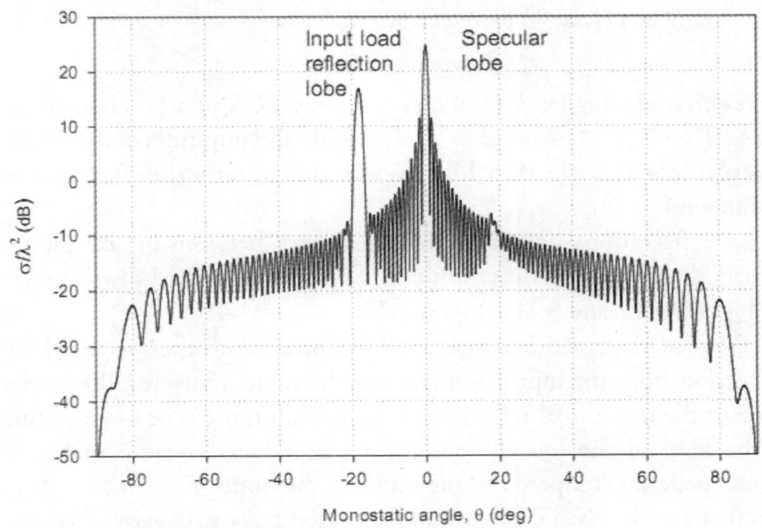

Figure 3.14 RCS of a series-fed linear array of $N = 64$, $\theta_s = 0°$, $\psi = \pi/4$, $d = 0.4\lambda$ and $l = 0.5\lambda$ with uniform amplitude distribution (unit amplitude)

Figure 3.14 shows the radar cross section (RCS) of a linear series-fed phased antenna array for $\psi = \pi/4$. The highest spike (specular lobe) is at $\theta = 0°$ while the spike due to the reflection from the input load is at about -20°. The RCS for $\psi = \pi/2$ is shown in Figure 3.15. All the inputs are remain same as shown in Figure 3.14. It can be observed that the location of the specular lobe in both the figures is the same. However, the lobe due to reflection from the input load shifts to the position of $\theta = -40°$ from $\theta = -20°$.

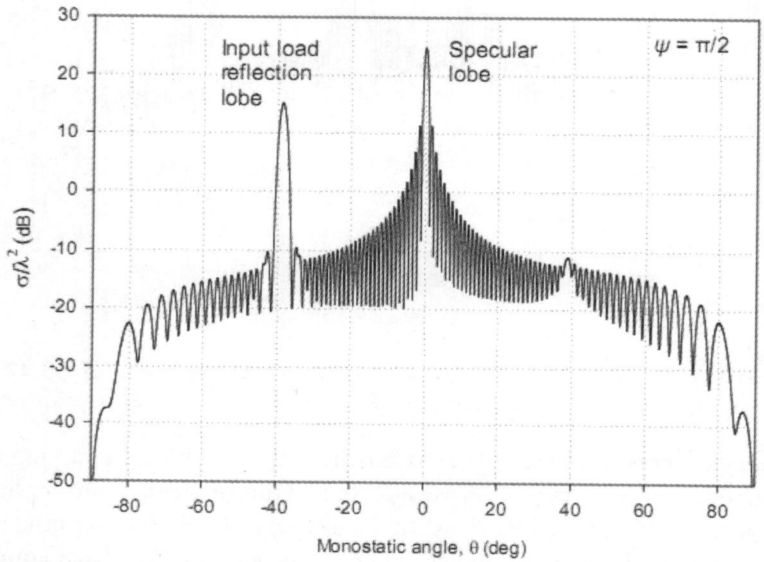

Figure 3.15 RCS of a series-fed linear array of $N = 64$, $\theta_s = 0°$, $\psi = \pi/2$, $d = 0.4\lambda$, $l = 0.5\lambda$ with uniform amplitude distribution (unit amplitude)

Coupling coefficients: Figure 3.16 shows the array RCS of a 50-element series-fed linear array ($l = 0.5\lambda$, $\theta_s = 0°$, $d = 0.4\lambda$ and $\psi = \pi/2$), with all coupling coefficients being equal to 0.25. The specular lobe is at $\theta = 0°$, while the lobe due to reflection from the input load is at –40° (approximately).

This is as per expectations since the electrical length between the couplers, $\psi = \pi/2$. The RCS for uniform distribution (unit amplitude) and cosine squared distribution on a pedestal are shown in Figures 3.17 and 3.18, respectively.

It may be observed that in either case, the location of both the specular lobe and the lobe due to the reflection from the input load remains the same. However, the level and broadness of the lobe due to the input load reflection varies based on the type of amplitude distribution considered. The level of the lobe is low and its broadness is more in the cosine squared distribution on a pedestal compared to the uniform distribution.

Thus to summarise, the RCS estimation of a phased array with series feed network is based on the hypothesis that an incident wave excites the forward and the backward travelling waves on the main line. In general, for a series-fed phased array, the spike (at $\theta = 0°$) in the RCS pattern can be attributed to the effect of specular reflection. Bragg diffraction is not taken into consideration owing to small inter-element spacing.

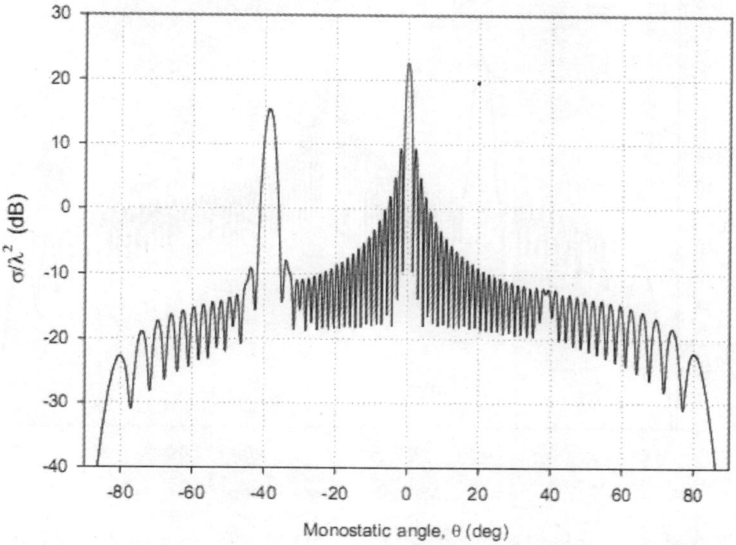

Figure 3.16 RCS of a series-fed linear array of $N = 50$, $\theta_s = 0°$, $\psi = \pi/2$, $d = 0.4\lambda$, $l = 0.5\lambda$ with uniform distribution (coupling coefficient = 0.25)

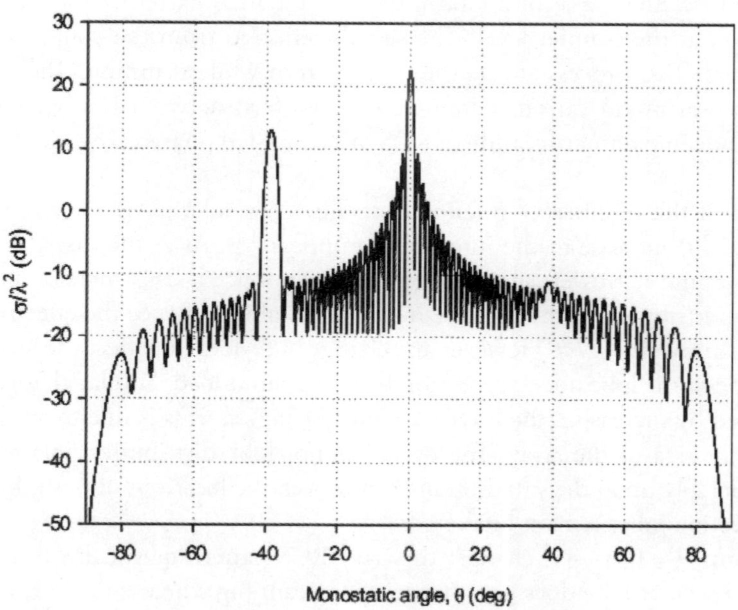

Figure 3.17 RCS of a series-fed linear array of N=50, θ_s=0°, ψ=$\pi/2$, d=0.4λ, l=0.5λ with uniform amplitude distribution (unit amplitude)

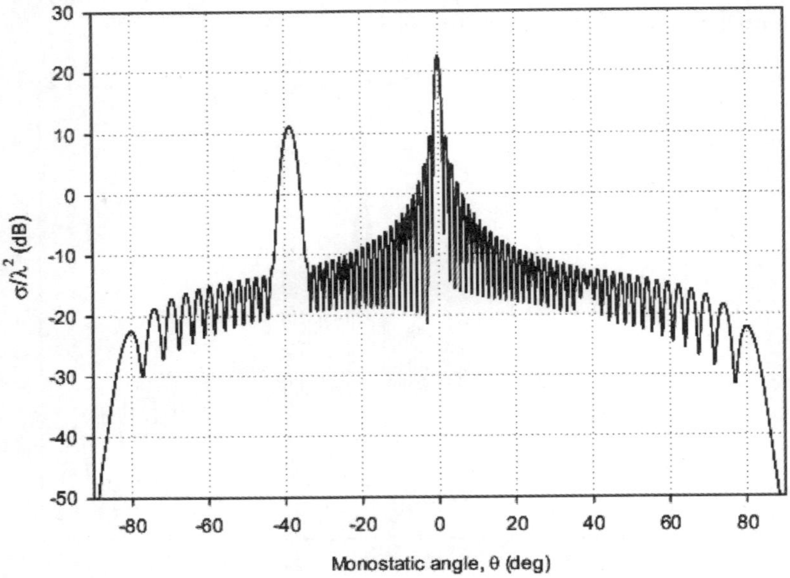

Figure 3.18 RCS of a series-fed linear array of $N = 50$, $\theta_s = 0°$, $\psi = \pi/2$, $d = 0.4\lambda$ and $l = 0.5\lambda$ with cosine squared on a pedestal amplitude distribution

This fact also holds good for the behaviour of reflections from the phase shifters. When the array beam is scanned, a second lobe appears in the RCS pattern. This lobe arises due to Bragg diffraction at the coupler level. The signals reflected from the couplers pass through the phase shifters. This process affects the RCS pattern while scanning. The dependence of RCS of a phased array on various parameters of the feed network has been analysed. The observations made in parametric study of RCS of a series-fed phased array can be summarised as follows:

The increase in the number of antenna elements, N, results in (i) the rise in the level of major lobes and (ii) increase in the number of minor lobes. As N increases, the lobes in the RCS pattern become narrower, sharper and well-pronounced. This increase in the levels of the lobes can be attributed to the increase in the physical (and hence the effective) area of the phased antenna array, $Ap = Ndl$. However, the change in N does not affect the location of either the specular lobe or the lobe due to reflection from the input load. Similar changes in the lobes are also observed by increasing the inter-element spacing, d. This is due to an increase in the effective aperture area of the array. The overall minor lobe distribution in the RCS pattern depends considerably upon the variation in d. However, the locations of both the specular and input load reflection lobes remain unaltered.

Scanning from $\theta_s = 0°$ to 45° changes the array RCS pattern drastically. It is observed that the location of specular lobe does not change upon scanning whereas the location of the lobe due to an input load is shifted. In other words, the lobes due to reflections from the points behind the phase shifters (e.g. the lobe due to the input load reflection) move either towards or away from the specular beam in the scanned RCS pattern. This creates a problem for an array on a stealthy platform if it is intended for tracking a threat. Moreover, additional lobes also appear due to Bragg diffraction and antenna beam location. A large lobe is observed in the RCS pattern due to the reflection from the feed input load. The location of this lobe depends on the electrical length between couplers, ψ. The change in ψ from $\pi/4$ to $\pi/2$ results in shifting the input load reflection lobe location by about 20°, while the location of all other lobes in the broadside RCS pattern remains the same. A similar shift may be observed in the scanned RCS pattern.

It is well known that the electrical gap between devices should be an odd multiple of quarter-wavelength for cancelling the reflections. However, it is difficult to predict the exact value of the electrical length between the couplers that can eliminate the input load reflection completely. Variation in coupling coefficients of the couplers in the feed network changes the level and broadness of the input load reflection lobe. However, the location of the lobe remains unchanged in the RCS pattern. For example, in the cosine squared distribution on a pedestal, the level of the input load reflection lobe in the RCS pattern is lower than that in the case of uniform distribution. Further it is observed that the broadness of the lobe in the former is more as compared to the latter (uniform) distribution. The location, level and the broadness of the specular lobe in the RCS pattern of both the distributions remain unaltered.

The possible errors in the RCS results may be due to the assumptions made in RCS estimation. The effects observed in the RCS pattern due to variation of different parameters are found to be independent. If one varies the parameters in combination, the combined effect is reflected in the RCS pattern. The lobes in the array RCS patterns are mainly due to the reflection of the feed network at different levels. The total control or suppression of RCS can only be achieved by matching all of the feed devices perfectly, which seems impractical.

3.4 Phased array with parallel feed network

A parallel feed network combines the signals coming from the antenna elements arranged along a particular direction (say, *x*-direction) using couplers. Parallel feeds are, in general, suited to rectangular element arrangements. A typical parallel feed network for a linear phased array is shown in Figure 3.19. It comprises phase shifters to facilitate beam scanning (Rudge et al. 1983) and the couplers to either divide (T_x mode) or combine (R_x mode) the power. Reflection and transmission coefficients of various antenna elements are also shown. Reflection coefficients ρ_r, ρ_p and ρ_c correspond to radiating element, phase-shifters, and couplers of the feed network, respectively.

Transmission coefficients τ_r and τ_p correspond to radiating element and phase shifters, respectively. The number of elements fed with the network N_x depends on the number of coupler levels, q, and is expressed as $N_x = 2^q$. The level of couplers considered will have its effect on the RCS of the target. Larger the number of couplers, more will be the lobes in the RCS/scattering pattern obtained.

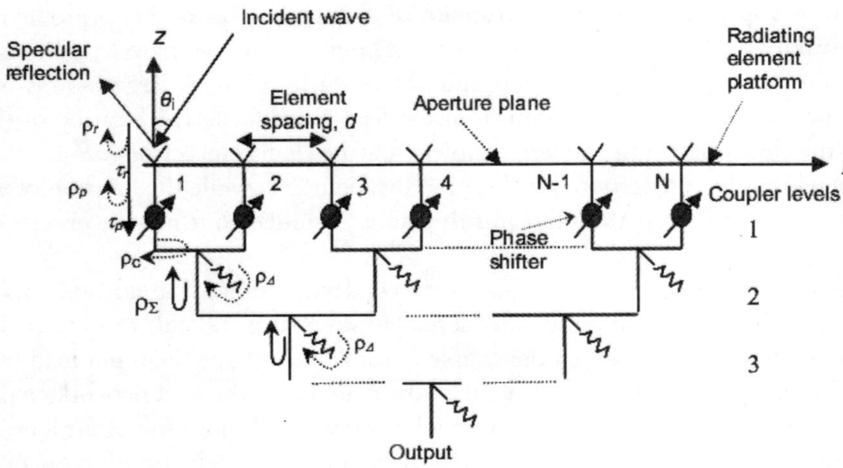

Figure 3.19 Parallel feed for a phased array

3.4.1 RCS formulation with isotropic array elements

The total RCS of the phased array is due to the scattering from all the signals, which enter the array and then return to the aperture after being reflected by the mismatches in the feed. The scattering sources in parallel feed network are antenna aperture, ρ_r, phase shifter inputs, ρ_p, input arms of first-level couplers, ρ_c, loads of sum and difference arm of the couplers at the first level of feed network, i.e. ρ_{Σ_1} and ρ_{Δ_1} and sum and difference arm loads of higher-level couplers, ρ_{Σ_i} and ρ_{Δ_i} Scattering at various mismatches in the feed network continues as the signal travels deeper into the feed network. The vector sum of all individual scattered fields that return to the aperture and re-radiate gives the total RCS of the antenna array. Assumptions made for an approximate model in parallel feed networks are the same as in series feed network, but with some additional features:

1. All couplers are represented by magic tees, which mean that the elements are uniformly excited. Conversely, it implies equal power splitting.
2. For an array with parallel feed network(s) along the *x*-direction, the components/devices along the *y*-direction are perfectly matched. Thus their contribution to the scattering is neglected.
3. Couplers in the fourth and higher levels of the feed are perfectly matched.

The array factors (Flokas 1994) for a linear array consisting of N_x elements along the *x*-axis and for other components (phase shifters/couplers) in the parallel feed network are the same as in the series network, with additional terms specific to the parallel feed. These are array factors of sum and difference arms of couplers at q^{th} level of the feed:

$$AF_{\Sigma_q} = \tau_r^2 \tau_p^2 \rho_\Sigma \prod_{l=1}^{q-1} \cos^2\left(\frac{2^{l-1}\zeta_x}{2}\right) \cos^2\left(\frac{2^{q-1}\zeta_x}{2}\right) \left(\frac{\sin(N_x\zeta_x)}{\left(N_x/2^q\right)\sin\left(2^q\zeta_x\right)}\right), \tag{3.71}$$

$$AF_{\Delta_q} = \tau_r^2 \tau_p^2 \rho_\Delta \prod_{l=1}^{q-1} \cos^2\left(\frac{2^{l-1}\zeta_x}{2}\right) \sin^2\left(\frac{2^{q-1}\zeta_x}{2}\right) \left(\frac{\sin(N_x\zeta_x)}{\left(N_x/2^q\right)\sin\left(2^q\zeta_x\right)}\right), \tag{3.72}$$

and for the planar array ($N_x \times N_y$),

$$\begin{aligned} AF_{\Sigma_q} = {} & \tau_r^2 \tau_p^2 \rho_\Sigma \prod_{l=1}^{q-1} \cos^2\left(\frac{2^{l-1}\zeta_x}{2}\right) \cos^2\left(\frac{2^{q-1}\zeta_x}{2}\right) \\ & \times \left(\frac{\sin(N_x\zeta_x)}{\left(N_x/2^q\right)\sin\left(2^q\zeta_x\right)}\right) \left[\frac{\sin(N_y\zeta_y)}{N_y \sin(\zeta_y)}\right] \end{aligned} \tag{3.73}$$

and

$$\begin{aligned} AF_{\Delta_q} = {} & \tau_r^2 \tau_p^2 \rho_\Delta \prod_{l=1}^{q-1} \cos^2\left(\frac{2^{l-1}\zeta_x}{2}\right) \sin^2\left(\frac{2^{q-1}\zeta_x}{2}\right) \\ & \times \left(\frac{\sin(N_x\zeta_x)}{\left(N_x/2^q\right)\sin\left(2^q\zeta_x\right)}\right) \left[\frac{\sin(N_y\zeta_y)}{N_y \sin(\zeta_y)}\right]. \end{aligned} \tag{3.74}$$

In the parallel feed network of a phased array, fraction of the signal entering the antenna array and returning to the aperture for re-radiation (Jenn and Flokas 1996) is given by

$$\Gamma_{mn}(\theta,\phi) \approx \underbrace{\rho_r e^{j\Delta_{mn}}}_{1^{\text{st}}\text{ term}} + \underbrace{\tau_r^2 \rho_p e^{j\Delta_{mn}}}_{2^{\text{nd}}\text{ term}} + \underbrace{\tau_r^2 \tau_p^2 \rho_c e^{j2\chi_{mn}} e^{j\Delta_{mn}}}_{3^{\text{rd}}\text{ term}} + \underbrace{\tau_r^2 \tau_p^2 \tau_c^2 e^{j\chi_{mn}} \left[(E_1')_{mn} + (E_2')_{mn} + \dots\right]}_{4^{\text{th}}\text{ term}}, \tag{3.75}$$

where, $\Delta_{mn} = (m-1)\alpha + (n-1)\beta$ and $\left(E_q'\right)_{mn}$ represents the reflected signals returning to the element (m,n), which originate at the qth level of couplers. The transmission coefficients are squared as the signal passes through each device twice. The factor of two is not explicitly included in case of the exponential factor multiplying the coupler terms since it is embedded within $\left(E_q'\right)_{mn}$ terms. The first and second terms of (3.75) represent the contributions of radiating elements and phase shifters to the scattered field. The contribution of couplers to the RCS is represented by the third and the fourth terms collectively. The total RCS of the phased array with a parallel feed network is expressed as

$$\sigma(\theta,\phi)=\frac{4\pi A_p^2\cos^2\theta}{\lambda^2}\left|\sum_{m=1}^{N_x}\sum_{n=1}^{N_y}\begin{bmatrix}\rho_r e^{j\Delta_{mn}}+\tau_r^2\rho_p e^{j\Delta_{mn}}+\tau_r^2\tau_p^2\rho_c e^{j2\chi_{mn}}e^{j\Delta_{mn}}\\ +\tau_r^2\tau_p^2\tau_c^2 e^{j\chi_{mn}}\left\{\left(E_1'\right)_{mn}+\left(E_2'\right)_{mn}+\ldots\right\}\end{bmatrix}e^{j\vec{k}.\vec{d}_{mn}}\right|^2.$$

Expanding the terms, one gets

$$\sigma(\theta,\phi)=\frac{4\pi A_p^2\cos^2\theta}{\lambda^2}\left|\begin{array}{l}\sum_{m=1}^{N_x}\sum_{n=1}^{N_y}\rho_r e^{j\Delta_{mn}}e^{j\vec{k}.\vec{d}_{mn}}+\sum_{m=1}^{N_x}\sum_{n=1}^{N_y}\tau_r^2\rho_p e^{j\Delta_{mn}}e^{j\vec{k}.\vec{d}_{mn}}\\ +\sum_{m=1}^{N_x}\sum_{n=1}^{N_y}\tau_r^2\tau_p^2\rho_c e^{j2\chi_{mn}}e^{j\Delta_{mn}}e^{j\vec{k}.\vec{d}_{mn}}\\ +\sum_{m=1}^{N_x}\sum_{n=1}^{N_y}\tau_r^2\tau_p^2\tau_c^2 e^{j\chi_{mn}}\left\{\left(E_1'\right)_{mn}+\left(E_2'\right)_{mn}+\ldots\right\}e^{j\vec{k}.\vec{d}_{mn}}\end{array}\right|. \tag{3.76}$$

Thus, the RCS equation, considering only the radiating elements can be written as

$$\sigma_r=\frac{4\pi A_p^2\cos^2\theta}{\lambda^2}\left|\sum_{m=1}^{N_x}\sum_{n=1}^{N_y}\rho_r e^{j\Delta_{mn}}e^{j\vec{k}.\vec{d}_{mn}}\right|^2, \tag{3.77}$$

where $\vec{k}=k\left(\hat{x}\sin\theta\cos\phi+\hat{y}\sin\theta\sin\phi+\hat{z}\cos\theta\right)$; $\vec{d}_{mn}=\hat{x}(m-1)d_x+\hat{y}(n-1)d_y$.

$$\text{Thus, } \vec{k}.\vec{d}_{mn}=(m-1)\alpha+(n-1)\beta=\Delta_{mn}. \tag{3.78}$$

Substituting (3.78) into (3.77) one gets

$$\sigma_r=\frac{4\pi A_p^2\cos^2\theta}{\lambda^2}\left|\sum_{m=1}^{N_x}\sum_{n=1}^{N_y}\rho_r e^{j\Delta_{mn}}e^{j\Delta_{mn}}\right|^2=\frac{4\pi A_p^2\cos^2\theta}{\lambda^2}\left|\sum_{m=1}^{N_x}\sum_{n=1}^{N_y}\rho_r e^{j2\Delta_{mn}}\right|^2 \tag{3.79}$$

$$=\frac{4\pi A_p^2\cos^2\theta}{\lambda^2}\left|\sum_{m=1}^{N_x}\sum_{n=1}^{N_y}\rho_r e^{j2[(m-1)\alpha+(n-1)\beta]}\right|^2.$$

Splitting the summation and removing the constant ρ_r out of the summation results in

$$\sigma_r=\frac{4\pi A_p^2\cos^2\theta}{\lambda^2}\left|\rho_r\sum_{m=1}^{N_x}e^{j2(m-1)\alpha}\sum_{n=1}^{N_y}e^{j2(n-1)\beta}\right|^2. \tag{3.80}$$

Using array theory, reduction of (3.80) to their closed form leads to

$$\sigma_r=\frac{4\pi A_p^2\cos^2\theta}{\lambda^2}\left|\rho_r e^{j(N_x-1)\alpha}\frac{\sin\left(N_x\alpha\right)}{\sin(\alpha)}e^{j(N_y-1)\beta}\frac{\sin\left(N_y\beta\right)}{\sin(\beta)}\right|^2$$

Normalisation by the total number of elements, N_x along the x-axis and N_y along the y-axis results in

$$\sigma_r = \frac{4\pi A_p^2 \cos^2\theta}{\lambda^2}\left|\rho_r \frac{\sin(N_x\alpha)}{N_x \sin(\alpha)}\frac{\sin(N_y\beta)}{N_y \sin(\beta)} e^{j(N_x-1)\alpha} e^{j(N_y-1)\beta}\right|^2.$$

Neglecting the phase terms form the above equation, one gets

$$\sigma_r \approx \frac{4\pi A_p^2 \cos^2\theta}{\lambda^2}\rho_r^2\left(\frac{\sin(N_x\alpha)}{N_x \sin(\alpha)}\right)^2\left(\frac{\sin(N_y\beta)}{N_y \sin(\beta)}\right)^2. \quad (3.81)$$

A similar mode of analysis yields the RCS, due to phase shifters only, as

$$\sigma_p \approx \frac{4\pi A_p^2 \cos^2\theta}{\lambda^2}\rho_p^2\tau_r^4\left(\frac{\sin(N_x\alpha)}{N_x \sin(\alpha)}\right)^2\left(\frac{\sin(N_y\beta)}{N_y \sin(\beta)}\right)^2. \quad (3.82)$$

Similarly, RCS due to the scattering at the coupler input ports (the third term of (3.75)) can be expressed as

$$\sigma_c = \frac{4\pi A_p^2 \cos^2\theta}{\lambda^2}\left|\sum_{m=1}^{N_x}\sum_{n=1}^{N_y}\tau_r^2\tau_p^2\rho_c e^{j2\chi_{mn}} e^{j\Delta_{mn}} e^{j\vec{k}.\vec{d}_{mn}}\right|^2. \quad (3.83)$$

Substituting (3.78) into (3.83) and simplifying the same leads to

$$\sigma_c = \frac{4\pi A_p^2 \cos^2\theta}{\lambda^2}\left|\sum_{m=1}^{N_x}\sum_{n=1}^{N_y}\tau_r^2\tau_p^2\rho_c e^{j2\chi_{mn}} e^{j2\Delta_{mn}}\right|^2$$

$$= \frac{4\pi A_p^2 \cos^2\theta}{\lambda^2}\left|\sum_{m=1}^{N_x}\sum_{n=1}^{N_y}\tau_r^2\tau_p^2\rho_c e^{j2(\chi_{mn}+\Delta_{mn})}\right|^2.$$

Expansion of terms χ_{mn} and Δ_{mn} results in

$$\sigma_c = \frac{4\pi A_p^2 \cos^2\theta}{\lambda^2}\left|\sum_{m=1}^{N_x}\sum_{n=1}^{N_y}\tau_r^2\tau_p^2\rho_c e^{j2[\{(m-1)\alpha_s+(n-1)\beta_s\}+\{(m-1)\alpha+(n-1)\beta\}]}\right|^2$$

$$= \frac{4\pi A_p^2 \cos^2\theta}{\lambda^2}\left|\sum_{m=1}^{N_x}\sum_{n=1}^{N_y}\tau_r^2\tau_p^2\rho_c e^{j2[(m-1)\zeta_x+(n-1)\zeta_y]}\right|^2$$

$$\sigma_c \approx \frac{4\pi A_p^2 \cos^2\theta}{\lambda^2}\tau_r^4\tau_p^4\rho_c^2\left(\frac{\sin(N_x\zeta_x)}{N_x \sin(\zeta_x)}\right)^2\left(\frac{\sin(N_y\zeta_y)}{N_y \sin(\zeta_y)}\right)^2. \quad (3.84)$$

The fourth term of (3.75) indicates the contribution of scattering at the coupler arms (sum and difference arms) to the antenna RCS. Mathematically, this can be expressed as

$$\sigma_{\Sigma\Delta} = \frac{4\pi A_p^2 \cos^2\theta}{\lambda^2}\left|\sum_{m=1}^{N_x}\sum_{n=1}^{N_y}\tau_r^2\tau_p^2\tau_c^2 e^{j\chi_{mn}}\left\{(E_1')_{mn}+(E_2')_{mn}+\ldots\right\}e^{j\vec{k}.\vec{d}_{mn}}\right|^2. \tag{3.85}$$

Substituting $\vec{d}_{mn} = \hat{x}(m-1)d_x + \hat{y}(n-1)d_y$, one gets

$$\sigma_{\Sigma\Delta} = \frac{4\pi A_p^2 \cos^2\theta}{\lambda^2}\left|\sum_{m=1}^{N_x}\sum_{n=1}^{N_y}\tau_r^2\tau_p^2\tau_c^2\left\{(E_1')_{mn}+(E_1')_{mn}+\ldots\right\}e^{j\left[(m-1)\zeta_x+(n-1)\zeta_y\right]}\right|^2.$$

The transmission coefficient of the couplers is taken to be one as they are assumed to be lossless. Thus, one gets

$$\sigma_{\Sigma\Delta} = \frac{4\pi A_p^2 \cos^2\theta}{\lambda^2}\left|\sum_{m=1}^{N_x}\sum_{n=1}^{N_y}\tau_r^2\tau_p^2\left\{(E_1')_{mn}+(E_2')_{mn}+\ldots\right\}e^{j\left[(m-1)\zeta_x+(n-1)\zeta_y\right]}\right|^2. \tag{3.86}$$

In order to simplify (3.86), the terms $(E_1')_{mn}$, for i =1, 2, …, q, are to be determined. For this, one need to understand the mode of propagation of signals via couplers in the feed network. In the parallel feed network, magic tees are used to combine the signals incident on the adjacent antenna elements. Figure 3.20 shows a first-level coupler used to combine the received signals, $(E_1)_{mn} = 1e^{j\delta_{mn}}$ and $(E_1)_{m'n} = 1e^{j\delta_{m'n}}$ from the antenna elements m and $m' = m+1$. This figure shows that the waves incident on coupler arms have unit amplitude with relative phase angles δ_{mn} and $\delta_{m'n}$. According to Flokas (1994), ($\delta_{mn} + \delta_{m'n}$) collectively represent (i) signal phase including insertion phases of devices between the first level of couplers and aperture, (ii) space path delay relative to the origin for the incident wave, given by

$$\delta_{pn} = (p-1)\zeta_x + (n-1)\zeta_y \text{ with } p = m \text{ or } m', \tag{3.87}$$

(iii) the signals incident on the coupler ($\tau_r\tau_p\tau_c$ on each side arm) are combined by the perfectly matched magic tee as per its scattering matrix κ given by

$$\kappa = \frac{1}{\sqrt{2}}\begin{bmatrix} 0 & 1 & 1 & 0 \\ 1 & 0 & 0 & 1 \\ 1 & 0 & 0 & -1 \\ 0 & 1 & -1 & 0 \end{bmatrix}.$$

This yields the combined signal at the sum arm of the magic tee as

$$(E_\Sigma)_{mn} = (E_1)_{mn}\kappa_{12} + (E_1)_{m'n}\kappa_{13} = \left(e^{j\delta_{mn}}\right)\left(\frac{1}{\sqrt{2}}\right) + \left(e^{j\delta_{m'n}}\right)\left(\frac{1}{\sqrt{2}}\right)$$

$$= \left(\frac{1}{\sqrt{2}}\right)\left(e^{j\delta_{mn}} + e^{j\delta_{m'n}}\right). \tag{3.88}$$

Splitting and multiplication by appropriate exponential factors yields

$$(E_\Sigma)_{mn} = \left(\frac{1}{\sqrt{2}}\right)\left(e^{j\frac{\delta_{mn}}{2}} e^{j\frac{\delta_{mn}}{2}} e^{j\frac{\delta_{m'n}}{2}} e^{-j\frac{\delta_{m'n}}{2}} + e^{j\frac{\delta_{m'n}}{2}} e^{j\frac{\delta_{m'n}}{2}} e^{j\frac{\delta_{mn}}{2}} e^{-j\frac{\delta_{mn}}{2}}\right),$$

$$(E_\Sigma)_{mn} = \sqrt{2}e^{j\frac{(\delta_{mn}+\delta_{m'n})}{2}} \cos\left(\frac{\delta_{m'n} - \delta_{mn}}{2}\right). \tag{3.89}$$

Similarly, the signal at the difference arm of the magic tee is given

$$(E_\Delta)_{mn} = (E_1)_{mn}\kappa_{42} + (E_1)_{m'n}\kappa_{43} = \left(1e^{j\delta_{mn}}\right)\left(\frac{\sqrt{2}}{2}\right) + \left(1e^{j\delta_{m'n}}\right)\left(-\frac{\sqrt{2}}{2}\right) = \left(\frac{\sqrt{2}}{2}\right)\left(e^{j\delta_{mn}} - e^{j\delta_{m'n}}\right).$$

Following the mode analysis similar to that for the case of the sum arm signal, one gets

$$(E_\Delta)_{mn} = \sqrt{2}\left(e^{j\frac{\delta_{mn}+\delta_{m'n}}{2}}\right)\left(\frac{e^{-j\frac{\delta_{m'n}-\delta_{mn}}{2}} - e^{j\frac{\delta_{m'n}-\delta_{mn}}{2}}}{2}\right) = j\sqrt{2}e^{j\frac{(\delta_{mn}+\delta_{m'n})}{2}} \sin\left(\frac{\delta_{m'n} - \delta_{mn}}{2}\right). \tag{3.90}$$

In practice, reflections inside the magic tee can originate by itself in the coupling region, at the junction of the sum port with the side arm of the next level of couplers, or at the difference port load. Other sources of reflection include the mismatches at the coupler input ports (ρ_c at port 2 and port 3) and the mismatches at the sum and difference output ports (ρ_Σ at port 1 and ρ_Δ at port 4).

The net effect of these internal scatterings (Zhang et al. 2011) can be modelled by two reflection coefficients: one for the sum-type of scattering sources (ρ_Σ) and the other for difference-type of scattering sources (ρ_Δ).

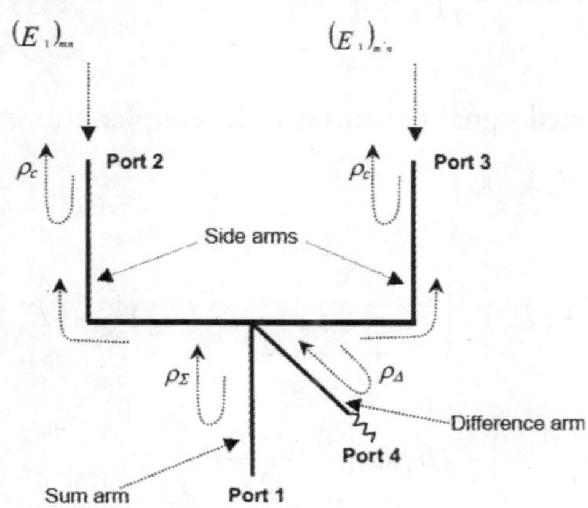

Figure 3.20 Schematic of magic tee with its difference port connected to the load

Thus, the reflected signals emerging from the sum and difference arms of the magic tee can be expressed as

$$(E'_\Sigma)_{mn} = (E_\Sigma)_{mn}\,\rho_\Sigma \tag{3.91a}$$

and

$$(E'_\Delta)_{mn} = (E_\Delta)_{mn}\,\rho_\Delta \tag{3.91b}$$

Likewise, the total reflected signal returning at the coupler arm input, port 2 is given by

$$(E'_1)_{mn} = (E'_\Sigma)_{mn}\,\kappa_{21} + (E'_\Delta)_{mn}\,\kappa_{24}.$$

The values of κ_{21} and κ_{24} can be obtained from the scattering matrix of the magic tee. Thus,

$$(E'_1)_{mn} = (E'_\Sigma)_{mn}\left(\frac{\sqrt{2}}{2}\right) + (E'_\Delta)_{mn}\left(\frac{\sqrt{2}}{2}\right) = \left(\frac{1}{\sqrt{2}}\right)\left[(E'_\Sigma)_{mn} + (E'_\Delta)_{mn}\right].$$

Substituting (3.91a) and (3.91b), one gets $(E'_1)_{mn} = \left(\frac{1}{\sqrt{2}}\right)\left[(E_\Sigma)_{mn}\,\rho_\Sigma + (E_\Delta)_{mn}\,\rho_\Delta\right]$ (3.92)

Substituting (3.89) and (3.90) into (3.92) yields

$$(E'_1)_{mn} = \left(\frac{1}{\sqrt{2}}\right)\begin{bmatrix} \left\{\sqrt{2}e^{j\frac{(\delta_{mn}+\delta_{m'n})}{2}}\cos\left(\frac{\delta_{m'n}-\delta_{mn}}{2}\right)\right\}\rho_\Sigma \\ +\left\{j\sqrt{2}e^{j\frac{(\delta_{mn}+\delta_{m'n})}{2}}\sin\left(\frac{\delta_{m'n}-\delta_{mn}}{2}\right)\right\}\rho_\Delta \end{bmatrix}. \tag{3.93}$$

Removing $\sqrt{2}e^{j\frac{(\delta_{mn}+\delta_{m'n})}{2}}$ as common leads to

$$(E'_1)_{mn} = \left[\rho_\Sigma\cos\left(\frac{\delta_{m'n}-\delta_{mn}}{2}\right) + j\rho_\Delta\sin\left(\frac{\delta_{m'n}-\delta_{mn}}{2}\right)\right]e^{j\frac{(\delta_{mn}+\delta_{m'n})}{2}}. \tag{3.94}$$

Similarly, the total reflected signal returning at the coupler arm input, port 3 is given by

$$(E'_1)_{m'n} = (E'_\Sigma)_{mn}\,\kappa_{31} + (E'_\Delta)_{mn}\,\kappa_{34}$$

$$= (E'_\Sigma)_{mn}\left(\frac{\sqrt{2}}{2}\right) + (E'_\Delta)_{mn}\left(-\frac{\sqrt{2}}{2}\right) = \left(\frac{1}{\sqrt{2}}\right)\left[(E'_\Sigma)_{mn} - (E'_\Delta)_{mn}\right],$$

$$(E'_1)_{m'n} = \left[\rho_\Sigma\cos\left(\frac{\delta_{m'n}-\delta_{mn}}{2}\right) - j\rho_\Delta\sin\left(\frac{\delta_{m'n}-\delta_{mn}}{2}\right)\right]e^{j\frac{(\delta_{mn}+\delta_{m'n})}{2}}. \tag{3.95}$$

The above expressions indicate that the net scattered field is not only dependent on the reflection coefficients, but also on the relative phases of the signals that enter the side arms of the coupler.

From (3.87), $\delta_{pn} = \Delta_{pn} + \chi_{pn} = (p-1)\zeta_x + (n-1)\zeta_y$, with $p = m$ or m'.

Therefore, for index m,

$$\delta_{mn} = (m-1)\zeta_x + (n-1)\zeta_y$$

$$\delta_{m'n} = (m'-1)\zeta_x + (n-1)\zeta_y = m\zeta_x + (n-1)\zeta_y \quad \because m' = m+1 \tag{3.96}$$

Thus, $$\frac{\delta_{m'n} - \delta_{mn}}{2} = \frac{\left[m\zeta_x + (n-1)\zeta_y\right] - \left[(m-1)\zeta_x + (n-1)\zeta_y\right]}{2} = \frac{\zeta_x}{2} \tag{3.97a}$$

and

$$\frac{\delta_{mn} + \delta_{m'n}}{2} = \frac{\left[(m-1)\zeta_x + (n-1)\zeta_y\right] + \left[m\zeta_x + (n-1)\zeta_y\right]}{2}$$

$$= \frac{\zeta_x}{2}(2m-1) + \frac{\zeta_y}{2}(2n-2). \tag{3.97b}$$

Substituting (3.97a) and (3.97b) into (3.94), one gets

$$\left(E_1'\right)_{mn} = \left[\rho_\Sigma \cos\left(\frac{\zeta_x}{2}\right) + j\rho_\Delta \sin\left(\frac{\zeta_x}{2}\right)\right] e^{j\left(\frac{\zeta_x}{2}(2m-1) + \frac{\zeta_y}{2}(2n-2)\right)}. \tag{3.98}$$

Thus, the total scattered field due to odd-numbered coupler ports (sum and difference ports) at the first level can be obtained by summing all the elements of the array. Mathematically,

$$\sum_{m=1,3,\ldots}^{N_x} \left(E_1'\right)_{mn} = \sum_{m=1,3,\ldots}^{N_x} \left[\rho_\Sigma \cos\left(\frac{\zeta_x}{2}\right) + j\rho_\Delta \sin\left(\frac{\zeta_x}{2}\right)\right] e^{j\left(\frac{\zeta_x}{2}(2m-1) + \frac{\zeta_y}{2}(2n-2)\right)}.$$

Removing constant terms out of summation leads to

$$\sum_{m=1,3,\ldots}^{N_x} \left(E_1'\right)_{mn} = \left[\rho_\Sigma \cos\left(\frac{\zeta_x}{2}\right) + j\rho_\Delta \sin\left(\frac{\zeta_x}{2}\right)\right] e^{j\frac{\zeta_y}{2}(2n-2)} \sum_{m=1,3,\ldots}^{N_x} e^{j\frac{\zeta_x}{2}(2m-1)}. \tag{3.99}$$

Similarly, for index m',

$$\delta_{mn} = (m-1)\zeta_x + (n-1)\zeta_y = (m'-2)\zeta_x + (n-1)\zeta_y \quad \because m = m'-1$$

$$\delta_{m'n} = (m'-1)\zeta_x + (n-1)\zeta_y.$$

Thus, $$\frac{\delta_{m'n} - \delta_{mn}}{2} = \frac{\zeta_x}{2}, \tag{3.100a}$$

$$\text{and } \frac{\delta_{mn}+\delta_{m'n}}{2}=\frac{\zeta_x}{2}(2m'-3)+\frac{\zeta_y}{2}(2n-2). \tag{3.100b}$$

By substituting (3.100a) and (3.100b) into (3.95), one gets

$$(E_1')_{m'n}=\left[\rho_\Sigma\cos\left(\frac{\zeta_x}{2}\right)-j\rho_\Delta\sin\left(\frac{\zeta_x}{2}\right)\right]e^{j\left(\frac{\zeta_x}{2}(2m'-3)+\frac{\zeta_y}{2}(2n-2)\right)}. \tag{3.101}$$

Thus, the total scattered field due to even-numbered coupler ports (sum and difference ports) at the first level can be obtained by summing over all the elements of the array. Mathematically,

$$\sum_{m'=2,4,\ldots}^{N_x}(E_1')_{m'n}=\left[\rho_\Sigma\cos\left(\frac{\zeta_x}{2}\right)-j\rho_\Delta\sin\left(\frac{\zeta_x}{2}\right)\right]e^{j\frac{\zeta_y}{2}(2n-2)}\sum_{m'=2,4,\ldots}^{N_x}e^{j\frac{\zeta_x}{2}(2m'-3)}. \tag{3.102}$$

Referring to (3.86), the RCS due to scattering at the first level of coupler ports (sum and difference arms) can be expressed as

$$\sigma_{\Sigma\Delta_1}=\frac{4\pi A_p^2\cos^2\theta}{\lambda^2}\left|\sum_{m=1}^{N_x}\sum_{n=1}^{N_y}\tau_r^2\tau_p^2(E_1')_{mn}\,e^{j[(m-1)\zeta_x+(n-1)\zeta_y]}\right|^2.$$

Interchanging the order of summation and the removal of the constant terms out of summation yields

$$\sigma_{\Sigma\Delta_1}=\frac{4\pi A_p^2\cos^2\theta}{\lambda^2}\tau_r^4\tau_p^4\left|\sum_{n=1}^{N_y}\sum_{m=1}^{N_x}(E_1')_{mn}\,e^{j[(m-1)\zeta_x+(n-1)\zeta_y]}\right|^2. \tag{3.103}$$

Substituting (3.99) and (3.102) into (3.103), one gets

$$\sigma_{\Sigma\Delta_1}=\frac{4\pi A_p^2\cos^2\theta}{\lambda^2}\tau_r^4\tau_p^4\left|\sum_{n=1}^{N_y}\left[\begin{aligned}&\left\{\rho_\Sigma\cos\left(\frac{\zeta_x}{2}\right)+j\rho_\Delta\sin\left(\frac{\zeta_x}{2}\right)\right\}e^{j\frac{\zeta_y}{2}(2n-2)}\\&\times\sum_{m=1,3,\ldots}^{N_x}e^{j\frac{\zeta_x}{2}(2m-1)}e^{j[(m-1)\zeta_x+(n-1)\zeta_y]}\\&+\left\{\rho_\Sigma\cos\left(\frac{\zeta_x}{2}\right)-j\rho_\Delta\sin\left(\frac{\zeta_x}{2}\right)\right\}e^{j\frac{\zeta_y}{2}(2n-2)}\\&\times\sum_{m'=2,4,\ldots}^{N_x}e^{j\frac{\zeta_x}{2}(2m'-3)}e^{j[(m'-1)\zeta_x+(n-1)\zeta_y]}\end{aligned}\right]\right|^2.$$

Rearranging and simplifying, one gets

$$\sigma_{\Sigma\Delta_1}=\frac{4\pi A_p^2\cos^2\theta}{\lambda^2}\tau_r^4\tau_p^4\left|\begin{aligned}&\left\{\rho_\Sigma\cos\left(\frac{\zeta_x}{2}\right)+j\rho_\Delta\sin\left(\frac{\zeta_x}{2}\right)\right\}\sum_{n=1}^{N_y}e^{j(n-1)\zeta_y}e^{j(n-1)\zeta_y}\sum_{m=1,3,\ldots}^{N_x}e^{j\frac{\zeta_x}{2}(2m-1+2m-2)}\\&+\left\{\rho_\Sigma\cos\left(\frac{\zeta_x}{2}\right)-j\rho_\Delta\sin\left(\frac{\zeta_x}{2}\right)\right\}\sum_{n=1}^{N_y}e^{j(n-1)\zeta_y}e^{j(n-1)\zeta_y}\sum_{m'=2,4,\ldots}^{N_x}e^{j\frac{\zeta_x}{2}(2m'-3+2m'-2)}\end{aligned}\right|^2,$$

$$\sigma_{\Sigma\Delta_1} = \frac{4\pi A_p^2\cos^2\theta}{\lambda^2}\tau_r^4\tau_p^4\left|\begin{array}{l}\left\{\rho_\Sigma\cos\left(\frac{\zeta_x}{2}\right)+j\rho_\Delta\sin\left(\frac{\zeta_x}{2}\right)\right\}\sum_{n=1}^{N_y}e^{j2(n-1)\zeta_y}\sum_{m=1,3,\ldots}^{N_x}e^{j\frac{\zeta_x}{2}(4m-3)}\\+\left\{\rho_\Sigma\cos\left(\frac{\zeta_x}{2}\right)-j\rho_\Delta\sin\left(\frac{\zeta_x}{2}\right)\right\}\sum_{n=1}^{N_y}e^{j2(n-1)\zeta_y}\sum_{m'=2,4,\ldots}^{N_x}e^{j\frac{\zeta_x}{2}(4m'-5)}\end{array}\right|^2. \tag{3.104}$$

In order to avoid complexity, the summation terms are evaluated separately. Thus,

$$\sum_{n=1}^{N_y}e^{j2(n-1)\zeta_y} = \sum_{n=0}^{N_y-1}\left(e^{j2\zeta_y}\right)^n = \frac{\sin\left(N_y\zeta_y\right)}{\sin\left(\zeta_y\right)}e^{j(N_y-1)\zeta_y}. \tag{3.105}$$

Considering $m = 2p+1$, for $p = 0, 1, \ldots, \left(N_x/2 - 1\right)$

$$\sum_{m=1,3,\ldots}^{N_x}e^{j\frac{\zeta_x}{2}(4m-3)} = \sum_{p=0}^{\left(\frac{N_x}{2}-1\right)}e^{j\frac{\zeta_x}{2}[4(2p+1)-3]} = \sum_{p=0}^{\left(\frac{N_x}{2}-1\right)}e^{j\frac{\zeta_x}{2}[8p+1]}$$

$$= e^{j\frac{\zeta_x}{2}}\sum_{p=0}^{\left(\frac{N_x}{2}-1\right)}\left(e^{j4\zeta_x}\right)^p = e^{j\frac{\zeta_x}{2}}\frac{\sin\left(\frac{N_x}{2}\frac{4\zeta_x}{2}\right)}{\sin\left(\frac{4\zeta_x}{2}\right)}e^{j\left\{\left(\frac{N_x}{2}-1\right)\frac{4\zeta_x}{2}\right\}},$$

$$\sum_{m=1,3,\ldots}^{N_x}e^{j\frac{\zeta_x}{2}(4m-3)} = \frac{\sin\left(N_x\zeta_x\right)}{\sin\left(2\zeta_x\right)}e^{j\frac{\zeta_x}{2}}e^{j\left\{\left(\frac{N_x}{2}-1\right)2\zeta_x\right\}}. \tag{3.106}$$

Similarly, by taking $m' = 2q$, for $q = 1, 2, \ldots, \left(N_x/2\right)$

$$\sum_{m'=2,4,\ldots}^{N_x}e^{j\frac{\zeta_x}{2}(4m'-5)} = \sum_{q=1}^{\frac{N_x}{2}}e^{j\frac{\zeta_x}{2}[4(2q)-5]} = \sum_{q=0}^{\frac{N_x}{2}-1}e^{j\frac{\zeta_x}{2}[4\{2(q+1)\}-5]} = \sum_{q=0}^{\frac{N_x}{2}-1}e^{j\frac{\zeta_x}{2}[8q+3]}$$

$$= e^{j\frac{3\zeta_x}{2}}\sum_{q=0}^{\frac{N_x}{2}-1}\left(e^{j4\zeta_x}\right)^q = e^{j\frac{3\zeta_x}{2}}\frac{\sin\left(\frac{N_x}{2}\frac{4\zeta_x}{2}\right)}{\sin\left(\frac{4\zeta_x}{2}\right)}e^{j\left\{\left(\frac{N_x}{2}-1\right)\frac{4\zeta_x}{2}\right\}},$$

$$\sum_{m'=2,4,\ldots}^{N_x}e^{j\frac{\zeta_x}{2}(4m'-5)} = \frac{\sin\left(N_x\zeta_x\right)}{\sin\left(2\zeta_x\right)}e^{j\frac{3\zeta_x}{2}}e^{j\left\{\left(\frac{N_x}{2}-1\right)2\zeta_x\right\}}. \tag{3.107}$$

Substituting (3.105) through (3.107) into (3.104) yields

$$\sigma_{\Sigma\Delta_1} = \frac{4\pi A_p^2 \cos^2\theta}{\lambda^2}\tau_r^4\tau_p^4 \left| \begin{array}{l} \left\{\rho_\Sigma \cos\left(\frac{\zeta_x}{2}\right) + j\rho_\Delta \sin\left(\frac{\zeta_x}{2}\right)\right\} \frac{\sin(N_y\zeta_y)}{\sin(\zeta_y)} e^{j(N_y-1)\zeta_y} \\ \times \frac{\sin(N_x\zeta_x)}{\sin(2\zeta_x)} e^{j\frac{\zeta_x}{2}} e^{j\left\{\left(\frac{N_x}{2}-1\right)2\zeta_x\right\}} \\ + \left\{\rho_\Sigma \cos\left(\frac{\zeta_x}{2}\right) - j\rho_\Delta \sin\left(\frac{\zeta_x}{2}\right)\right\} \frac{\sin(N_y\zeta_y)}{\sin(\zeta_y)} e^{j(N_y-1)\zeta_y} \\ \times \frac{\sin(N_x\zeta_x)}{\sin(2\zeta_x)} e^{j\frac{3\zeta_x}{2}} e^{j\left\{\left(\frac{N_x}{2}-1\right)2\zeta_x\right\}} \end{array} \right|^2 .$$

Simplification of exponential terms in the above equation gives

$$\sigma_{\Sigma\Delta_1} = \frac{4\pi A_p^2 \cos^2\theta}{\lambda^2}\tau_r^4\tau_p^4 \left| \begin{array}{l} \left\{\rho_\Sigma \cos\left(\frac{\zeta_x}{2}\right) + j\rho_\Delta \sin\left(\frac{\zeta_x}{2}\right)\right\} \frac{\sin(N_y\zeta_y)}{\sin(\zeta_y)} e^{j(N_y-1)\zeta_y} \\ \times \frac{\sin(N_x\zeta_x)}{\sin(2\zeta_x)} e^{j\left(N_x-\frac{3}{2}\right)\zeta_x} \\ + \left\{\rho_\Sigma \cos\left(\frac{\zeta_x}{2}\right) - j\rho_\Delta \sin\left(\frac{\zeta_x}{2}\right)\right\} \frac{\sin(N_y\zeta_y)}{\sin(\zeta_y)} e^{j(N_y-1)\zeta_y} \\ \times \frac{\sin(N_x\zeta_x)}{\sin(2\zeta_x)} e^{j\left(N_x-\frac{1}{2}\right)\zeta_x} \end{array} \right|^2 .$$

Expanding these terms and then regrouping them results in

$$\sigma_{\Sigma\Delta_1} = \frac{4\pi A_p^2 \cos^2\theta}{\lambda^2}\tau_r^4\tau_p^4 \left| \begin{array}{l} \frac{\sin(N_y\zeta_y)}{\sin(\zeta_y)} e^{j(N_y-1)\zeta_y} \frac{\sin(N_x\zeta_x)}{\sin(2\zeta_x)} \\ \times \left\{ \begin{array}{l} \rho_\Sigma \cos\left(\frac{\zeta_x}{2}\right)\left(e^{j\left(N_x-\frac{3}{2}\right)\zeta_x} + e^{j\left(N_x-\frac{1}{2}\right)\zeta_x}\right) \\ -j\rho_\Delta \sin\left(\frac{\zeta_x}{2}\right)\left(e^{j\left(N_x-\frac{1}{2}\right)\zeta_x} - e^{j\left(N_x-\frac{3}{2}\right)\zeta_x}\right) \end{array} \right\} \end{array} \right|^2 .$$

Removing the common factor $e^{-j\zeta_x/2}$ then the above equation yields

$$\sigma_{\Sigma\Delta_1} = \frac{4\pi A_p^2 \cos^2\theta}{\lambda^2}\tau_r^4\tau_p^4 \left| \begin{array}{l} \dfrac{\sin(N_y\zeta_y)}{\sin(\zeta_y)} e^{j(N_y-1)\zeta_y} \dfrac{\sin(N_x\zeta_x)}{\sin(2\zeta_x)} \\ \times \left\{ \begin{array}{l} \rho_\Sigma \cos\left(\dfrac{\zeta_x}{2}\right) e^{j\left(N_x\zeta_x - \frac{\zeta_x}{2} - \frac{\zeta_x}{2}\right)} \left(e^{-j\frac{\zeta_x}{2}} + e^{j\frac{\zeta_x}{2}}\right) \\ -j\rho_\Delta \sin\left(\dfrac{\zeta_x}{2}\right) e^{j\left(N_x\zeta_x - \frac{\zeta_x}{2} - \frac{\zeta_x}{2}\right)} \left(e^{j\frac{\zeta_x}{2}} - e^{-j\frac{\zeta_x}{2}}\right) \end{array} \right\} \end{array} \right|^2,$$

$$\sigma_{\Sigma\Delta_1} = \frac{4\pi A_p^2 \cos^2\theta}{\lambda^2}\tau_r^4\tau_p^4 \left| \begin{array}{l} \dfrac{\sin(N_y\zeta_y)}{\sin(\zeta_y)} e^{j(N_y-1)\zeta_y} \dfrac{\sin(N_x\zeta_x)}{\sin(2\zeta_x)} e^{j(N_x-1)\zeta_x} \\ \times \left\{ 2\rho_\Sigma \cos^2\left(\dfrac{\zeta_x}{2}\right) + 2\rho_\Delta \sin^2\left(\dfrac{\zeta_x}{2}\right) \right\} \end{array} \right|^2.$$

Normalisation by the total number of elements, N_x and N_y leads to

$$\sigma_{\Sigma\Delta_1} = \frac{4\pi A_p^2 \cos^2\theta}{\lambda^2}\tau_r^4\tau_p^4 \left| \begin{array}{l} \dfrac{\sin(N_y\zeta_y)}{N_y\sin(\zeta_y)} e^{j(N_y-1)\zeta_y} \dfrac{\sin(N_x\zeta_x)}{N_x\sin(2\zeta_x)} e^{j(N_x-1)\zeta_x} \\ \times \left\{ 2\rho_\Sigma \cos^2\left(\dfrac{\zeta_x}{2}\right) + 2\rho_\Delta \sin^2\left(\dfrac{\zeta_x}{2}\right) \right\} \end{array} \right|^2.$$

Neglecting the phase terms from the above equation, one gets

$$\begin{aligned} \sigma_{\Sigma\Delta_1} \approx{}& \frac{4\pi A_p^2 \cos^2\theta}{\lambda^2}\tau_r^4\tau_p^4 \left(\frac{\sin(N_y\zeta_y)}{N_y\sin(\zeta_y)}\right)^2 \left(\frac{\sin(N_x\zeta_x)}{N_x\sin(2\zeta_x)}\right)^2 \left\{2\rho_\Sigma\cos^2\left(\frac{\zeta_x}{2}\right)\right\}^2 \\ &+ \frac{4\pi A_p^2 \cos^2\theta}{\lambda^2}\tau_r^4\tau_p^4 \left(\frac{\sin(N_y\zeta_y)}{N_y\sin(\zeta_y)}\right)^2 \left(\frac{\sin(N_x\zeta_x)}{N_x\sin(2\zeta_x)}\right)^2 \left\{2\rho_\Delta\sin^2\left(\frac{\zeta_x}{2}\right)\right\}^2. \end{aligned}$$

Rearranging these terms, one gets

$$\begin{aligned} \sigma_{\Sigma\Delta_1} \approx{}& \frac{4\pi A_p^2 \cos^2\theta}{\lambda^2}\tau_r^4\tau_p^4\rho_\Sigma^2\cos^4\left(\frac{\zeta_x}{2}\right) \left(\frac{\sin(N_x\zeta_x)}{N_x/2\,\sin(2\zeta_x)}\right)^2 \left[\frac{\sin(N_y\zeta_y)}{N_y\sin(\zeta_y)}\right]^2 \\ &+ \frac{4\pi A_p^2 \cos^2\theta}{\lambda^2}\tau_r^4\tau_p^4\rho_\Delta^2\sin^4\left(\frac{\zeta_x}{2}\right) \left(\frac{\sin(N_x\zeta_x)}{N_x/2\,\sin(2\zeta_x)}\right)^2 \left[\frac{\sin(N_y\zeta_y)}{N_y\sin(\zeta_y)}\right]^2 \\ \approx{}& \sigma_{\Sigma_1} + \sigma_{\Delta_1}, \end{aligned} \qquad (3.108)$$

where σ_{Σ_1} and σ_{Δ_1} represent the RCS due to the scattering contribution from the sum and difference arms of the couplers at the first level of the feed, respectively.
Substituting the array factors for these arms into the general RCS equation for an array of *x*-polarised elements, one gets

$$\sigma_{\Sigma_1} = \frac{4\pi A_p^2 \cos^2\theta}{\lambda^2} \tau_r^4 \tau_p^4 \rho_\Sigma^2 \cos^4\left(\frac{\zeta_x}{2}\right) \left(\frac{\sin(N_x\zeta_x)}{\left(N_x/2\right)\sin(2\zeta_x)}\right)^2 \left[\frac{\sin(N_y\zeta_y)}{N_y\sin(\zeta_y)}\right]^2 \tag{3.109}$$

and

$$\sigma_{\Delta_1} = \frac{4\pi A_p^2 \cos^2\theta}{\lambda^2} \tau_r^4 \tau_p^4 \rho_\Delta^2 \sin^4\left(\frac{\zeta_x}{2}\right) \left(\frac{\sin(N_x\zeta_x)}{\left(N_x/2\right)\sin(2\zeta_x)}\right)^2 \left[\frac{\sin(N_y\zeta_y)}{N_y\sin(\zeta_y)}\right]^2 . \tag{3.110}$$

It can be noted that expressions (3.109) and (3.110) are similar to the two terms of (3.108). This indicates that the RCS due to the sum and difference arms of the couplers at any level can be obtained by substituting appropriate array factors into the general RCS equation. Thus the RCS contributions from the second-level couplers, sum-type and difference-type, respectively, are as follows:

$$\sigma_{\Sigma_2} = \frac{4\pi A_p^2 \cos^2\theta}{\lambda^2} \tau_r^2 \tau_p^2 \rho_\Sigma{}^2 \cos^4\left[\frac{\zeta_x}{2}\right] \cos^4[\zeta_x] \left(\frac{\sin(N_x\zeta_x)}{\left(N_x/4\right)\sin(4\zeta_x)} \frac{\sin(N_y\zeta_y)}{(N_y)\sin(\zeta_y)}\right)^2 , \tag{3.111}$$

$$\sigma_{\Delta_2} = \frac{4\pi A_p^2 \cos^2\theta}{\lambda^2} \tau_r^2 \tau_p^2 \rho_\Delta{}^2 \cos^4\left[\frac{\zeta_x}{2}\right] \sin^4[\zeta_x] \left(\frac{\sin(N_x\zeta_x)}{\left(N_x/4\right)\sin(4\zeta_x)} \frac{\sin(N_y\zeta_y)}{(N_y)\sin(\zeta_y)}\right)^2 . \tag{3.112}$$

In general, the expressions for q level of couplers is written as

$$\sigma_{\Sigma_q} = \tau_r^4 \tau_p^4 \rho_\Sigma^2 \prod_{l=1}^{q-1} \cos^4\left(\frac{2^{l-1}\xi_x}{2}\right) \cos^4\left(\frac{2^{q-1}\xi_x}{2}\right) \times \left[\left(\frac{\sin(N_x\zeta_x)}{\left(N_x/2^q\right)\sin(2^q\zeta_x)} \frac{\sin(N_y\zeta_y)}{N_y\sin(\zeta_y)}\right)^2\right] \tag{3.113}$$

and

$$\sigma_{\Delta_q} = \tau_r^4 \tau_p^4 \rho_\Delta^2 \prod_{l=1}^{q-1} \cos^4\left(\frac{2^{l-1}\xi_x}{2}\right) \sin^4\left(\frac{2^{q-1}\xi_x}{2}\right)$$
$$\times \left[\left(\frac{\sin(N_x \zeta_x)}{\left(N_x/2^q\right)\sin(2^q \zeta_x)} \frac{\sin(N_y \zeta_y)}{N_y \sin(\zeta_y)}\right)^2\right]. \tag{3.114}$$

The equations (3.109) through (3.114) lead to the inference that in order to include the effect of one or more coupler levels, two terms, σ_Σ and σ_Δ, for each level are to be added to the existing RCS expression. Thus, considering q-levels of couplers the overall RCS of the phased array is given by

$$\sigma \approx \sigma_r + \sigma_p + \sigma_c + \sum_{i=1}^{q} \sigma_{\Sigma_i} + \sum_{i=1}^{q} \sigma_{\Delta_i}. \tag{3.115}$$

3.4.2 RCS pattern analysis

The RCS of a parallel fed phased array is computed for linear and planar arrays. For linear arrays, all the elements are aligned along the x-axis with equal inter-element spacing, d_x. The z-axis is taken as the broadside of the array. In planar arrays, all the elements are assumed to be in the x–y plane with uniform spacing of d_x and d_y between the elements N_x and N_y, respectively. The radar and operating frequencies of the antenna array are assumed to be the same. All reflection coefficients are assumed to be constant. The RCS of the phased array is computed for both broadside and scanned beams. The level of couplers is chosen to be q, i.e. for $n > q$, the terms ρ_Σ and ρ_Δ vanish. These sub-arrays of q elements are coupled at the aperture.

Table 3.2 Symbols used in the plots and their significance

Symbol	Cause for the appearance of the lobe
♦	Lobe due to specular scattering
○	Lobe due to scattering from the first level of couplers
●	Lobe due to scattering from the second level of couplers
□	Lobe due to scattering from the third level of couplers
▲	Lobe due to Bragg diffraction
◇	Antenna beam location lobe
■	Lobes due to beam scan of $\theta_s = 45°$

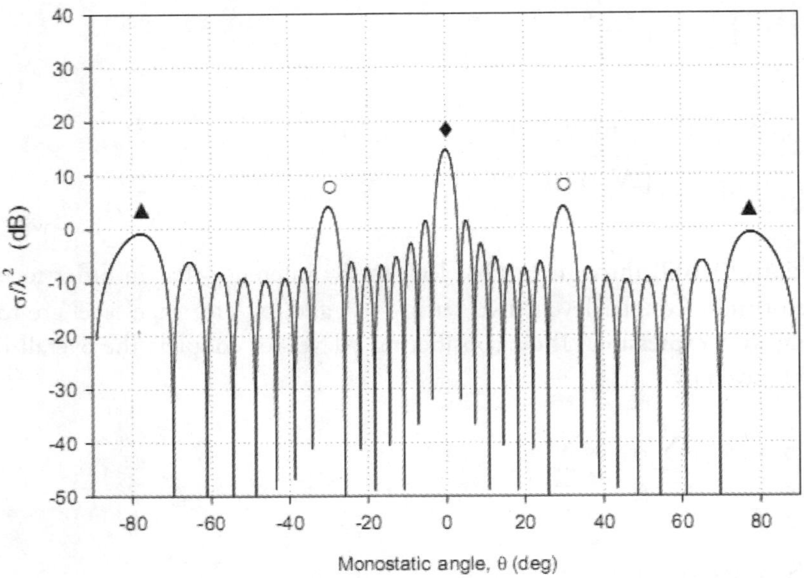

Figure 3.21 RCS of a parallel-fed linear array with $N_x = 16$, $d_x = 0.5\lambda$, $h = 0.5\lambda$, $\theta_s = \phi_s = \phi = 0°$, and level of couplers, $q = 1$

Any large phased arrays can be decomposed into arrays of q-element sub-arrays with the appropriate phase-shifted excitation for analysis. However, any change in the parameter ψ, the electrical path length between the couplers, does not seem to affect the RCS value. This is because the approximate model used here, considers only the non-coherent sum of individual scattering contributions from the devices and components.

Linear array: The RCS of a parallel-fed linear antenna array of identical elements with $N_x \times 1$ (i.e. for $N_y = 1$) is analysed by varying the parameters, *viz.* the number of antenna element, the inter-element spacing, the antenna effective height, beam scan angle and the level of couplers. Figure 3.21 shows the broadside RCS for a parallel-fed linear array of 16 elements with a single coupler level ($q = 1$) and $d_x = 0.5\lambda$, $h = 0.5\lambda$ and $\theta_s = \phi_s = \phi = 0°$. The conventions used to identify these lobes in the RCS pattern is given in Table 3.2. The highest lobe at $\theta = 0°$ represents a specular lobe due to scattering from the antenna aperture. The lobes near $\theta = \pm 80°$ arise due to Bragg diffraction. Spikes at $\theta = \pm 30°$ correspond to the mismatches between the first-level couplers.

Figure 3.22 presents the broadside RCS of a 128-element linear phased array for similar arrangements. The location of major lobes remain unaltered (neglecting a slight shift of about 5° in the Bragg diffraction lobe) while their levels increase in magnitude as the number of antenna elements increases. It is also evident that the lobes in the pattern become narrower, sharper and hence more pronounced with the increase in the number of antenna elements. This is because of the increase in the physical (and hence effective) area of the phased antenna array, $A_p = N_x N_y d_x h$. It may also be observed that the increase in the number of antenna elements in the array increases the number of sidelobes.

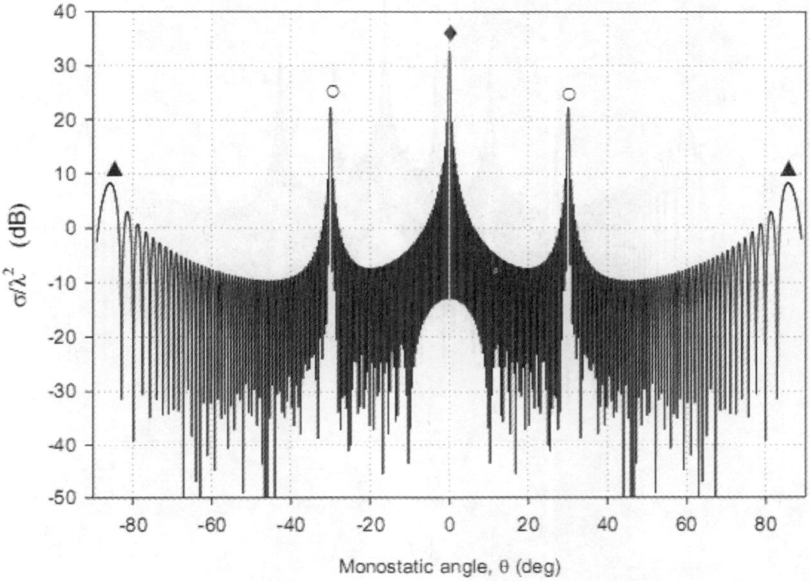

Figure 3.22 RCS of a parallel-fed linear array with $N_x = 128$, $d_x = 0.5\lambda$, $h = 0.5\lambda$, $\theta_s = \phi_s = \phi = 0°$, and level of couplers, $q = 1$

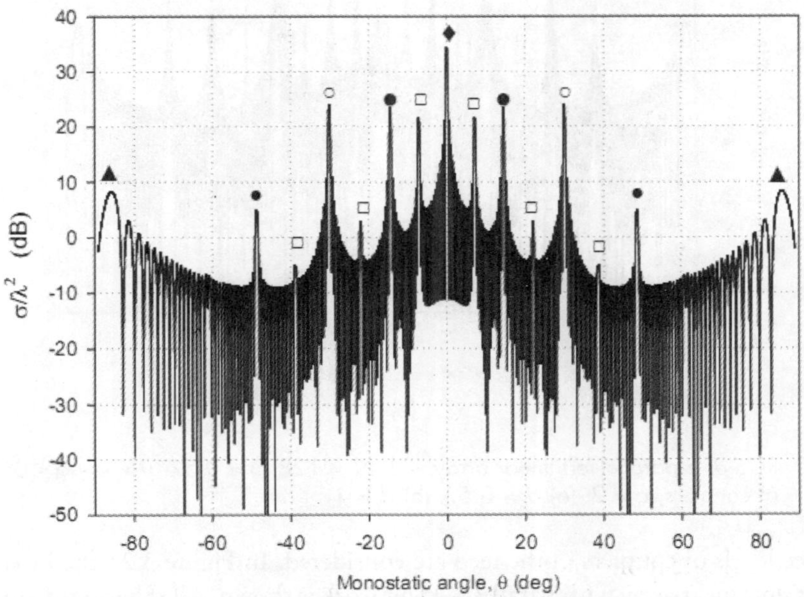

Figure 3.23 RCS of a parallel-fed linear array with $N_x = 128$, $d_x = 0.5\lambda$, $h = 0.5\lambda$, $\theta_s = \phi_s = \phi = 0°$, and level of couplers, $q = 3$

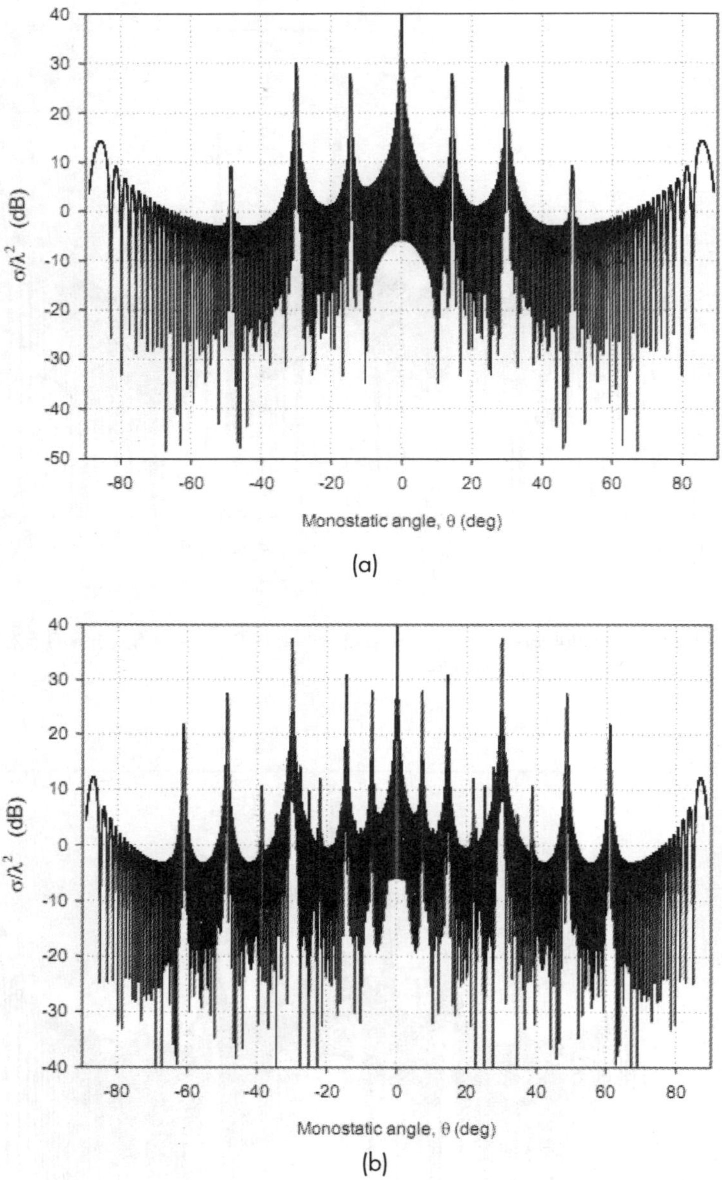

Figure 3.24 RCS of a parallel-fed linear array with $N_x = 128$, $h = 0.5\lambda$, $\theta_s = \phi_s = \phi = 0°$, and level of couplers, $q = 2$: (a) $d_x = 0.5\lambda$, (b) $d_x = \lambda$

Next, three levels of couplers in the feed are considered. In Figure 3.23 the broadside RCS of a 128-element linear array with parallel feed network is shown. All other inputs are the same as shown in Figure 3.22. The lobes at $\theta = 0°$, ±30° and ±80° (approximately) can be attributed to specular scattering, Bragg diffraction and the mismatches between the first-level couplers, respectively. However, additional lobes at θ = ±15° and ±45° arise due to the mismatches between the second-level couplers while those at θ=±7° and ±21° are due to the mismatches between the third-level couplers. It may be inferred from the above fact that identification of lobes due to the scattering from specific levels of couplers is difficult when small sized arrays are considered, e.g. it is difficult to identify the lobes, arising due to the scattering of couplers above the first level, in a 16-element linear array (Figure 3.21). However, this is not true for large sized arrays, e.g. the lobes arising due to the specific scattering sources may be easily identified in a 128-element array.

The variation of element spacing along the axis of the array, d_x, affects the RCS pattern more prominently than the change in element spacing along the direction broadside to the array. This is evident from Figure 3.24, which shows many additional lobes appearing in the pattern due to the variation of d_x from 0.5λ (Figure 3.24a) to λ (Figure 3.24b), keeping all other parameters constant. The levels and location of the lobes due to specular scattering and Bragg diffraction remain the same. This is because the area of the antenna array remains the same, i.e. $d_x \times h = 0.5\lambda^2$. However, the lobes arising due to the mismatches between couplers differ significantly. It may be due to the fact that the change in d_x may affect the pattern directly, while the change in h would only change the amplitude of the lobes.

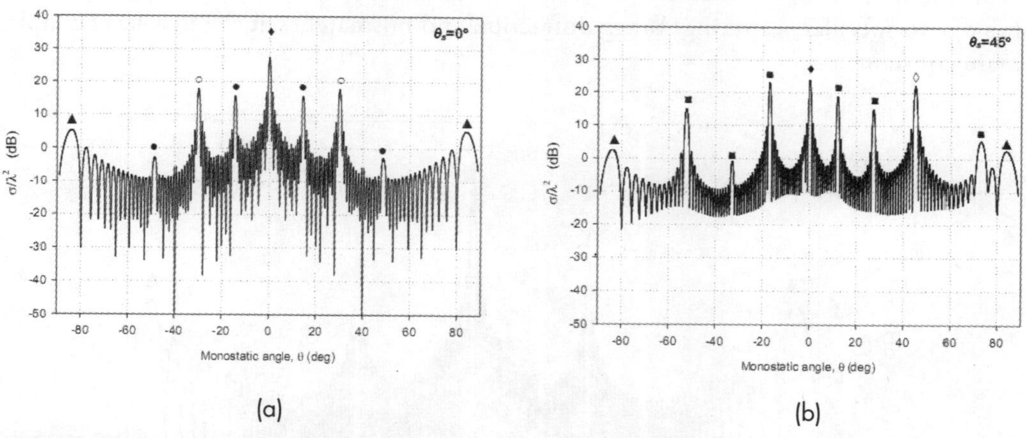

Figure 3.25 RCS of a parallel-fed linear array with N_x = 64, d_x = 0.5λ, h = 0.5λ, $\phi_s = \phi$ = 0°, and level of couplers, q=2 (a) θ_s = 0°, (b) θ_s = 45°

The dependence of RCS on the beam scan angle, θ_s is considered in the following. A linear phased array with a parallel feed network, with N_x = 64, d_x = 0.5λ, h = 0.5λ, $\phi_s = \phi$ = 0° and q = 2, is considered. Figure 3.25 shows the broadside RCS for a parallel-fed linear array with θ_s = 0° and θ_s = 45°. The specular lobe at θ = 0° represents the scattering from the antenna aperture (Figure 3.25a). The lobes close to θ = ±85° arise from Bragg diffraction; the lobes at θ = ±30° are due to the mismatches between the couplers at first level, while those at θ = ±15° and at θ = ±45° are due to the coupler mismatches at level two. It may be observed that even after scanning (Figure 3.25b), the position of the specular lobe and

Bragg diffraction lobe remains unaltered. However, the lobes due to the mismatches at first level of couplers get shifted to positions at $\theta = +10°$ and $\theta = -50°$ from their initial position at $\theta = \pm 30°$.

Similarly, the lobes associated with the mismatches at the second and third levels of couplers also get shifted from their initial positions. The large lobe, at $\theta = 45°$, is due to in-phase addition of the scattered signals passing through the phase shifters. It is evident that the lobes due to mismatches beyond phase shifters of the feed network scan along with the antenna beam. This is because the factor χ_{mn} of the phase shifters, due to the terms ζ_x and ζ_y, depends on the beam scan angle, θ_s. However, the RCS components arising from the reflections due to radiating elements and the components before phase shifters are exempted from the χ_{mn} term and thus remain fixed as illustrated by the specular lobe at $\theta = 0°$.

The effect of the number of couplers levels in the feed network on the array RCS is discussed in the following. Let us consider the RCS of a 64-element linear array with $dx = 0.5\lambda$, $h = \lambda$, $\theta_s = \phi_s = \phi = 0°$ and parallel feed network. Figure 3.26 shows the RCS considering the single level of couplers. The highest spike at $\theta = 0°$ can be attributed to scattering from the antenna aperture while the ones close to $\theta = \pm 85°$ arise due to Bragg diffraction. Spikes at the position of $\theta = \pm 30°$ correspond to the mismatches between the first-level couplers.

For one more coupler level, i.e. for $q = 2$, (Figure 3.27), the location and levels of the lobes due to specular scattering, Bragg diffraction and mismatches at the first-level couplers remain constant.

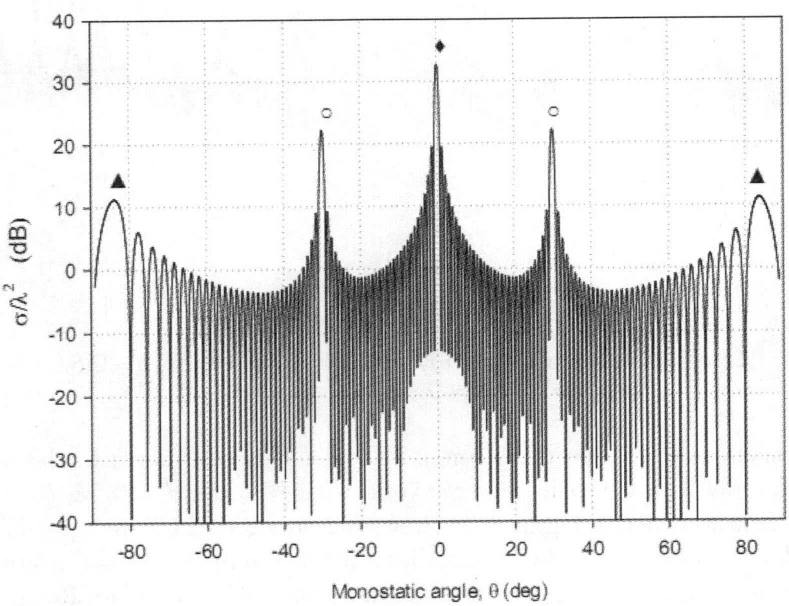

Figure 3.26 RCS of a parallel-fed linear array with $N_x = 64$, $d_x = 0.5\lambda$, $h = \lambda$, $\theta_s = \phi_s = \phi = 0°$, and level of couplers, $q = 1$

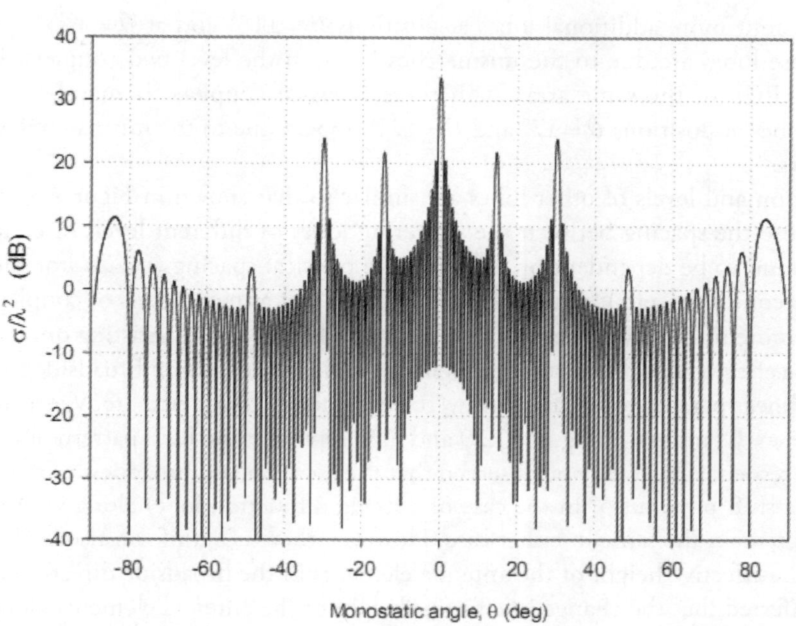

Figure 3.27 RCS of a parallel-fed linear array with $N_x = 64$, $d_x = 0.5\lambda$, $h = \lambda$, $\theta_s = \phi_s = \phi = 0°$, and level of couplers, $q = 2$

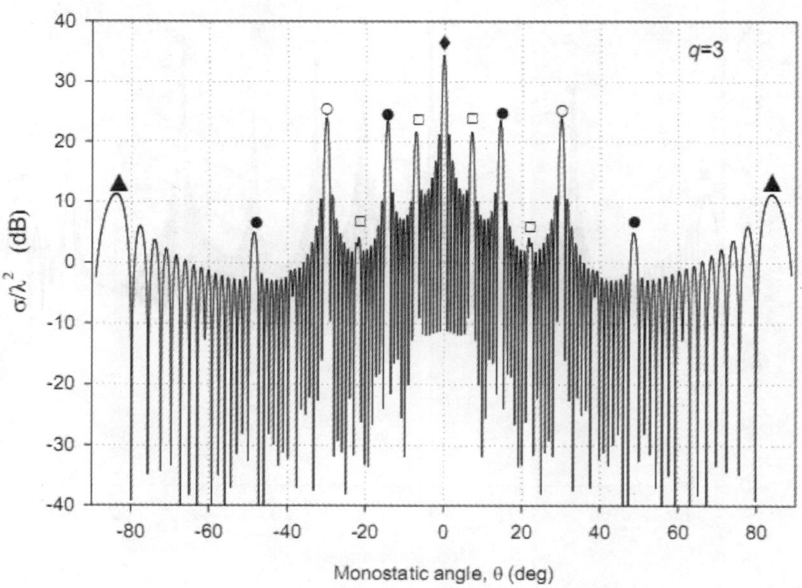

Figure 3.28 RCS of a parallel-fed linear array with $N_x = 64$, $d_x = 0.5\lambda$, $h = \lambda$, $\theta_s = \phi_s = \phi = 0$, and level of couplers, $q = 3$

However, four more additional lobes at positions $\theta = \pm 15°$ and at $\theta = \pm 45°$ appear in the pattern. These lobes are due to the mismatches between the level-two couplers. Figure 3.28 presents the RCS of the same array with three levels of couplers. It may be seen that the additional lobes at positions $\theta = \pm 7°$ and $\theta = \pm 21°$ appear due to the mismatches at the third-level couplers.

The location and levels of other lobes are similar to that shown in Figures 3.26 and 3.27. This is because the spacing between the scattering lobes of different levels of couplers in the pattern is found to be dependent on their relative physical spacing (i.e. $2d_x$ for the first level, $4d_x$ for the second level, etc.). Thus, it can be inferred that as more levels of couplers are added to the feed, more number of lobes would appear amidst the already existing ones.

Lastly, the effective height of antenna elements along the direction broadside to the array is varied for a linear parallel-fed phased array. The parameters considered are $N_x = 128$, $d_x = 0.5\lambda$, $\theta_s = 45°$, $\phi_s = \phi = 0°$ and $q = 2$. Figures 3.29 and 3.30 represent the RCS pattern obtained for $h = 0.5\lambda$ and λ, respectively. It may be observed that the location and broadness of the major lobes (neglecting a shift of about 5° in the case of a Bragg diffraction lobe) along with those of the minor lobes remain the same in either case. However, there is a uniform rise in the level of all lobes with the effective height of the antenna elements in the broadside direction of the array. It may be inferred that the change in effective height of the antenna elements along the y-axis changes only the levels of the lobes but not their location, no matter whether it is broadside or the scanned RCS pattern.

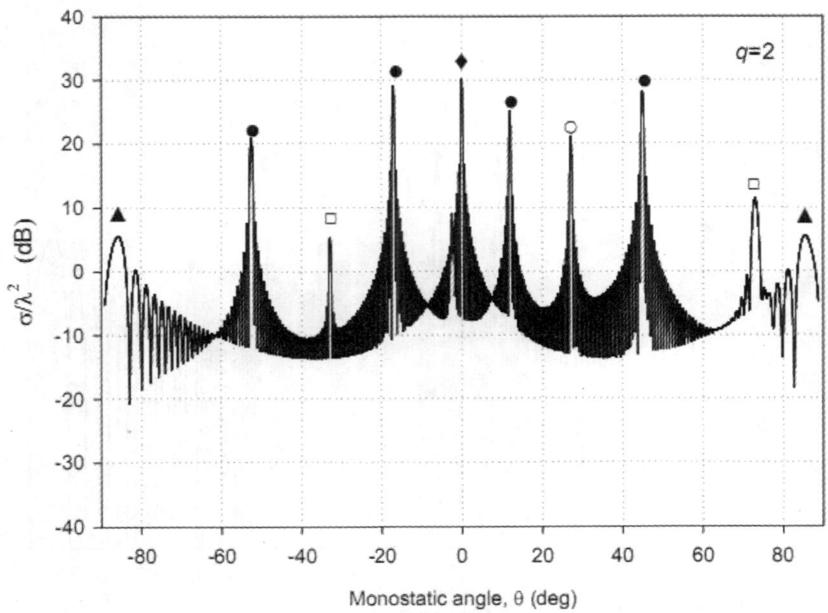

Figure 3.29 RCS of a parallel-fed linear array with $N_x = 128$, $d_x = 0.5\lambda$, $h = 0.5\lambda$, $\theta_s = 45°$, $\phi_s = \phi = 0°$, and level of couplers, $q = 2$

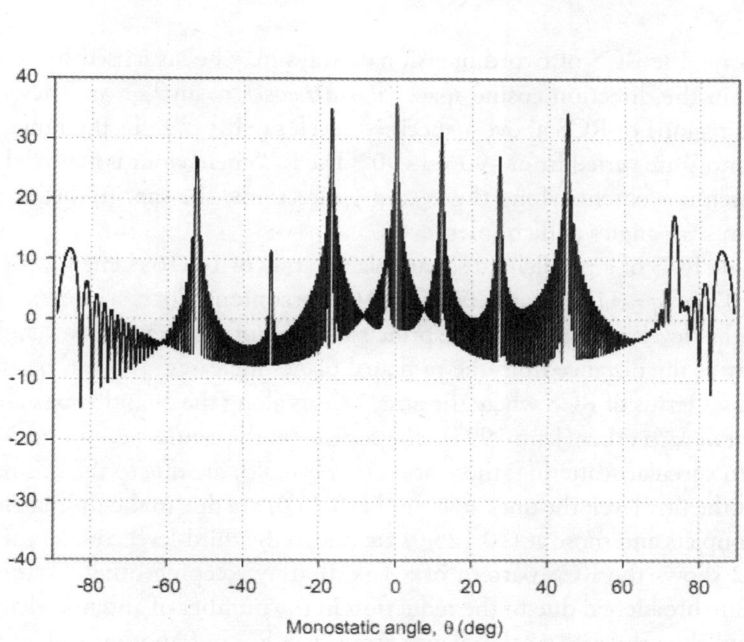

Figure 3.30 RCS of a parallel-fed linear array with $N_x = 128$, $d_x = 0.5\lambda$, $h = \lambda$, $\theta_s = 45°$, $\phi_s = \phi = 0°$, and level of couplers, $q = 2$

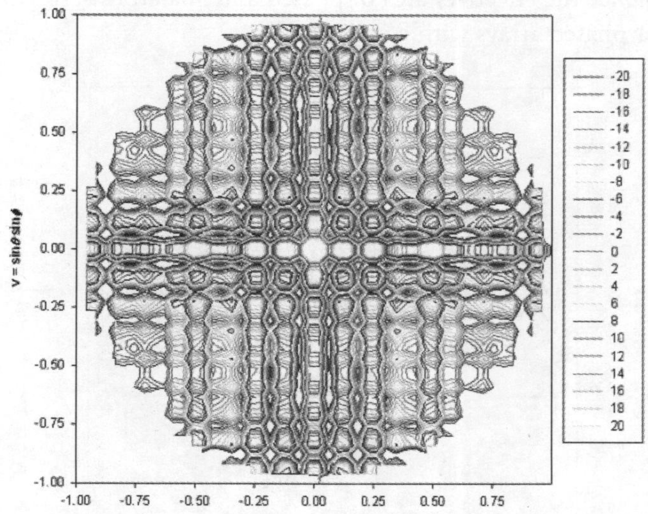

Figure 3.31 RCS of a parallel-fed planar array with $N_x = 16$, $N_y = 16$, $d_x = d_y = 0.5\lambda$, $\theta_s = \phi_s = 0°$, and level of couplers, $q = 3$

Planar array: The RCS of two-dimensional arrays may be visualised by plotting constant level contours in the direction cosine space (u=sinθ cosϕ; v=sinθ sinϕ). These RCS contours enclose spatial regions of RCS above a specified level, say for σ/λ^2 in the range -20 to 20 dB. Here, both θ and ϕ are varied from –90° to +90°. The RCS behaviour is analysed by varying the number of antenna elements along the x- and y-directions, the spacing between the antenna elements, beam scan angles and coupler levels.

The broadside RCS of a parallel-fed square planar array of 16×16 elements, for $d_x = 0.5\lambda$, $d_y = 0.5\lambda$, $\theta_s = \phi_s = 0°$ and $q = 3$ is shown in Figure 3.31. The contour plot is symmetric about the axis $v = 0$ and has higher levels of RCS along the principal planes of the array. However, their level drops off very quickly as the distance from the principal planes increases. This can be attributed to the separable product terms of RCS where the array factors along the x- and y-axes are considered to be independent of each other (Jenn 1995). The contour at the centre, i.e. around $(u,v) = (0,0)$, can be attributed to specular scattering; those at around (±0.5,0) are due to the mismatches between the couplers at the first level; the ones at around (±0.25,0) are due to the mismatches between the second-level couplers and those at (±0.125,0) are due to the third-level coupler mismatches.

Figure 3.32 shows the RCS pattern of a 16×10 array, keeping other parameters constant. The contours are broadened due to the reduction in the number of antenna elements along the y-direction, and thus the area of the planar array. The RCS of another square planar array of size 64×64 for a similar kind of arrangement is shown in Figure 3.33. The contours obtained in the pattern are very similar to those in case of a 16×16 array except that they are narrower and hence distinct. This indicates that as the number of antenna elements increases, the contours in the pattern become narrower, sharper and more pronounced. This is due to the increase in the physical (and hence the effective) area of phased antenna array, $A_p = N_x N_y d_x d_y$, as observed in the case of linear phased arrays with parallel feed.

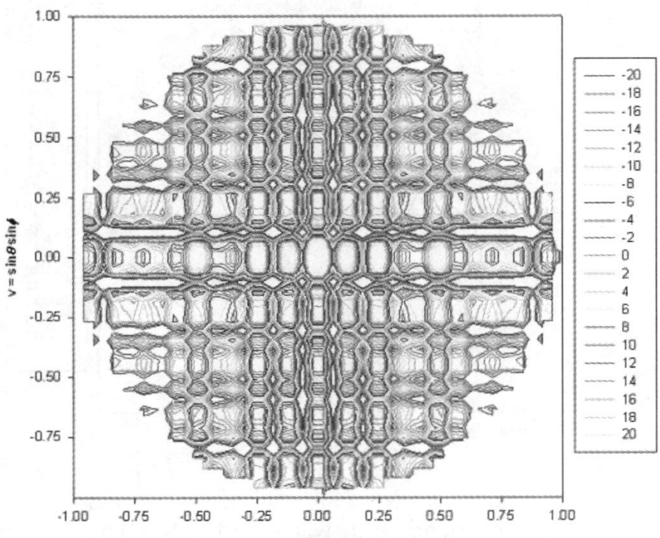

Figure 3.32 RCS of a parallel-fed planar array with N_x=16, N_y=10, d_x=d_y=0.5λ, θ_s=ϕ_s=0°, and level of couplers, q = 3

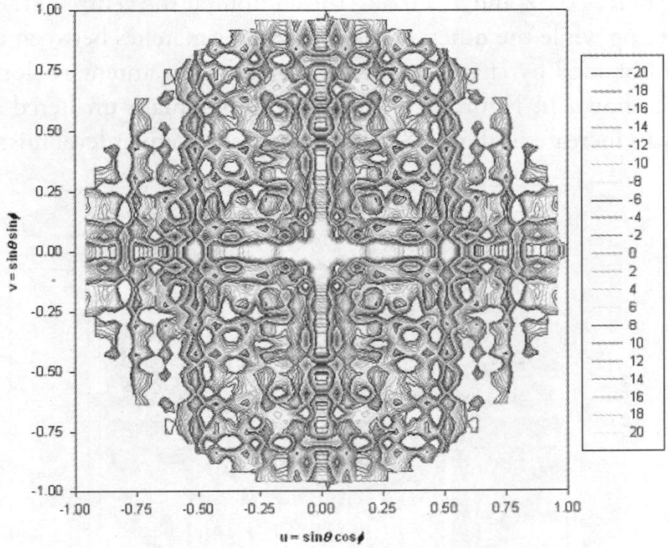

Figure 3.33 RCS of a parallel-fed planar array with N_x=64, N_y=64, d_x=d_y=0.5λ, θ_s=ϕ_s=0°, and level of couplers, q=3

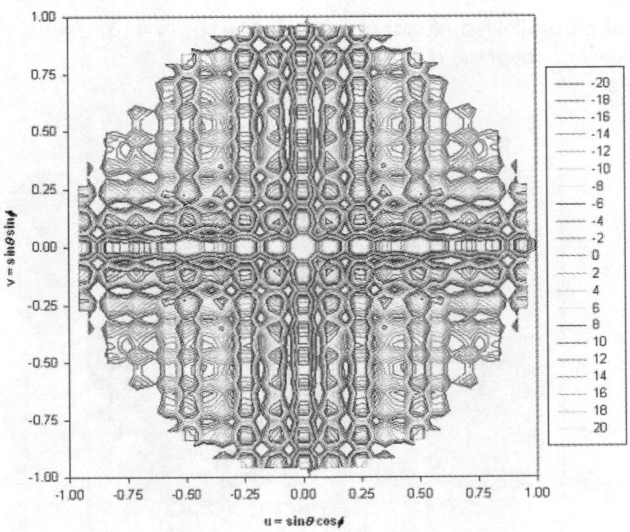

Figure 3.34 RCS of a parallel-fed planar array with N_x=N_y=16, d_x=0.5λ, d_y=0.5λ, θ_s=ϕ_s=0°, and level of couplers, q = 2

The effect of spacing between antenna elements on the RCS pattern of a planar phased array is shown in Figures 3.34 through 3.36. Figure 3.34 presents the RCS of the16×16 array for an elemental spacing of $d_x = 0.5\lambda$ and $d_y = 0.5\lambda$. The contour at the centre of the pattern is due to the specular scattering while the others are due to the mismatches between the coupler levels. The RCS pattern obtained by changing the spacing between antennas along the *x*-direction from 0.5λ to λ is shown in Figure 3.35. The contours remain unaltered in their location. However, there is an increase in the RCS value, due to the coupler level mismatches.

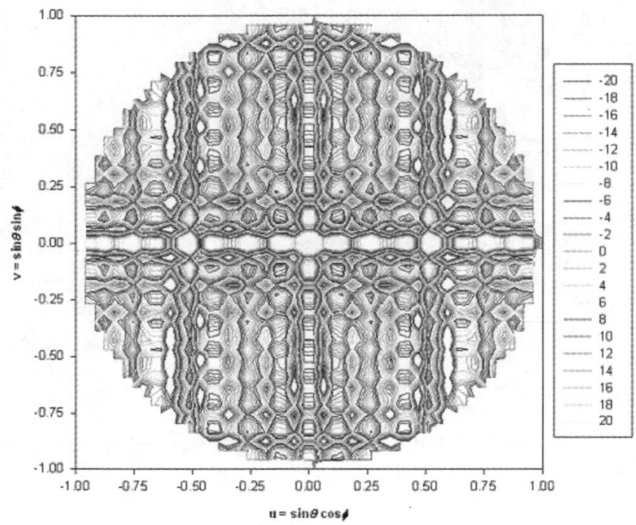

Figure 3.35 RCS of a parallel-fed planar array with N_x=16, N_y=16, d_x=1λ, d_y=0.5λ, θ_s=ϕ_s=0°, and level of couplers, q = 2

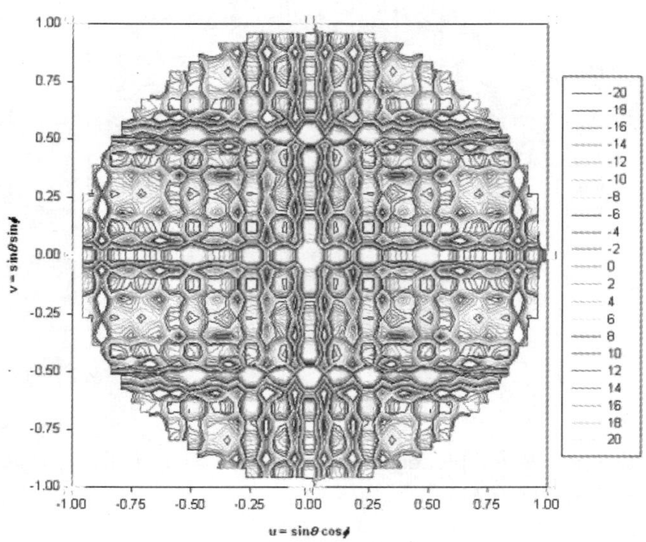

Figure 3.36 RCS of a parallel-fed planar array with N_x=16, N_y=16, d_x=0.5λ, d_y=1λ, θ_s=ϕ_s=0°, and level of couplers, q = 2

Moreover, the additional contours [those seen beyond $(u,v) = (\pm 0.5,0)$] appear due to the increase in the inter-elemental spacing along the x-axis. These changes may be due to the increase in the physical (and hence the effective) area of the phased antenna array, $A_p = N_x N_y d_x d_y$ with the increase in d_x.

Figure 3.36 shows the RCS of same phased array with $d_x = 0.5\lambda$ and $d_y = \lambda$. The RCS pattern along the axis $v = 0$, once again, is similar to that shown in Figure 3.34. However, it may be observed that the pattern similar to the one centred on the axis of v=0 gets repeated at both v = +0.5 and v = -0.5. This change in the RCS pattern is along the expectation lines.

The effect of the scan angle is considered in the following section. The broadside RCS of a 16×16 planar array ($d_x = d_y = 0.5\lambda$ and $\theta_s = \phi_s = 0°$) with a single level of couplers is shown in Figure 3.37a. The contours at the centre and at $(u,v) = (\pm 0.5,0)$ can be attributed to specular scattering and to the mismatches between the couplers at the first level, respectively. Scanning the antenna beam, for θ_s and ϕ_s from 0° to 45°, while keeping all other parameters same, results the change in RCS pattern as shown in Figure 3.37b. The location of the lobe due to specular scattering remains unchanged; however, six more prominent lobes at (u,v)=(+0.5,+0.5), (+0.5, -0.5), (-0.5,+0.5), (-0.5, -0.5), (0,+0.5) and (0, -0.5), appear in the pattern. These lobes correspond to the specific antenna beam location, and to the scattering due to the mismatches between the coupler levels. In other words, the lobes associated with the mismatches behind the phase shifters are observed to scan along with the antenna beam. This is because the factor χ_{mn} of the phase shifters, due to the terms ζ_x and ζ_y, depends on the beam scan angles, θ_s and ϕ_s. However, the RCS components arising from reflections due to radiating elements and the components before phase shifters are independent of the χ_{mn} term and thus remain fixed as illustrated by the specular lobe at $\theta = 0°$.

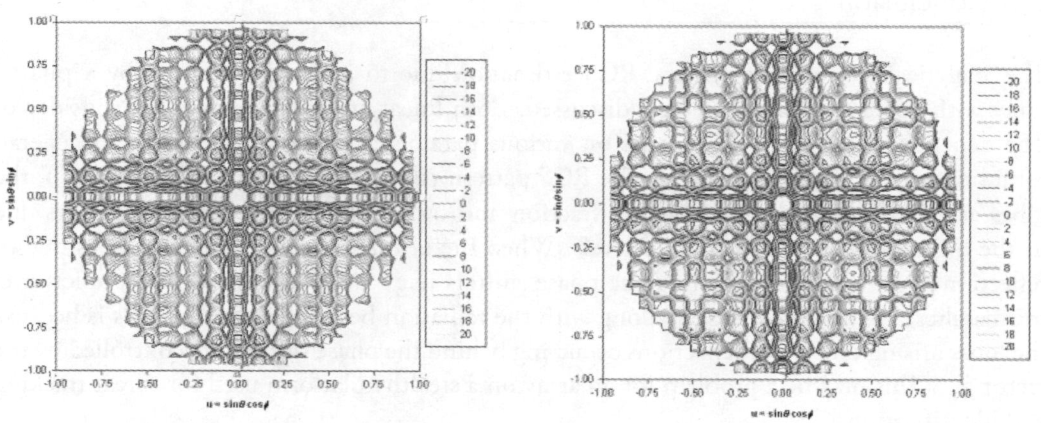

(a) (b)

Figure 3.37 RCS of a parallel-fed planar array with N_x=16, N_y=16, d_x=d_y=0.5λ, and level of couplers, q=1. (a) θ_s=ϕ_s=0°, (b) θ_s=ϕ_s=45°

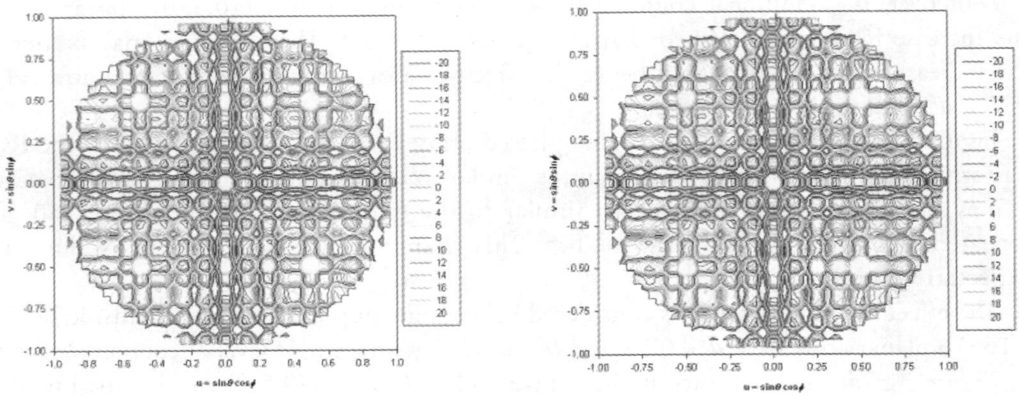

(a) (b)

Figure 3.38 RCS of a parallel-fed planar array with $N_x = 16$, $N_y = 16$, $d_x = d_y = 0.5\lambda$, $\theta_s = \phi_s = 45°$: (a) $q = 2$ and (b) $q = 3$

Figures 3.38a and 3.38b show the scanned RCS patterns for the same array with two and three levels of couplers, respectively. It may be observed that the location of contours due to the specular scattering and due to the single level of couplers remain unaltered. However, more lobes appear in the RCS pattern are due to the additional coupler levels considered. This shows that the change in the pattern observed due to the scan of θ_s and ϕ_s from 0° to 45° is same as in Figures 3.34 through 3.36. It is clear that even in the planar arrays, the effect of beam scanning is independent of the level of couplers.

3.5 Conclusion

The analytical formulation for the RCS estimation due to in-band scattering by a phased array with series and parallel feed is discussed. Both linear and planar arrays are considered. The dependence of the RCS pattern on various parameters has been analysed. In general, for a linear array with parallel feed, the RCS pattern consists of a specular lobe ($\theta = 0°$), the lobes at around ±80° due to Bragg diffraction and other lobes in between these lobes due to the mismatches between coupler levels. When the beam scans the array, the lobes due to reflections from the points behind the phase shifters (e.g. the lobes due to the coupler level mismatches) are observed to scan along with the radiation beam of the array. This is because the lobes arising due to the reflections occurring behind the phase shifters are controlled by the factor χ_{mn}. This may be a problem for an array on a stealthy platform used for threat tracking and identification.

Similarly, for a planar array with parallel feed, the RCS pattern shows contours with high value at its centre, which corresponds to the specular scattering. All other prominent contours in the pattern are due to the coupler mismatches at various levels. The observations made in the parametric study of the RCS pattern of a parallel-fed phased array can be summarised as follows:

The increase in the number of antenna elements (N in a linear array; N_x and N_y in a planar array) results in the rise of the level of major lobes (in linear array) or contours (in planar array). As the number of elements in the array increases, the lobes or contours in the RCS pattern become narrower, sharper and well-pronounced. The increase in the level of the lobes or contours may be attributed to the increase in the physical (and hence the effective) area of the phased array. However, the change in the number of antenna elements does not affect the location of major lobes or contours in the RCS pattern. In case of a linear array, the number of sidelobes increases with increase in the number of elements. However, in a planar array, the changes in contour plot are not clear for analysing the number of sidelobes.

There is no change in the location of specular lobes and major lobes upon increasing the inter-element spacing for either the linear or the planar array. This may be due to the increase in the effective aperture area of the array. Similar changes in the lobes or contours are observed with increasing the inter-element spacing d_x and h (in the linear array) or d_x and d_y (in the planar array). The RCS pattern changes significantly with the change in d_x, h (in the linear array) and d_x, d_y (in the planar array). As the spacing increases, the number of lobes (in the linear array) or contours (in the planar array) obtained in the pattern increases. This may be due to the dependence of lobe or contour spacing on the relative physical spacing between antenna elements.

In the scanned mode, the RCS pattern of both linear and planar arrays changes significantly. It is observed that the lobes or the contours representing the scattering beyond the phase shifters in the feed network are controlled by the factor χ_{mn}. This factor is dependent on the phase shifter settings. The lobes or the contours arising due to the scattering from elements below phase shifters scan along the radiation beam. However, the RCS components arising from the reflections due to radiating elements and the components before phase shifters remain fixed. This is evident from the specular lobe at $\theta = 0°$ in the linear array and from the contour at $(u,v)=(0,0)$ in the planar array.

As the level of couplers considered increases, the number of lobes (in linear array) or the contours (in planar array) obtained in the RCS pattern increases. This is because in the pattern the spacing between the lobes for different levels of couplers is found to be dependent on their relative physical spacing (i.e. $2d_x$ for the first level, $4d_x$ for the second level and so on). Thus it may be inferred that as more levels of couplers are introduced to the feed, more number of lobes or contours appear amidst the already existing ones.

The effects observed on the RCS pattern due to the variation of different parameters are found to be independent. For instance, the number of antenna elements varies and then only the levels of the major lobes and the number of minor lobes appearing in the RCS pattern vary. On the other hand, if the number of coupler levels varies, then there will be an increase in the number of lobes, which correspond to the coupler mismatches. However, there will be no increase in the level of the lobes already present in the pattern. This indicates that the level of the major lobes is a function of the number of antenna elements only. Conversely, it also establishes that the number of major lobes, i.e. lobes due to the coupler mismatches, is a function of the levels of couplers only. However, if the parameters are varied in combination, the combined effect is reflected in the RCS pattern.

The results obtained in both series and parallel fed-phased arrays may not be error free due to the assumptions made in RCS estimation. A lossless feed network and identical reflection coefficients are assumed. Practically all devices will have non-zero dB loss associated with them. Moreover, reflection coefficients vary randomly in their amplitude and phase components due to fabrication imperfections.
Secondly, equal power dividers have been assumed. However, low sidelobe design of an antenna requires a feed with tapered amplitude distribution, which results in further variation of coupler distribution. This once again reduces the sidelobe RCS level as compared to the case of equal power splitters.

Thirdly, the scattered signals might travel through the device for more than once, indicating the term τ_p^4 for the reflections that occur behind the phase shifters, and this would significantly reduce the RCS. However, the approximate method outlined here is computationally simple and flexible. It may be easily extended to an arbitrary number of elements and coupler levels. In general, contributions from each scattering source may be segregated and analysed individually. The total suppression of RCS can only be achieved by perfectly matching all the feed devices. It is noted that the results are independent of the changes in electrical length between the couplers. In other words, any change in the electrical path difference between the couplers does not affect the RCS value.

References

Chu, R. S. 1991. 'Analysis of an infinite phased array of dipole elements with RAM coating on ground plane and covered with layered radome.' *IEEE Transactions on Antennas and Propagation* 39: 164–76.

Flokas, V. 1994. 'In-band radar cross section of phased arrays with parallel feeds.' *MS Thesis Report* Naval Postgraduate School, Monterey, California, 58.

Green, R. B. 1966. 'Scattering from conjugate matched antennas.' *IEEE Transactions on Antennas and Propagation* 14: 17–21.

Gustafsson, M. 2006. 'RCS reduction of integrated antenna arrays with resistive sheets.' *Journal of Electromagnetic Waves and Applications* 20: 27–40.

Hansen, R. C. 1989. 'Relationships between antennas as scatterers and as radiators.' *Proceedings of IEEE* 77: 659–62.

Jenn, D. C. 1995. *Radar and Laser Cross Section Engineering.* Washington, DC: AIAA Education Series, 476

Jenn, D. C. and S. Lee. 1995. 'In-band scattering from arrays with series feed networks.' *IEEE Transactions on Antennas and Propagation* 43: 867–73.

Jenn, D. C. and V. Flokas. 1996. 'In-band scattering from arrays with parallel feed networks.' *IEEE Transactions on Antennas and Propagation* 44: 172–78.

Kahn, W. K. and H. Kurss. 1965. 'Minimum scattering antennas.' *IEEE Transactions on Antennas and Propagation* 13: 67–75.

Knott, E. F., J. F. Shaeffer, and M. T. Tuley. 1985. *Radar Cross Section.* Dedham, MA: Artech House Inc., 462.

Kuhn, J. 2011. *AlGaN/GaN-HEMT Power Amplifiers with Optimized Power-added Efficiency for X-band Applications*. Karlsruhe, Germany: KIT Scientific Publishing, 230.

Lee, E. Y.-C. 2008. 'Electromagnetically transparent feed networks for antenna arrays.' Ph.D. Dissertation, Ohio State University, 127.

Lee, S. 1994. 'In-band scattering from arrays with series feed networks.' MS Thesis Report, Naval Postgraduate School, Monterey, California, 45.

Lo, Y. T. and S. W. Lee. 1993. *Antenna Handbook*. Vol. 1. New York, USA: Van Nostrand Reinhold, 913.

Lu, B., S.-X. Gong, S. Zhang, and J. Ling. 2009. 'A new method for determining the scattering of linear polarised element arrays.' *Progress in Electromagnetics Research M* 7: 87–96.

Mallioux, R. J. 1994. *Phased Array Antenna Handbook*. Second edition Boston: Artech House, 496.

Rudge, A. W., K. Milne, A. D. Olver, and P. Knight. 1983. *The Handbook of Antenna Design*. Vol. 2. London, UK: Peter Peregrinus Ltd., 945.

Schindler, J. K., R. B. Mack, and P. Blacksmith. 1965. 'The control of electromagnetic scattering by impedance loading.' *Proceedings of IEEE* 53: 993–1004.

Volakis, J. L. 2007. *Antenna Engineering Handbook*. Fourth edition New York, USA: McGraw Hill, 1773.

Wang, W. T., Y. Liu, S. X. Gong, Y. J. Zhang, and X. Wang. 2010. 'Calculation of antenna mode scattering based on method of moments.' *PIERS Letters* 15: 117–26.

Zhang, S., S.-X. Gong, Y. Guan, and B. Lu. 2011. 'Optimized element positions for prescribed radar cross section pattern of linear dipole arrays.' *International Journal of RF and Microwave Computer-Aided Engineering* 21(6): 622–28.

Zhang, S., S.-X. Gong, Y. Guan, J. Ling, and B. Lu. 2010. 'A new approach for synthesizing both the radiation and scattering patterns of linear dipole antenna array.' *Journal of Electromagnetic Waves and Applications* 24: 861–70.

4

Active RCS Reduction in Phased Arrays

4.1 Introduction

Active radar cross section reduction (RCSR), i.e. active stealth technology relates to phased arrays and involves rejection of unwanted signals incident on aircrafts or missiles from different angles and simultaneous maintenance of distortion-less gain towards the signals from desired sources. This has additional advantages over the existing passive techniques for signature reduction, *viz.* shaping and coatings (Vinoy and Jha 1996). Moreover, it does not face difficulties in rejection of wideband and multiple targets incident from hostile sources.

Phased arrays consist of a large number of antenna elements, such as dipoles or waveguides, each of which is fed from a distribution network (Mailloux 1994). Depending on the spacing between elements and the relative phase of their excitation, the net radiation pattern can be steered in any direction without any mechanical motion of the array itself. Since this is done electronically, very high scan rates can be achieved. The main advantages of phased array are rapid electronic-beam steering and adaptive control of radiation pattern. In antenna arrays, effect of the probing signals can be removed/cancelled if a sufficient deep null is created in its radiation pattern towards the impinging angle of hostile radar. If this is done automatically, then such an array is referred to as adaptive array.

An adaptive array is an array of antennas and receiver with a real-time adaptive signal processor. The array follows the beam-steering command, maintaining the gain in the desired directions (Elliot 2005). Moreover, it adapts the weights to control the magnitude and phase of the feed current of each array element towards optimisation. Adaptive arrays may be used for active cancellation of probing effect of hostile radars. The adaptive algorithms employed for weight adaptation play an important role in controlling the performance indices of phased array.

An array comprises a set of sensors, the outputs of which are combined in some way to produce a desired effect. The sensors may be of many forms, for example, acoustic transducers for sonar, monopulse antennas for HF reception and microwave horns in a radar system. The antenna beam can be electronically steered in the desired directions and sufficiently deep nulls can be placed towards the unwanted directions, by adjusting the phase and amplitude of the current at each array element (Krim and Viberg 1996), (Veen and Buckley 1988). In other words, an adaptive array can alter its beam pattern according to signal environment by employing an efficient adaptive optimisation algorithm. An alternate use of such adaptive array is one in which the impinging signals are considered as the probing ones and the angles at which the nulls are placed are utilised to extract the information regarding direction of arrival (DOA). Since the angular width of a null appears to be smaller than that of the lobe formed by the same antenna aperture, such an approach may have superior resolution than the conventional approaches (Chandran 2004), (Fenn 2008).

It is known that the receiving array antenna can steer its main beam toward any direction by adjusting the complex weight in each element. However, it is difficult to steer the beam and null simultaneously towards two pre-specified directions, with a single set of weights. This ability of a phased array depends on several factors, *viz.* element positions, orientation of elements, antenna patterns of elements, polarisation of signals and direction of the beams and/or nulls (Podilchak et al. 2009), (Lin 1982). The coefficient called spatial correlation includes these factors and completely characterises the array beam pointing and nulling. Thus, by choosing these parameters intelligently to reduce the spatial correlation between the antenna elements, the performance of antenna array may be enhanced.

In a phased array, the beam pattern is optimised using adaptive algorithm so that the main lobes are maintained in the directions of desired signals, while deep nulls are placed in specific directions corresponding to the undesired probing signals. In simple words, the signal is fed from each element of an array to the weighting network in which each weight has been chosen such that any unwanted signals are cancelled out. This simple principle is effective only when one is dealing with narrow-band signals such that they can be adequately characterised by single phase and single amplitude term. For broadband signals, care must be taken to equalise the matched amplitude and frequency responses of each channel, including the group delay in the paths to a summing point. The lack of path matching leads to amplitude and phase distortions across the cancellation band of the wanted signal. This requires some form of frequency-dependent weighting (Hung and Kaveh 1990).

A wideband adaptive beamformer comprises a multiple input and single output combiner along with an efficient adaptive algorithm to obtain optimum weights (Veen and Buckley 1988), (Hung and Kaveh 1990). Broadband beamforming is employed when the impinging signals are wideband. One advantage of the narrowband system is that the array can be steered simply by phase changes, whereas broadband system requires physical delay. Thus, the broadband system requires simultaneous processing of a large number of samples from each sensor and is more complex.

An alternative approach to obtain optimum solution in adaptive processing is the use of constraints. For example, if one constraints the gain of the system to be equal to unity for the impinging signals and minimises the total output from the array subject to the constraint, then the ratio of wanted to unwanted signals is enhanced. An adaptive beamformer selects the weight vector as a function of the data to optimise the performance subjected to various constraints. However, they are quite sensitive to errors, such as the array steering vector errors, mismatch errors, etc. One way to curtail the uncertainty in the response of phased array is to impose unity-gain-based constraints over a small angular spread in the region of the desired direction. These constraints are referred to as main-beam constraints (Lee and Cho 2004), (Lee and Hsu 2000). For better response, additional constraints may be included in the formulation.

One type of constraint is the derivative main beam constraint in which the derivative of the optimal array output obtained is made zero at the desired angle (Huarng and Yeh 1992), (Thng et al. 1993). Imposing a constraint is associated with the loss of one of the available degrees of freedom for cancelling the probing signals. Unfortunately, it is still not clear as to which is the best way of picking the additional constraints. It is also difficult to predict the cost of imposing additional constraints on the array design specifications and, hence, the performance (Lee et al. 2006).

Most of the techniques for adaptive array processing are based on the conjecture that the array response towards the impinging signals is known *a priori*. Further, these techniques involve several assumptions related to the signal environment, for example, *a priori* knowledge of signal arrival direction, uncorrelated sources and absence of desired signals. This may results in drastic degradation of the performance if the assumptions about the environment, antenna array and/or sources are wrong or inaccurate (Lee and Wu 2005). A distinctive reason for degradation in array performance is the divergence of the real response from the presumed response for a given signal scenario. Such disparity may occur due to the look direction/pointing errors, deformed shape of array, distortions in the wavefront, fading of signal, local scattering, etc. Conventional beamformers are known to be too sensitive to even a small disparity of such kind. In such situations of disparity, the beamformers have the tendency to misinterpret the desired signal as a probing signal and, consequently, nullify the desired signal as well instead of maintaining main lobe towards it. This phenomenon is referred to as self-nulling in the literature related to adaptive array processing (Trees 2002).

This chapter presents a brief outline of the adaptive algorithms used in array processing. The performance of algorithms is discussed in view of the convergence rate, flexibility and output signal-to-interference-noise ratio (SINR). The efficacy of these algorithms can be exploited towards achieving an accurate adapted pattern and, hence, active RCSR in different radar/RF scenario. The radar environment may consist of narrowband/wideband sources, single/multiple desired and probing sources, or correlated sources. The adaptive phased array is expected to cater to such radar signal environment, nullifying the probing effect.

4.2 Adaptive algorithms

The optimal weights of an antenna array, which maximises the output SINR in the absence of errors, are in general computed using the array correlation matrix (ACM) of noise and the steering vector in the look direction. In many situations where the estimation of noise correlation matrix is not feasible, the total ACM is used to obtain the optimal adaptive weights of the beamformer. In the absence of the knowledge of the ACM, various adaptive algorithms that involve iterative weight adaptation are used to estimate optimum antenna weights. Various weight adaptation algorithms have been proposed. In general, such algorithms are made for signal environment in which both the desired and probing signals impinge on a beamformer. The function of the algorithm is to minimise the interference and maximise the desired signal at the array output.

There are two main approaches leading to three widely used families of algorithms for the adaptive estimation of optimum mean square error (MSE) parameters on the basis of the available data set (Haykin 1996). The first one is based on the stochastic approximation of the method of steepest descent and is known as the least mean square (LMS) family. These are gradient-based optimisation algorithms employing a noisy estimate of the required gradient vector.

The second ones are matrix inversion algorithms, based on the inversion of the sample correlation matrix of the array processor. The third one is based on the stochastic approximation of the Gauss–Newton method and is known as the recursive least square (RLS) family. They utilise the matrix inversion lemma to develop sample-by-sample updating of the inverse of the sample correlation matrix.

Pioneering work in the area of algorithms was done by (Shor 1966), (Applebaum 1976), (Widrow et al. 1967). The Shor algorithm (Shor 1966) maximises the signal-to-noise ratio (SNR) at the array output where noise includes the interference. Since both the desired signal and interference ought to be known, the Shor's algorithm is somewhat difficult to implement. The Applebaum algorithm (Applebaum 1976) too maximises SNR, but under the assumption that the desired signal is absent most of the time as in a pulsed radar. The Applebaum array is more practical to implement and has been used in various problems of clutter and interference rejection in radar systems. The adaptive array concept due to Widrow et al. (1967), which is known as the LMS algorithm, minimises MSE between the actual array output and the desired array output called the reference signal. Thus, the LMS algorithm minimises waveform distortion (where the reference signal is the replica of the desired signal), whereas the Applebaum algorithm maximises SNR. The LMS algorithm has been used to achieve the interference suppression in various scenarios.

The difference between the weights estimated by the adaptive algorithm and the optimal weights is further characterised by the ratio of the average excess steady-state MSE and the minimum mean square error (MMSE). This is referred to as the misadjustment. It is a dimensionless parameter that determines the performance and efficacy of the adaptive algorithm. It is essentially an error due to the random estimate of the gradient for the weight adaptation. This is termed as the misadjustment noise (Evans et al. 1993). The misadjustment noise increases with the step-size. On the contrary, it leads to faster convergence. Moreover, instead of having a single step-size for an entire weight vector, one may select a variable step-size for each weight separately, leading to an increased convergence of the algorithm.

Signal Representation: Consider an adaptive antenna array system with L elements (Figure 4.1). The product of the signal at each array element and the antenna weight when summed gives an array output.

The array output is given by

$$y(t) = \sum w_l^* x_l(t) \tag{4.1}$$

where, w_l is the antenna weight. The signal received by antenna array is given by

$$x(t) = [x_1(t), x_2(t), \ldots, x_L(t)]^T \tag{4.2}$$

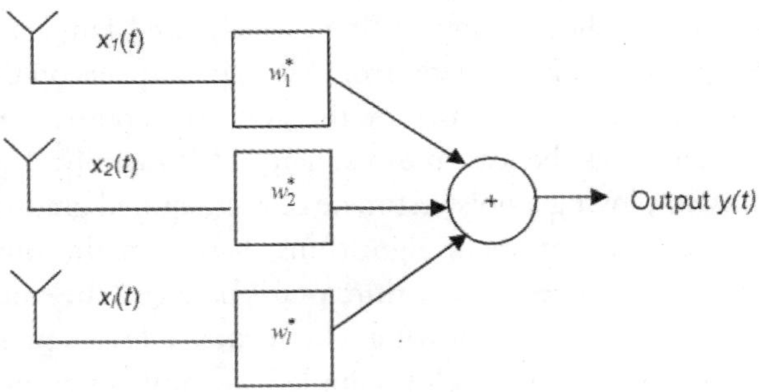

Figure 4.1 Adaptive antenna array system

where T is the matrix transpose. Thus, the output of array becomes

$$y(t) = w^H x(t) \tag{4.3}$$

where w^H denotes Hermitian of w. The power $P(t)$ is expressed as

$$P(t) = y(t)y^*(t) \tag{4.4}$$

Substituting the value of $x(t)$ in (4.4), we get

$$P(t) = w^H Rw \tag{4.5}$$

where R is the signal correlation matrix given by $R = E[x(t)x^H(t)]$, in which E is the expectation operator.

It is assumed that the antenna array is placed in the far-field region of impinging sources. Thus the signal impinging on the array has a planar wavefront. The coordinate origin is taken as reference. The time taken by an impinging wave coming from the kth source to reach l^{th} antenna element is expressed as

$$\tau_l(\theta_k) = \frac{d}{c}(l-1)\cos\theta_k \tag{4.6}$$

The signal received due to the kth source is expressed as $m_k(t)e^{j2\pi f_o t}$, where m_k is a modulating function and f_o is the frequency. The modulating function provides the details of the modulation employed, e.g. $m_k(t) = A_k e^{j\xi_k(t)}$ represents a frequency division multiple access (FDMA) system. Here, A_k denotes the amplitude and $\xi(t)$ is the message. On the contrary, for time division multiple access (TDMA), $m_k(t) = \sum d_k(n)p(t - n\Delta)$, where $d_k(n)$ is the message symbol, $p(t)$ is the sampling pulse and Δ is the sampling interval.

Steering Vector Representation: The steering vector (S_k) is the response of L antenna elements towards the kth source. For an array of identical L antenna elements, it is expressed as

$$S_k = [\exp(j2\pi f_0 \tau_l(\theta_k, \phi_k)), ..., \exp(j2\pi f_0 \tau_L(\theta_k, \phi_k))]^T \quad (4.7)$$

Now, the received signal vector is written as

$$x(t) = \sum_k m_k(t) S_k + n(t) \quad (4.8)$$

where $n(t)$ is the $L \times 1$ vector representing additive white gaussian noise (AWGN) and

$$y(t) = \sum_k m_k(t) w^H S_k + w^H n(t) \quad (4.9)$$

White noise is a randomly generated signal with constant power spectral density. The first term is due to all the impinging sources and the second term is the random noise. The ACM for uncorrelated impinging sources is given by

$$R = \sum_k p_k S_k S_k^H + \sigma_n^2 I_d \quad (4.10)$$

where p_k is the variance of modulating function, S_k is the steering vector and $\sigma^2 I_d$ denotes the correlation matrix due to thermal noise in which I_d denotes the identity matrix.

4.2.1 Least mean square algorithm

The original version of LMS algorithm (Griffiths 1969), (Widrow and McCool 1976) does not require any constraint on the weights and is, thus, referred to as the unconstrained LMS algorithm. The constrained LMS algorithm has been applied to adaptive beamforming for various applications (Frost 1972), (Godara and Cantoni 1986), (Godara 1986). The analyses of both the constrained and unconstrained LMS algorithms include the transient behaviour of the weights and the covariance matrix, convergence, and misadjustment analyses.

Adaptive array processing employing a constrained LMS algorithm requires an unbiased estimate of the gradient of the output power with respect to the array weights. There are a number of schemes for determining the unbiased estimate of this gradient. Though the gradient is unbiased in every scheme, the covariance of the estimated gradient with each method is different and, thus, the transient and the steady-state behaviour of the constrained algorithm are different in each case.

Standard LMS: The LMS algorithm updates the weights at each iteration by estimating the gradient of the quadratic surface and then moving the weights in the negative direction of the gradient by a small amount. The constant determining of this amount is referred to as the step-size. When this step-size is small enough, the process leads the estimated weights to approach the optimal value. If the outputs from all the receivers are available, a required gradient can be determined from the product of the array output and the outputs from each receiver. This is the principle of standard LMS algorithm (Widrow and Stearns 1985). The weight updation equation is given as

$$W(t+1) = P[W(t) - \mu\, g(W(t))] + \frac{S_o}{S_o^H S_o} \tag{4.11}$$

where, $S_o = S(\theta_o)$, is the steering vector corresponding to the desired signal impinging at θ_o. Similarly, P, the projection operator is expressed as

$$P = I - \frac{S_o S_o^H}{S_o^H S_o} \tag{4.12}$$

and $g[W(t)] = 2X(t+1)X^H(t+1)W(t)$ (4.13)

Structured Gradient LMS: An alternative method, referred to as the structured gradient method (Godara and Gray 1989), exploits the Toeplitz structure of the ACM for estimating the gradient. The noisy estimate of ACM used in the standard algorithm is not constrained to have this structure. A simple way to implement such constraint is to compute the average of the unconstrained estimate of the diagonal elements (Godara 1990). The gradient (4.13) takes the form

$$\hat{g}[W(t)] = 2\hat{R}(t+1)W(t) \tag{4.14}$$

where, $\hat{R}(t)$ is the ACM at the t instant of time,

$$\hat{R}(t) = \begin{bmatrix} \hat{r}_o(t) & \hat{r}_1(t) & \cdots & \hat{r}_{L-1}(t) \\ \hat{r}_1^*(t) & . & & \\ \cdots & \cdots & & \\ \cdots & \cdots & & \\ \hat{r}_{L-1}^*(t) & & & \hat{r}_o(t) \end{bmatrix} \tag{4.15}$$

with

$$\hat{r}_l(t) = \frac{1}{N_l}\sum_i x_i(t)x_{i+1}^*(t),\ l = 0, 1, \ldots, L-1 \tag{4.16}$$

where $N_l = L - l$ for a linear array.

The convergence rate of the standard algorithm and the required step-size are found to be sensitive to the signal power in the look direction. This signal sensitivity is greatly reduced by the structured gradient method. Further, in the presence of a strong signal in the look direction, the standard algorithm requires a smaller step-size as compared to the structured gradient adaptive algorithm. The smaller step-size makes the convergence rate slower. In addition, the signal sensitivity increases with the number of array elements, while the converse is true for the structured gradient method.

Normalised LMS: For certain conditions, the convergence time of the adaptive array, while using conventional LMS algorithm, can be longer than the radar dwell time (Godara 2004). It is found that because of long convergence time, a single low-power jammer can be more effective than higher power jammers. Although the techniques based on various types of covariance matrix are available, their implementation cost is quite high. The normalised least mean square (NLMS) algorithm (Nitzberg 1985) can substantially decrease the convergence time with the same implementation cost as of the conventional LMS.

It is shown that by neglecting the thermal noise, the convergence behaviour of NLMS algorithm becomes independent of jammer power for the single jammer case. For multiple jammers, the convergence time is unchanged when each jammer power changes by the same factor (Homer et al. 2007). Another form of NLMS algorithm, proposed by Bershad (1986), is a variation of the constant-step-size LMS algorithm. It uses the data-dependent step-size at each iteration and avoids the need for estimating the eigenvalues of the correlation matrix or its trace. It has a better convergence speed and is less sensitive to signal compared to the standard LMS (Givens 2009).

Recursive LMS: The standard LMS algorithm is sensitive to the received signal due to the usage of a signal correlation matrix in the gradient estimation. This signal sensitivity could be avoided by computing the correlation matrix for all the samples. On the contrary, recursive LMS algorithm utilises each and every previous sample and updates the signal correlation matrix for each new sample. This matrix is then used for estimating the required gradient (Widrow and Stearns 1985). The gradient is expressed as

$$g_{\text{recursive}}[W(t)] = 2R(t+1)W(t) \tag{4.17}$$

where $$R(t+1) = \frac{1}{t+1}\left[tR(t) + X(t+1)X^H(t+1)\right] \tag{4.18}$$

Alternately, the signal sensitivity may be avoided via spatial averaging in place of sample averaging. This approach is followed in weight estimation and adaptation in structured gradient LMS algorithm (Godara and Gray1989).

Improved LMS: In this algorithm, the weight adaptation utilises the Toeplitz structure (Godara 1990) of ACM. The gradient is expressed as

$$g_I[W(t)] = 2\tilde{R}(t+1)W(t) \tag{4.19}$$

where $$\tilde{R}(t+1) = \frac{1}{t+1}\left[t\tilde{R}(t) + \hat{R}(t+1)\right] \tag{4.20}$$

Other forms of LMS: An LMS algorithm automatically tracks a desired signal and suppresses the interferences. Thus, it is one of the best choices for radar systems in which the arrival angle of the desired signal is unknown at the receiver. However, the LMS adaptive array requires a reference signal to control each weight (Brennan and Reed 1982). If one assumes that there is a deterministic component in the signal impinging from the desired direction and is well-known to the receiver, then this deterministic component may be used as a reference signal. This fully known component can be an amplitude-modulated signal and a pilot signal added to the transmitted communication signal. Ogawa et al. (1985) analysed the steady-state performance of the LMS adaptive array in which a pilot signal is used as a reference signal. The LMS array accepts all the signals except the reference signal/pilot signal, which is taken to be interference. To separate the pilot signal easily from other components, its frequency is taken to be different from that of the information signal. This avoids the problem of cancellation of the desired signal.

Griffiths (1969) proposed a modified LMS algorithm for the case of correlated signals. Khatib and Compton (1978) developed an algorithm that maximised the gain of an antenna array on an incoming signal. The method was based on the maximisation of steepest ascent of the output of antenna array subjected to the constraint so as to avoid infinitely large weights. Other algorithms that have been proposed for adaptive arrays include recursive algorithms, *viz.* least square estimation techniques, additional modified versions of LMS algorithms, covariance matrix inversion methods, etc.

Performance analysis: The efficiency of an algorithm is evaluated in terms of its rate of convergence and output SINR. Figure 4.2 shows the output noise power for a uniform 10-element linear array with $\lambda/2$ inter-element spacing. Results are presented for the case of two probing sources from two distinct directions (70°, 100; 100°, 1) and one desired signal having the power ratio of 100. The power levels of the probing sources are different (100 and 1). It may be observed that the improved LMS algorithm has better performance as compared to standard and recursive LMS, in view of the output noise power level and the convergence rate. Moreover, the improved LMS algorithm has better weight covariance properties than the standard LMS.

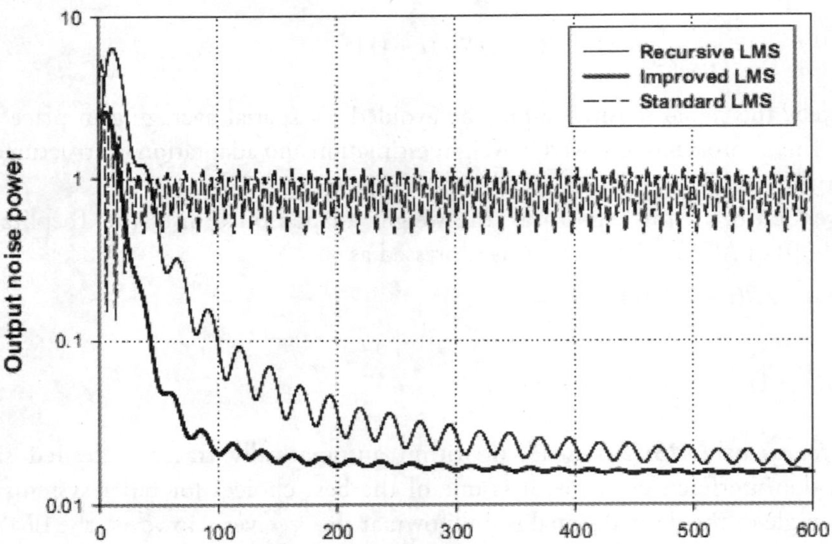

Figure 4.2 Mean output noise power of a 10-element uniform linear array, $\lambda/2$ spacing, ϕ = 90°; 2 jammers: 70°, 100; 100°, 1

If the power of incident signal is increased, the output noise power in case of recursive LMS algorithm increases. However, this trend does not exist for improved LMS algorithm. Rather the performance of algorithm is independent on signal power level and shows improvement with increase in the signal power. If a very strong signal having power of 30 dB or more is incident, the improved LMS algorithm performs much better than the recursive LMS algorithm. This is true for both the convergence rate and the output noise power of the algorithm.

The steady-state performance of phased array in the radar environment consisting of three probing sources (36°, 48° and 66°; 30 dB) and one desired source (0°; 0 dB) is shown in Figure 4.3. The improved LMS algorithm shows better performance as compared to standard LMS and structured LMS algorithm in view of the output SINR and the rate of convergence.

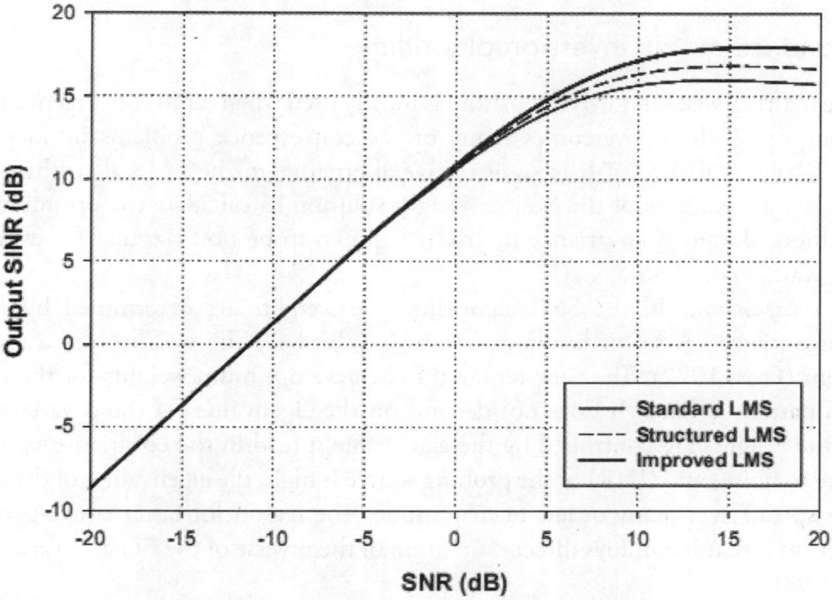

Figure 4.3 Steady-state performance of 10-element linear array in different SNR environments

4.2.2 Recursive least square algorithm

In this adaptive algorithm, the array gain matrix replaces the gradient step-size in weight adaptation. The RLS algorithm offers better convergence rate, steady-state MSE and parameter-tracking capability over the LMS-based algorithms. The simplest LMS algorithm performs adaptive filtering at the cost of double the count of unknown parameters. The rate of convergence of the LMS algorithm mainly depends on the spread of eigenvalues of the signal correlation matrix. On the contrary, the RLS algorithm is free from such problem but is complex in nature as weight adaptation depends on the square of the number of the unknowns. Thus, a numerically stable RLS algorithm, called the QR-decomposition–RLS (QRD–RLS), has been proposed (Lueng and Haykin 1989; Najm 1990). In this algorithm, QRD of the input data matrix using Givens Rotation is computed and the least square (LS) weight vector is solved by back substitution.

The back substitution, however, to be performed in the array structure is a costly operation. Hence, an inverse QRD–RLS algorithm was proposed where the LS weight vector is computed without the back substitution (Alexander and Ghirnikar 1993), (Chern and Chang 2002). The algorithm, which has better convergence performance than RLS algorithm, is the improved LMS, in the presence of the strong look-direction signal. It uses the structured method to estimate the correlation matrix using all the samples. Many other forms of RLS algorithms have been proposed that improve the computational efficiency (Cioffi and Kailath 1984), (Fabre and Gueguen 1986).

4.2.3 Standard matrix inversion algorithm

The sample matrix inversion (SMI) algorithm is mostly used when rapid convergence is required for high-output SINR. It overcomes many of the convergence problems in adaptive array processing (Reed et al. 1974). It is an attractive alternative to the LMS algorithm. The SMI algorithm uses an estimate of the Weiner weight solution to calculate the optimum weights. In this method, the noise covariance matrix is required to be non-singular for estimation of inversion (Ward et al. 1986).

Weight adaptation: In the SMI algorithm, the weights are determined by estimation of covariance matrix. A set of linear equations is solved for the weights, for a given signal environment (Frost 1972). The time required to achieve optimum weights for the generation of antenna pattern is fixed. It does not depend on the eigenvalues of the covariance matrix. However, the weights are controlled by the eigenvalue spread in the covariance matrix. If the interference to noise ratio (INR) of the probing source is high, the eigenvalues of the covariance matrix can spread over many orders of magnitude. The reason for faster convergence rate in SMI algorithm is that it employs direct estimation of the inverse of the signal covariance matrix (Carlson 1988).

For a received signal, $x(t)$, the covariance matrix R and the correlation matrix, c, are expressed as

$$R = E\left[x(t)x^H(t)\right] \tag{4.21}$$

$$c = E\left[d(t)x(t)\right] \tag{4.22}$$

Here, $d(t)$ is the reference signal. If *a priori* information about impinging desired and the probing signals is known, then the optimum weights may be directly obtained from the Weiner solution (Godara 2004) as

$$w_{\text{optimal}} = R^{-1}c \tag{4.23}$$

In a practical situation, the information regarding the incident signals is not known, and the signal environment varies dynamically. Thus, the optimum antenna weights are determined from the signal covariance matrix after time averaging the blocks of received data. The matrices of a block size $N_2 - N_1$ are expressed as

$$\hat{R} = \sum_{m=N_1}^{N_2} x(m)x^H(m) \tag{4.24}$$

$$\hat{c} = \sum_{m=N_1}^{N_2} d^*(m)x^H(m) \tag{4.25}$$

where N_1 and N_2 are the lower and the upper limits of the receiving block, respectively. Thus, the weight vector can be calculated as

$$\hat{} \quad \frac{\hat{}}{\hat{}} \tag{4.26}$$

For each incoming block of input data, covariance matrix keeps on changing and accordingly the weights are updated. However, in estimation-based algorithm such as the SMI algorithm, there always exists a residual error, which is usually larger than the LMS error. The estimate-based error is obtained as

$$e = \hat{R}w_{optimal} - \hat{c} \tag{4.27}$$

There are modified versions of the SMI algorithm, which generates the adapted weight vectors if the noise covariance matrices become singular. For example, one such method adds a factor to each diagonal matrix element before its inversion (Hudson 1981). Another technique is based on Buehring's orthogonal projection method (Buehring 1976). This involves iterative estimation of the weight vectors and has an accelerated convergence rate.

The stability of the SMI algorithm relies on its capability to invert large-size signal covariance matrix. For avoiding a singular covariance matrix, a zero-mean white Gaussian noise is included in the steering vector. This approach adds a strong component to each diagonal element of the covariance matrix, facilitating the matrix inversion. The singularity problem occurs when the number of signals to be resolved is less than the degrees of freedom, i.e. the number of antenna array elements. SMI algorithm is also termed as an exact algorithm. This algorithm can be employed for (a) the synthesis of radiation pattern of uniform/non-uniform linear phased array and (b) the active cancellation of the probing sources impinging the antenna array. This requires the estimation of optimal weights used for calculating the output of the beamformer.

Pattern synthesis: For a uniform linear array with N isotropic elements, the signal vector at t snapshot is given by

$$x(t) = s_d(t)a(\theta_d) + \sum_{i=1}^{L} s_i(t)a(\theta_i) + \sigma(t) \tag{4.28}$$

Here, s_d is the desired signal arriving at the angle θ_d, s_i are the probing signals arriving at angle θ_i, L is the number of probing signals impinging at the array and σ is the thermal noise. a(θ_d) and $a(\theta_i)$ are the steering vectors for the desired and probing signals, respectively. The steering vector is given by

$$a(\theta) = \left[1, e^{-j\frac{2\pi}{\lambda}d\sin(\theta)}, \ldots, e^{-j\frac{2\pi}{\lambda}d(N-1)\sin(\theta)} \right] \tag{4.29}$$

In the SMI algorithm, the signal covariance matrix is determined by sampling the incoming data (Yasin et al. 2010). It is given by

$$R(t+1) = \frac{tR(t) + x(t+1)x^H(t+1)}{t+1} \tag{4.30}$$

Once the covariance matrix is known, the weight vector is determined as

$$w(t) = R^{-1}(t)a(\theta) \tag{4.31}$$

To calculate R^{-1}, a matrix inversion lemma (Ganz et al. 1990) is used as

$$R^{-1}(t) = R^{-1}(t-1) - \frac{R^{-1}(t-1)x(t)x^H(t)R^{-1}(t-1)}{1 + x^H(t)R^{-1}(t-1)x(t)} \tag{4.32}$$

where $R^{-1}(0) = \frac{I}{\varepsilon_o}, \quad \varepsilon_o > 0$

The number of samples grows with the snapshots, and the covariance matrix approaches its true value. This in turn makes the estimated weights the optimal one. The array output is given by

$$y = w^H x \tag{4.33}$$

Table 4.1 Effect of number of array elements on the pattern

Number of elements	Peak sidelobe level (PSLL)	3 dB Beamwidth
8	–6.2 dB	17°
10	–6.5 dB	14°
16	–6.4 dB	8°
32	–5.5 dB	4°

It is assumed that the desired signal incident on an array is expressed as

$$S(t) = A\cos(\omega t), \ t > 0 \tag{4.34}$$

The radiation pattern of uniform linear array with $\lambda/2$ inter-element spacing is obtained for a different number of elements (Figure 4.4). As the number of array elements is increased, the main beam of the pattern gets narrower or improves in directivity. However, the number of sidelobes increases with the number of array elements. Table 4.1 demonstrates the role of the number of array elements on peak sidelobe level (PSLL) and 3 dB beamwidth of the pattern.

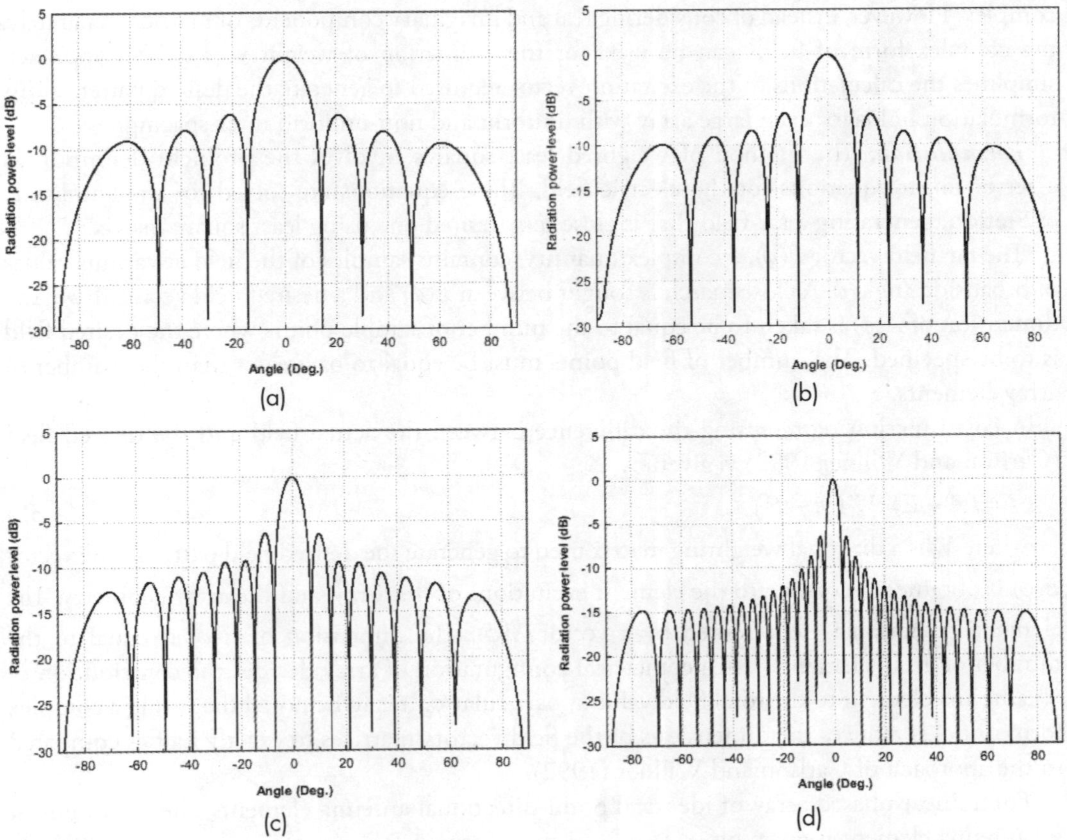

Figure 4.4 Radiation pattern of uniform linear array using SMI algorithm, (a) $N = 8$ (b) $N = 10$ (c) $N = 16$ (d) $N = 32$

4.2.4 Weighted least square algorithm

The method of weighted least squares can be considered as an extension of the unweighted matrix inversion method (Hirasawa and Strait 1971). This method is comparably more flexible and has greater control over the actual pattern due to the weighting matrix, which keeps on updating until the desired pattern is obtained. This method can also be applied to non-uniformly spaced antenna arrays, and it can be easily extended to non-linear arrays, to arrays with different element patterns on each element or for near-field pattern synthesis. Carlson and Willner (1992) employed the method of weighted least squares in pattern synthesis by segregating the real and imaginary components of the complex electric field, i.e. the complex excitation vectors.

This approach doubles the dimension of all the matrices corresponding to the desired far-field, excitation field and weights. The inversion of these large real matrices becomes computationally complex. However, instead of considering real and imaginary components of matrices separately, the complex form of fields can be used for the estimation of weight vector. This approach simplifies the calculations of the excitation vector required to generate the desired pattern. This formulation holds for even large array with uniform and non-uniform inter-spacing.

Formulation: The method of weighted least squares involves the problem definition as a set of linear equations for the electric field. These equations are solved for the amplitude excitation, generating a far-field that is close to a desired one using least square matrix.

The far-field vector $E(\phi)$, a complex quantity, contains samples of the field at various values of broadside angle ϕ. A close match is sought between $E(\phi)$ and a desired field vector $A(\phi)$. The dimension of $A(\phi)$ is taken to be equal to the number of sample points where the desired field is to be specified. The number of field points must be equal to or greater than the number of array elements.

A cost function representing the difference between the actual field and the desired field (Carlson and Willner 1992) is given by

$$J = (A - E)^T W (A - E) \tag{4.35}$$

where W is a diagonal weighting matrix used to generate the desired field pattern. The vector $E(\phi)$ is obtained according to the element excitations of the array, and the array geometry. The element excitation is represented by a vector $X(\phi)$. The dimension of $X(\phi)$ is equal to the number of array elements. The geometrical configuration of array decides the contribution of nth antenna element to the total field value at particular ϕ. It is to be noted that using a complex form of fields reduces the dimensions of the field vectors/matrices to exactly half as compared to the approach of Carlson and Willner (1992).

For a linear phased array of identical omni-directional antenna elements, the field value at n^{th} antenna element is given by

$$e_n(\phi) = x_n \exp(j\theta_n) \tag{4.36}$$

where x_n is the element excitation and

$$\theta_n = \frac{2\pi}{\lambda} d_n \sin(\phi) \tag{4.37}$$

and d_n is the location of the nth antenna element with respect to the phase centre of array, and λ is the wavelength. The field vector is related to the excitation vector as

$$E(\phi) = H(\phi)X \tag{4.38}$$

where the matrix H relates each antenna element with the corresponding sample field point. In matrix form, $(E)_{m\times 1} = (H)_{m\times n} \times (X)_{n\times 1}$

where m and n denote the field point and the antenna element, respectively. Hence, the cost function becomes

$$J = (A - HX)^T W (A - HX) \tag{4.39}$$

For the minimum cost function, the derivative of J with respect to X must be 0. This gives the excitation vector as

$$X = \left(H^T W H\right)^{-1} H^T W A \tag{4.40}$$

If the weight vector is proper, this excitation vector would provide the desired field pattern. Since (4.40) is an exact matrix equation, the result is obtained within a few iterations. The H matrix can also be modified for arrays with different element patterns and non-uniform inter-element spacing. Further, through the weight matrix, W, a greater control over the antenna pattern is possible. This makes the method of weighted least squares very efficient, flexible and fast.

For pattern synthesis, the real part of the desired field vector is set equal to 1 (in the main lobe region) and the imaginary part to zero. The weight matrix is a square matrix with dimension m. It is initialised as an identity matrix. An error weighting is obtained from (4.39) and inserted in the weight matrix W.

Accordingly, a new excitation vector X is calculated. Once optimum X is obtained, the pattern is generated using

$$P = X^H . U \tag{4.41}$$

where U is the steering vector and superscript H denoted the Hermitian, i.e. transpose of the complex conjugate.

Uniform array: Figure 4.5 shows the pattern of a linear array of 16 elements with half-wavelength spacing. A pencil beam in the radiation pattern is desired in the broadside direction, which ranges from –2° to +2°.

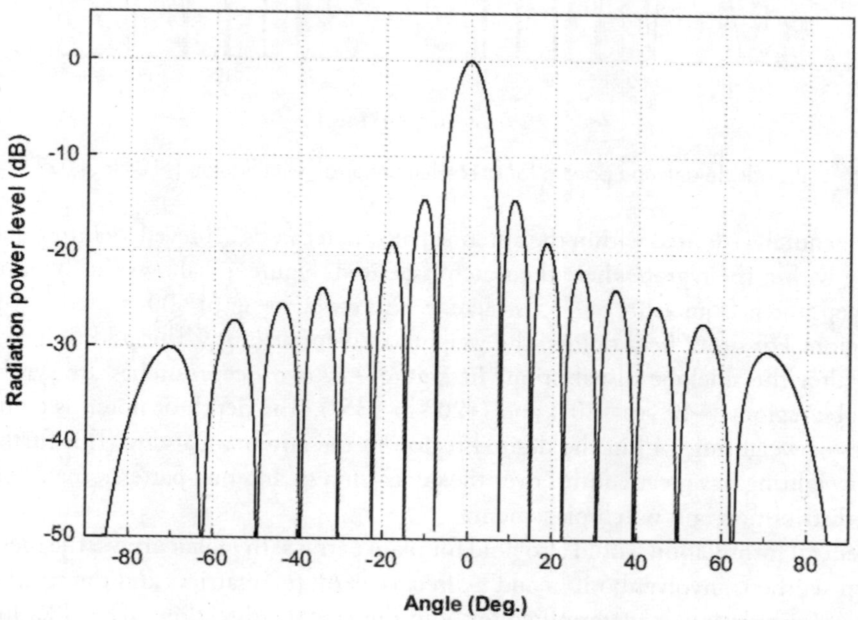

Figure 4.5 Broadside antenna pattern for a 16-element linear array (Main lobe: –2° to +2°)

The desired field is chosen to be 1 in the region from –2° to +2°, and 0 in the sidelobe region with sample points chosen at every degree from –90° to +90°. The weighting matrix being set equal to the identity matrix is updated at each step based on the algorithm. Once the optimum excitation vector is obtained using the corrected weighting matrix, the pattern is generated.

Figure 4.6 shows the pattern of a 16-element phased array having a desired 20° flat top main lobe. In this case, the desired field is chosen to be unity for region –10° to +10°. This establishes the flexibility of the method employed for the pattern synthesis. The sidelobe distribution of the antenna pattern can be further reconfigured by varying the weight matrix or the desired field vector.

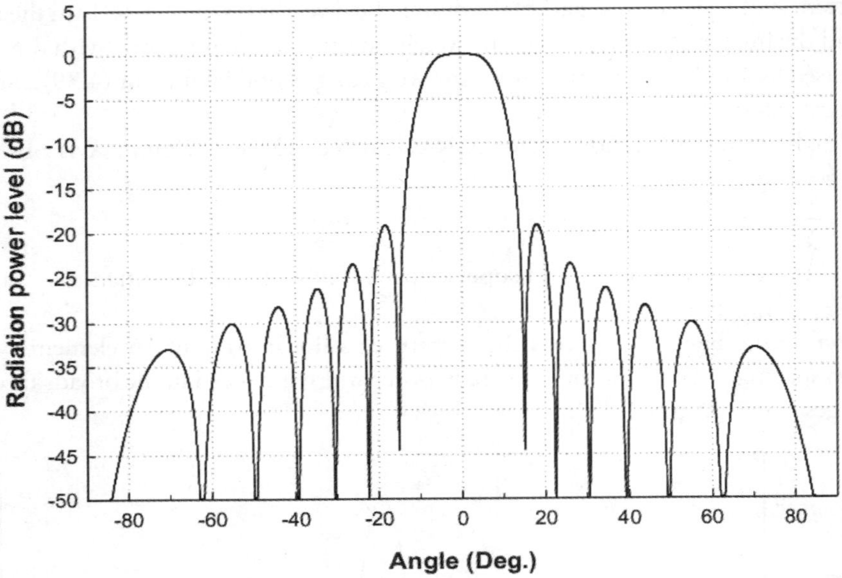

Figure 4.6 Broadside antenna pattern for a 16-element array with flat top (–10° to +10°)

If a deep notch is desired within the antenna pattern, it can be achieved by introducing heavy weighting within the region where the notch is desired. Figure 4.7 shows the antenna pattern with a deep notch from +20° to +35°. Here, a heavy weighting of 500 is used in the desired notch region. For other field points, the weights are kept at unity. This additional constraint tends to alter the sidelobe distribution. In Figure 4.8, two deep notches are synthesised in the sidelobe regions (–40° to –20°) and (+20° to +35°). The depth of notch is controlled by the values of weighting within the desired region in the antenna pattern. This further proves that the weighting has great control over the generation of antenna pattern, main lobe width, sidelobe distribution and notch placement.

The general formulation would also hold for planar arrays. In planar arrays, the methodology of pattern synthesis involves both *x*- and *y*-directions. All the matrices and the steering vectors used in the formulation are determined for both the *x*- and *y*-directions. The radiation pattern is obtained by multiplying the space factors corresponding to *x*- and *y*-directions (Elliot 2005).

Figure 4.9 presents the radiation pattern of 16 × 10 planar phased array with uniform spacing (0.5λ) and main beam steered to 10°. It is apparent that the steering of main beam can be controlled efficiently using weighted least square algorithm. Figure 4.10 presents the pattern of the same 16 × 10 planar array with a notch in the region from –30° to –25°. It may be observed that due to the notch in the sidelobe region, the overall sidelobe level comes down. Figure 4.11 shows that the main beam steering (10°) and the position of the notch (–30° to –25°) can be effectively and simultaneously controlled for a planar array (16 × 10, 0.5λ spacing) using weighted least square algorithm.

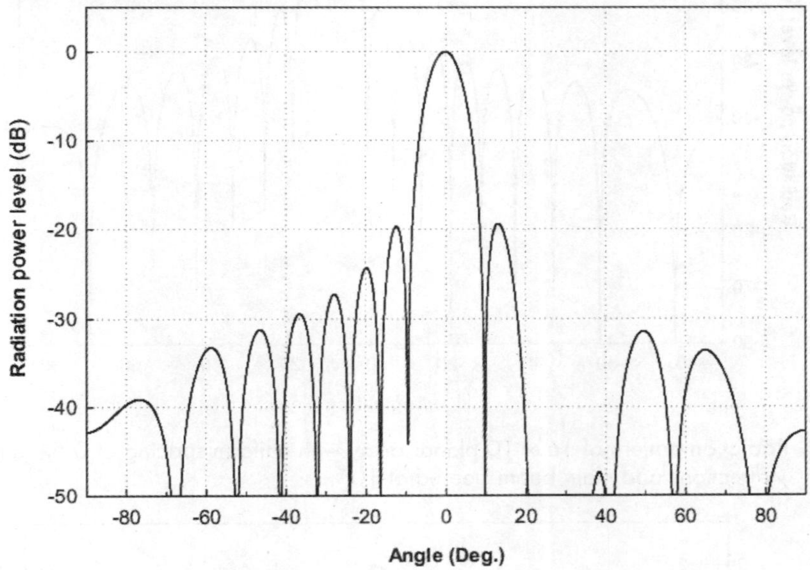

Figure 4.7 Radiation pattern for a 16-element linear array with notch from +20° to +35°

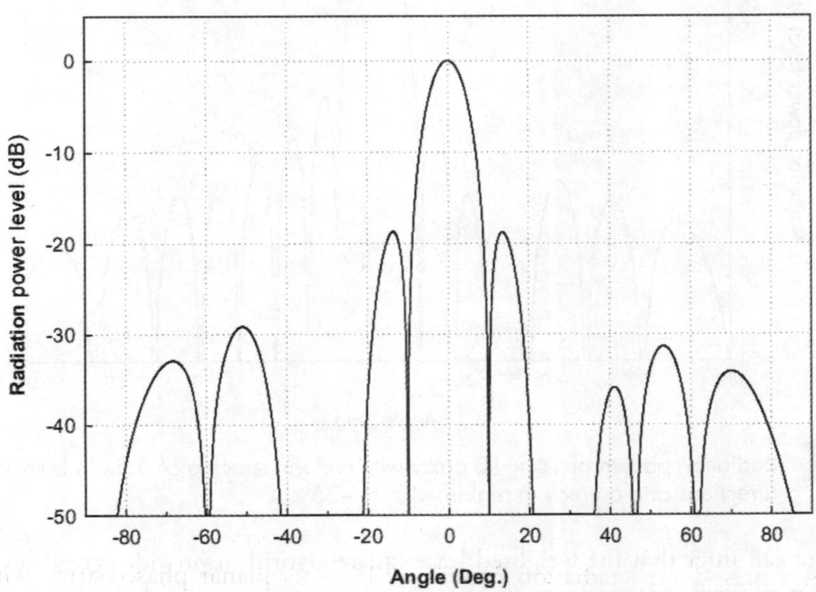

Figure 4.8 Radiation pattern for a 16-element uniform linear array with notches from –40° to –20° and +20° to +35°

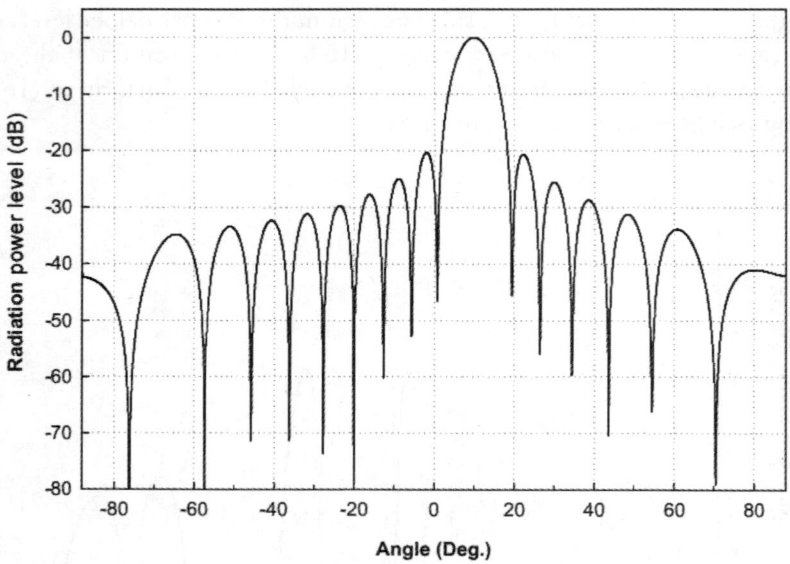

Figure 4.9 Radiation pattern of 16 × 10 planar array with uniform spacing of 0.5λ in both *x* and *y* directions and main beam steered at 10°

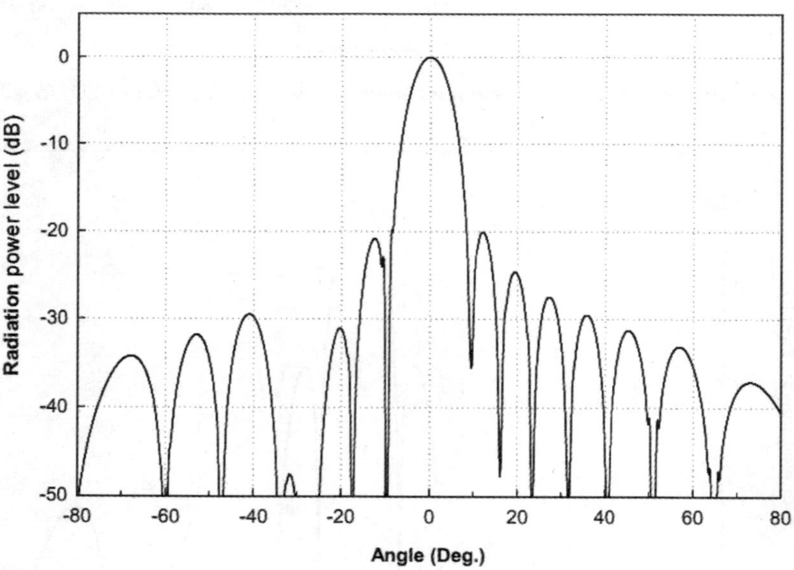

Figure 4.10 Radiation pattern of 16 × 10 array with uniform spacing of 0.5λ in both *x* and *y* directions and a notch in region –30° to –25°

Thus, one can infer that the weighted least square algorithm provides great flexibility and accuracy over unweighted matrix inversion methods in generation of desired antenna pattern. This is achieved by controlling the desired field vector and the weighting matrix used to generate the antenna pattern. Since the pattern synthesis is done using complex matrices, one does not encounter the problem of the inversion of large matrices. This method provides the exact and accurate solution within a few iterations.

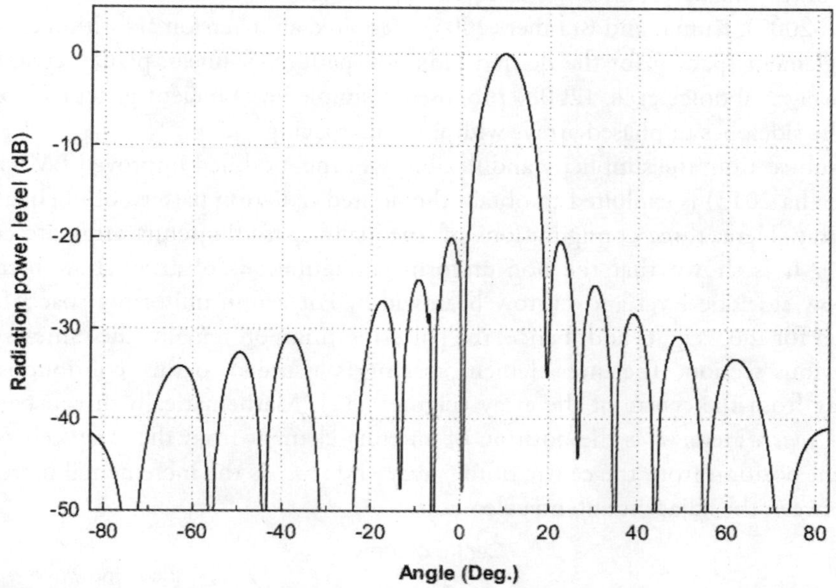

Figure 4.11 Radiation pattern of 16 × 10 array with uniform spacing of 0.5λ. Main beam steered at 10°; notch in region –30° to –25°

The simulation results demonstrate the versatility of the method. The antenna patterns with different main lobe widths and the notches at different regions within the antenna pattern can be accurately generated through proper weighting. The weighted least square algorithm proves to be very efficient even for the case of planar arrays. The algorithm has significant control over the sidelobe level, the main lobe width and the position of the notch.

Non-uniform array: When the array elements are kept at unequal or random spacing, the steering vector, i.e. response of the array towards the impinging signal is different from that of uniformly spaced antenna array. Moreover, unequal inter-element spacing alters the optimum weighting of the antenna elements and, thus, the synthesised pattern.

The antenna pattern synthesis of non-uniform array has been carried out since last five decades. This study is known to originate with the research work of Unz (1960), who proposed a matrix-based method for the current distribution and, hence, the desired radiation characteristics of linear phased array with unequal spacing. Extending the Unz's approach, Harrington (1961) proposed a way to reduce the first sidelobe level of a non-uniform linear phased array. King et al. (1961) used unequally spaced antenna elements to reduce the grating lobes in the pattern. Andreasen (1962) computed various possibilities of unequally spaced linear arrays using the digital techniques. The 3 dB beam width of the pattern is reported to depend mainly on the array length and the sidelobe level is controlled by size and geometry of the array. Ishimaru (1962) introduced a new function (source position function) based on Poisson sum expansion, which made the design of unequally spaced arrays with the possibly desired radiation characteristics. A successive approximation method was proposed for controlling the amplitude of the radiation pattern of unequally spaced arrays (Gallaudet and Moustier 2000), (Chen et al. 2007). Kumar and Branner (2005) presented an inversion algorithm to obtain the optimum element spacing for the desired radiation pattern of linear, planar, cylindrical and spherical arrays. Abdolee et al. (2008) proposed a simple and efficient genetic algorithm for reducing the sidelobes in phased arrays with unequal spacing.

In this sub-section, the simplicity and flexibility of the modified improved LMS algorithm (Singh and Jha 2013) is exploited to obtain the desired radiation pattern of unequally spaced antenna array. Here, three configurations of unequally spaced antenna array are considered (Table 4.2). It is shown that the non-uniform configurations of array show better results with the low sidelobe level and narrow beamwidth. For a non-uniformly spaced array, the formulation for the weights and, hence, the pattern estimation remains the same as discussed in the previous section. The inter-element spacing is assumed to vary continuously as one moves away from the centre of the array (Figure 4.12). Mathematically, it can be expressed as $d_n = d + space$. Here, d_n is the position of antenna element from the array centre, d is the first element position from the centre of the array and *space* is the incremental increase in the spacing between the adjacent antenna elements.

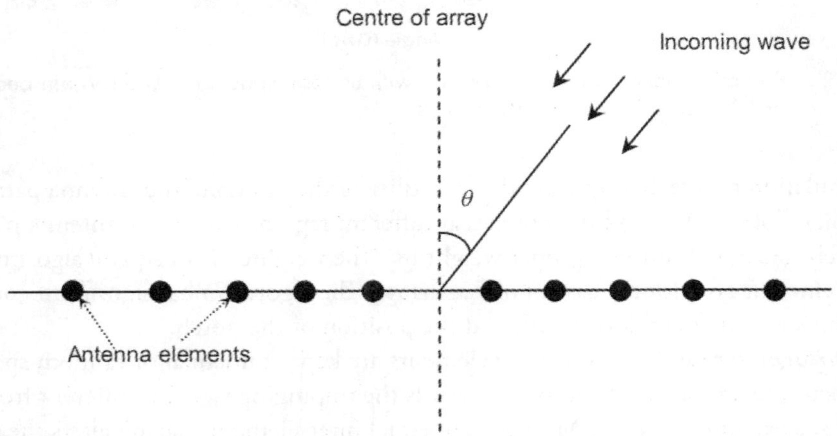

Figure 4.12 Schematic of a non-uniform array

Three cases considered for N-element non-uniform array are:

Case 1: $space = 0.12\,n$, where $n = 0, 1, 2, \ldots N/2$,

Case 2: $space = 0.1\,n$, where $n = 0, 1, 2, \ldots N/2$, and

Case 3: Random spacing with length of array equal to 8λ

Table 4.2 provides the position of elements from the centre of the array for these inter-element spacing. Due to non-uniform spacing between elements, d_n in (4.37) takes different values. This changes the value of θ_n as given by (4.37) and, hence, new steering vector and H matrix are obtained.

The steering vector $U(\theta)$ is the array response to the incident angle impinging the array at θ. It is given by

$$U(\theta) = \left[1 \;\; e^{-j\phi} \;\; e^{-2j\phi} \dots \;\; e^{-j(N-1)\phi}\right]^T \tag{4.42}$$

where T is the matrix transpose and ϕ is the phase shift in between the elements along the array. If the spacing between the antenna elements is non-uniform, the steering vector is expressed as

$$U(\theta) = \left[1 \;\; e^{-j\Phi} \;\; e^{-2j\Phi} \dots \;\; e^{-j(N-1)\Phi}\right]^T \tag{4.42a}$$

where, $\Phi = \dfrac{2\pi\{S(p+1,\,1) - S(p,1)\}}{\lambda}\sin\theta;\; p = 1, 2, \dots N$

Here, S is the element position along the axis.

The matrix H now takes the form

$$H = \cos\theta + j\sin\theta$$

with

$$\theta = \frac{2\pi}{\lambda} S(p,1) \times \sin\phi \tag{4.43}$$

Table 4.2 Position of non-uniform array elements with respect to the centre of the array

Position	Uniform array	Non-uniform array (1)	Non-uniform array (2)	Non-uniform array (3)
1	–3.75	–1.76	–4.0	–3.6
2	–3.25	–1.52	–3.1	–2.8
3	–2.75	–1.30	–2.3	–2.1
4	–2.25	–1.10	–2.1	–1.5
5	–1.75	–0.92	–0.9	–1.0
6	–1.25	–0.76	–0.4	–0.6
7	–0.75	–0.62	–0.3	–0.3
8	–0.25	–0.50	–0.1	–0.1
9	0.25	0.50	0.1	0.1
10	0.75	0.62	0.3	0.3
11	1.25	0.76	0.4	0.6
12	1.75	0.92	0.9	1.0
13	2.25	1.10	2.1	1.5
14	2.75	1.30	2.3	2.1
15	3.25	1.52	3.1	2.8
16	3.75	1.76	4.0	3.6

Figures 4.13–4.15 present the pattern of a 16-element non-uniform array (NU1 of Table 4.2) with different main lobe widths. It may be seen that the modified improved LMS algorithm is efficient enough to generate the array pattern with the desired main lobe width. Moreover, the main lobe width is increased while the sidelobe level is suppressed further. This can be explained in terms of the principle of conservation of energy. Since the input power fed to the antenna array is constant, for broader main lobe, energy distributed among the sidelobes will be less as compared to the narrower main lobe.

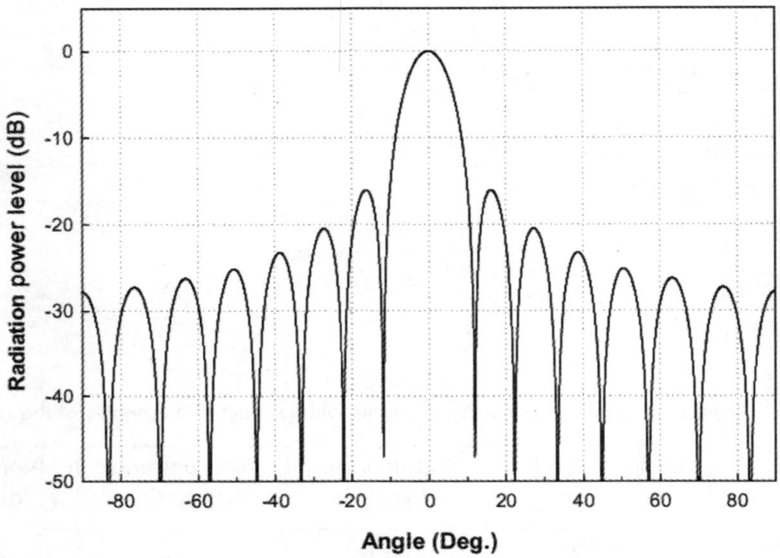

Figure 4.13 Radiation pattern of 16-element non-uniform linear array (NU1), main lobe = – 5° to + 5°

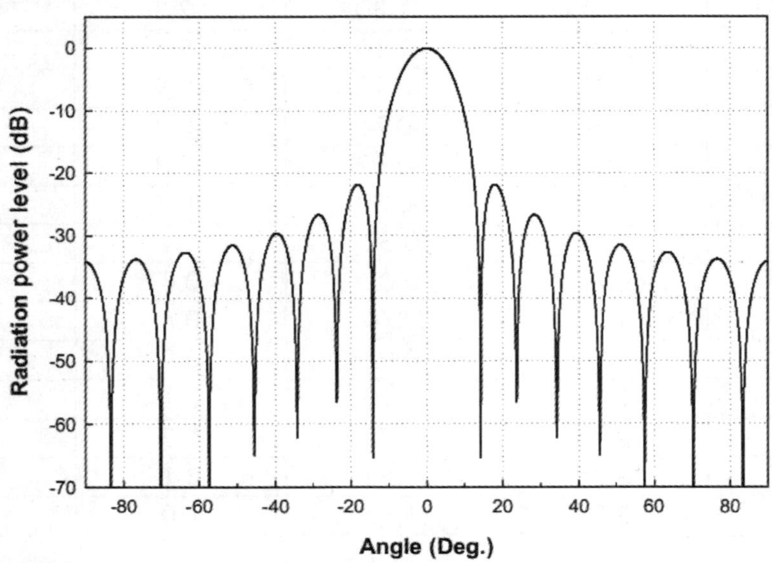

Figure 4.14 Radiation pattern of 16-element non-uniform linear array (NU1), main lobe = –8° to +8°

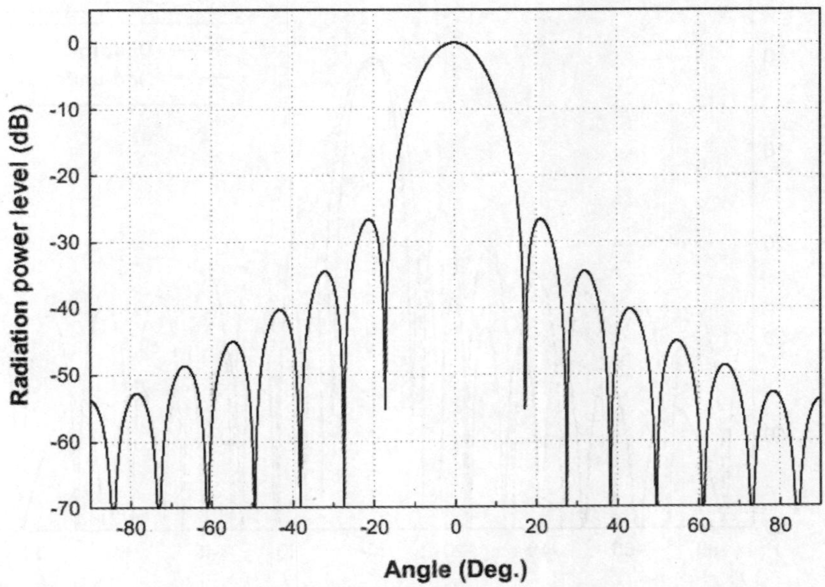

Figure 4.15 Radiation pattern of 16-element non-uniform linear array (NU1), main lobe = –10° to +10°

Figures 4.16 through 4.18 show the comparison between the pattern of uniform and non-uniform 16-element arrays (NU1) for the three different main lobe widths (–5° to +5°; –8° to +8°; –10° to +10°). It may be observed that for the main lobe width of –5° to +5° and –8° to +8°, SLL is higher in non-uniform array as compared to uniform array. Further, as the desired main lobe width is increased, the directivity of the array is improved. Figure 4.18 shows that unlike uniform array pattern, for a broadside region from –10° to +10°, the sidelobe level is lower, and instead of flat top, pointed main lobe is generated. This implies that the algorithm generates an improved radiation pattern for the prescribed main lobe region. The observations are summarised in Table 4.3.

Table 4.3 Observations made from comparison of radiation pattern of 16-element linear phased array with uniform and non-uniform spacing

Desired main lobe width	Uniform PSLL (in dB)	Non-uniform array PSLL (in dB)	Main lobe
–5° to +5°	–19.21	–26.73	Wider for non-uniform array
–8° to +8°	–23.64	–21.88	Equal
–10° to +10°	–21.12	–16.15	Wider for uniform array

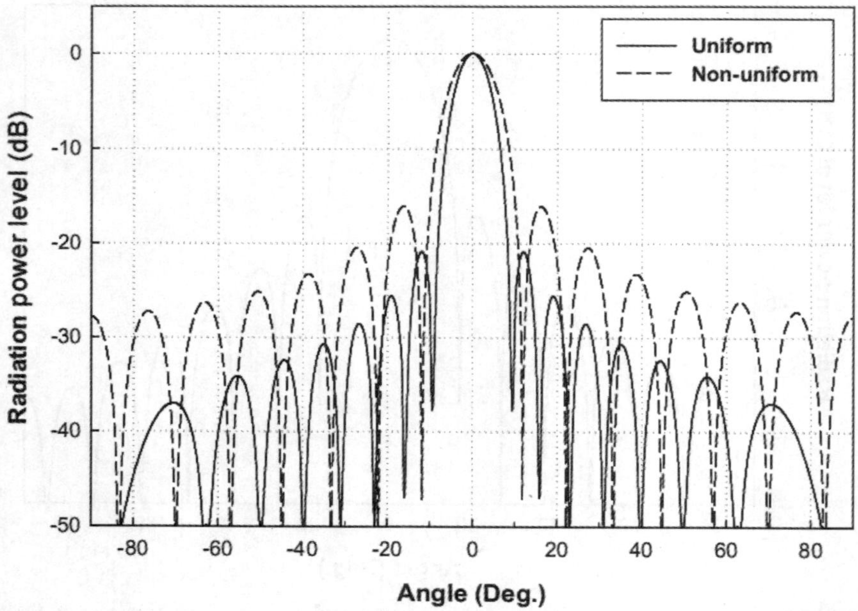

Figure 4.16 Comparison of radiation pattern of 16-element array for uniform and non-uniform inter-element spacing, main lobe = –5° to +5°

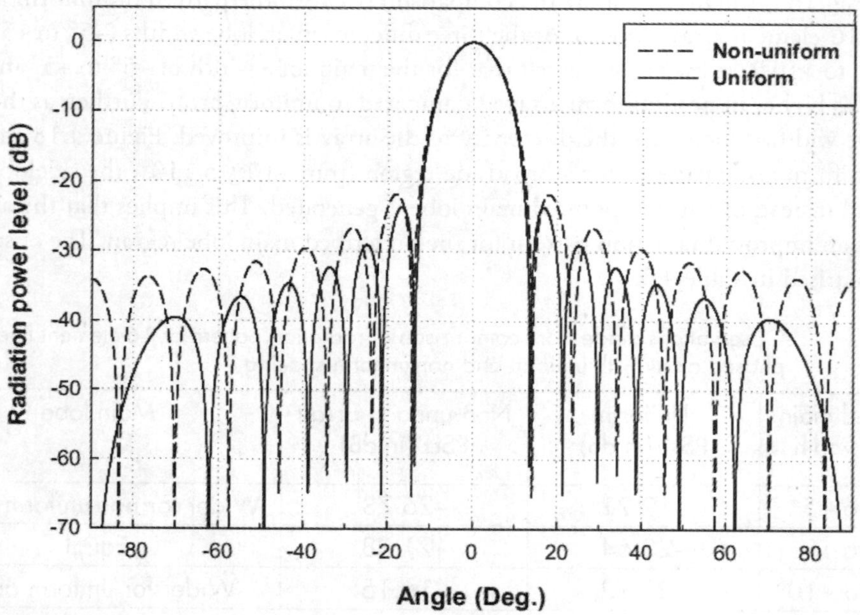

Figure 4.17 Comparison of radiation patterns of non-uniform and uniform linear array, N = 16, main lobe = –8° to +8°

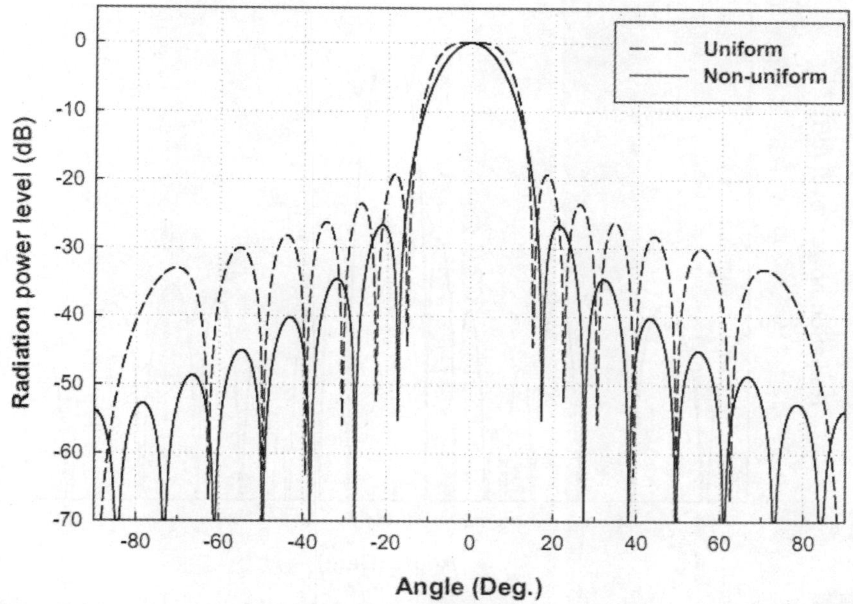

Figure 4.18 Comparison of radiation patterns of non-uniform and uniform linear array, $N = 16$, main lobe = –10° to +10°

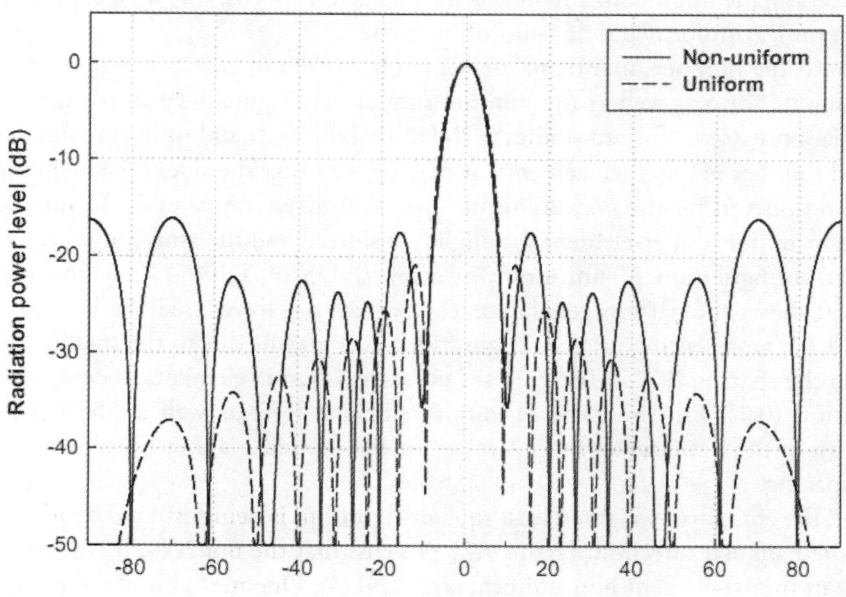

Figure 4.19 Comparison of radiation patterns of non-uniform linear array with uniform linear array, $N = 16$, main lobe = –5° to +5°, length of array = 8 λ

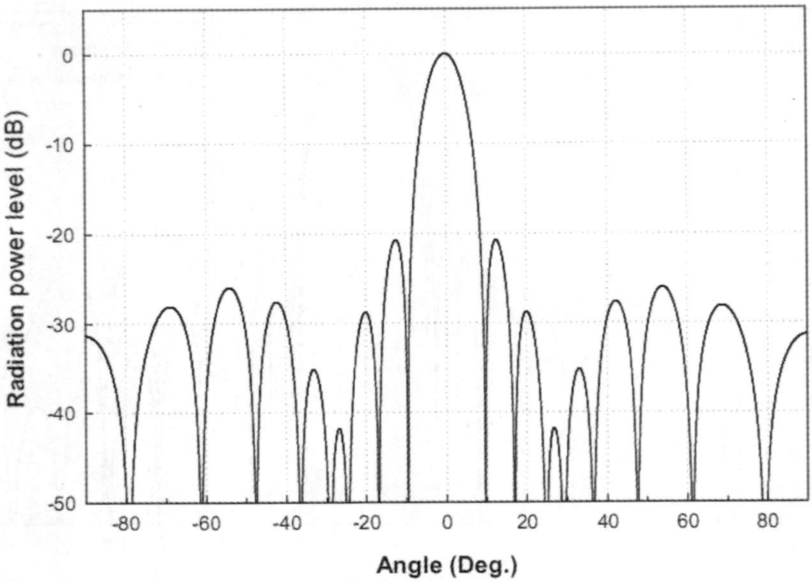

Figure 4.20 Radiation pattern of NU3 non-uniform array, N = 16, main lobe region = –5° to +5°

So far, the results discussed are for the array with uniform and non-uniform inter-element spacing, keeping the number of antenna elements fixed (N = 16). The total length of the array is different for the uniform and non-uniform arrays.

Now, both the number of antenna elements (N = 16) and size of the array (8λ) are kept same for the uniform as well as the non-uniform arrays. Figure 4.19 shows the comparison of the radiation patterns for non-uniform (NU2 in Table 4.2) and uniform arrays with equal length and number of antenna elements. It may be seen that the sidelobe distribution is very different and uneven for the non-uniformly spaced array as compared to Figure 4.13, where only the non-uniform inter-element spacing is considered and the array length is ignored.

Another configuration of non-uniform array (NU3 of Table 4.2) is considered next. Figure 4.20 shows the pattern generated. The pattern has lower sidelobe level as compared to Figure 4.13, representing NU1 configuration. This may be due to the smaller incremental increase in the spacing ($0.1\ \lambda$) between the adjacent antenna elements as compared to NU1 configuration ($0.12\ \lambda$). Thus, the maximum sidelobe level as well as the entire sidelobe distribution in the pattern can be controlled in the desired manner by adjusting the inter-element spacing.

Further, the efficacy of weighted least squares algorithm is demonstrated by placing a notch in the desired angular direction. Figure 4.21 presents that the notch (+20° to +50°) is placed accurately in the 16-element non-uniform array (NU3). One may also observe that although notch is accurately generated in the pattern, the entire sidelobe level (on the side of notch) is reduced. This was not observed when the uniform spacing was considered. Thus, one can infer that the weighted least square algorithm is efficient for the synthesis of radiation patterns of non-uniform arrays.

The radiation pattern with low sidelobe level is obtained for the non-uniform array than the uniform array. However, this is not true if the spacing is made non-uniform, keeping the length of array fixed. The flexibility of the algorithm is established by the capability of controlling the main lobe width, the steering and the position of the notch in the desired sidelobe region. The placement of notch can be exploited towards active cancellation of probing in phased arrays.

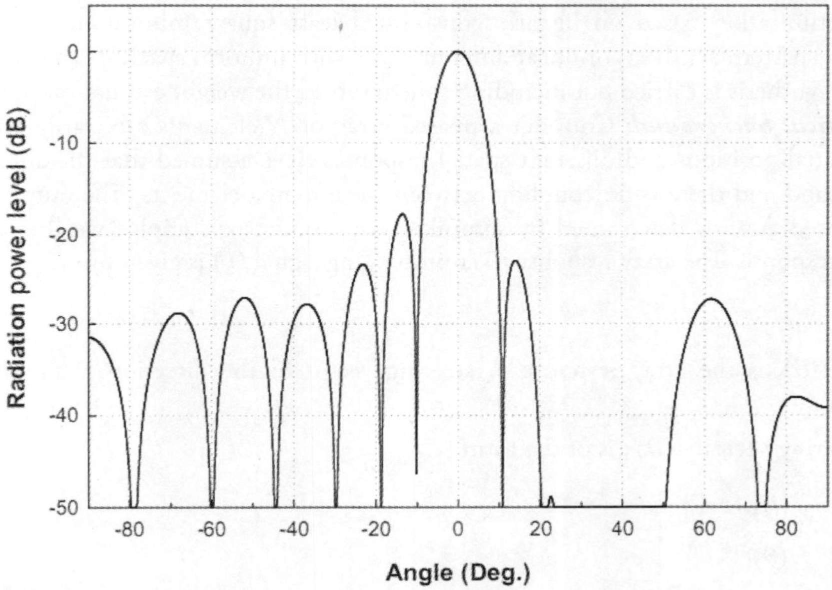

Figure 4.21 Radiation pattern of 16-element non-uniform array (NU3) with notch in the region from +20° to +50°

4.2.5 Linearly constrained least square algorithm

Adaptive algorithms rely on the estimation of optimum weights to generate the desired radiation characteristics. Apart from the generation of desired radiation pattern, efforts have been made towards optimising the process so as to get the optimal solution within fewer iterations. Researchers have proposed optimisation techniques based on mathematical methods like linear programming, cone programming, convex optimisation (Lebret and Boyd 1997) and linear least square minimisation (Tseng and Griffiths 1992).

The methods so far proposed for weight estimation (Fuchs and Fuchs 2010) can be divided into two classes. One class of algorithms uses an iterative re-weighted least squares and the other class uses methods based on adaptive array algorithms. Both the classes of algorithms use iterative method for pattern synthesis. The disparity between the desired and the computed radiation pattern is determined and the weights are adjusted accordingly to generate an adapted pattern.

The estimation of array weighting coefficients is done to produce a prescribed beam pattern response of non-adaptive deterministic arrays. Further, by imposing an appropriate linear constraint, a quiescent adaptive pattern can be pre-specified (Griffiths and Buckley 1987).

The disadvantage of these methods is slow convergence. There is no guarantee that the desired pattern will be obtained. It is difficult to draw an exact inference about the achievable array performance, *a priori*. The potential of Convex Optimisation in eliminating these disadvantages has been extensively explored (Lebret 1996), (Isernia et al. 2000). In this section, convex optimisation based on linearly constrained least square minimisation is discussed towards the pattern synthesis of linear antenna array with uniform spacing. The optimisation of pattern synthesis is carried out including constraints in the weight estimation process.

Theoretical background: Consider a phased array of N elements arbitrarily placed with known spatial positions and different spatial responses. It is assumed that the antenna array is narrowband and there is no coupling between the antenna elements. The output response of the phased array is determined by summing the product of complex weights w and the element's response. The array response to an impinging signal (θ) is given by

$$p(\theta) = w^H v(\theta) \tag{4.44}$$

where $v(\theta)$ is the array response or steering vector in the direction θ and H denotes Hermitian.

The steering vector, $v(\theta)$, is of the form

$$v(\theta) = \begin{bmatrix} g_1(\theta)e^{j\omega\tau_1(\theta)} \\ g_2(\theta)e^{j\omega\tau_2(\theta)} \\ g_3(\theta)e^{j\omega\tau_3(\theta)} \\ \vdots \\ g_N(\theta)e^{j\omega\tau_N(\theta)} \end{bmatrix} \tag{4.45}$$

where $g_l(\theta)$ is the complex pattern gain of thee l^{th} element in the angle of arrival θ, ω is the operating frequency and τ_l is the time delay of the element with respect to a spatial reference point.

To generate a pattern (Figure 4.22) with a main lobe towards a desired direction and sidelobes at a pre-specified level over two disjointed areas on both sides of the mainlobe, the optimisation algorithm requires the input parameters, *viz.* look direction θ_s and edges of the sidelobe regions (θ_a, θ_b) and (θ_c, θ_d), and the desired sidelobe level ε. The width of the main beam is defined by the transition regions between the sidelobe edges on both sides of the main lobe.

The algorithm for pattern synthesis consists of two stages, (i) initialisation and (ii) iteration. The convex optimisation problem may be expressed as:

$$\min_w w^H A w \tag{4.46}$$

subjected to $M \le N$ constraints, $C^H w = f$ (4.47)

Here, the matrix A (dimension $N \times N$) is a signal covariance matrix R_{xx}. The matrix C is a $N \times M$ full rank matrix. A closed-form solution of such problem is given by (Tseng 1992) as

$$w = A^{-1}C(C^H A^{-1}C)^{-1} f \tag{4.48}$$

which provides optimum weights for the generation of the array pattern.

The algorithm starts with weights w such that a minimum power gain over the sidelobe regions and unity gain in the desired direction is maintained. The antenna weights are initialised by solving the least-square problem, i.e. $\min w^H Aw$.

The constraints for maintainin a unity peak in the desired direction are expressed as

$$v_s^H w = 1 \tag{4.49}$$

$$\text{and } \operatorname{Re}\{v_d^H w\} = 0 \tag{4.50}$$

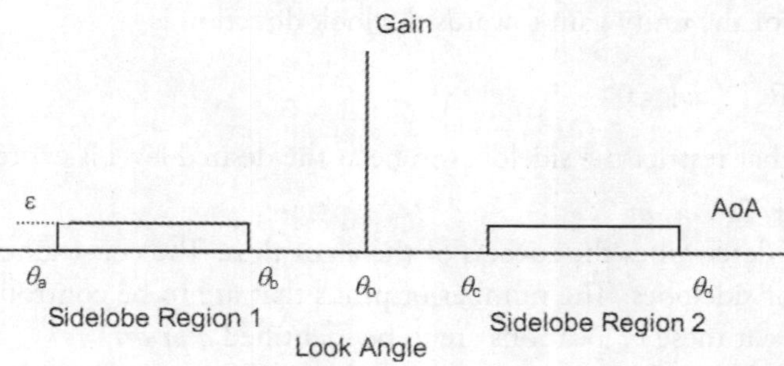

Figure 4.22 A typical specification of a desired antenna array pattern

v_s *and* v_d are the steering vectors in the direction of θ_s and the derivative of the same with respect to θ, respectively. The matrix A simulates the covariance matrix and assumes the form given by

$$A = \frac{1}{2}\sum_{l=1}^{L} v(\theta_l)v^H(\theta_l) \tag{4.51}$$

where L depends on the angles at which one wishes to define the steering vectors depending on the accuracy required. The solution of the above problem has low sidelobe response over the specified regions. The second constraint in (4.50) restricts the real part of $v_d^H w$ to be equal to zero.

This combination of constraints, steering vector and its derivative ensures that the resulting pattern has a unity peak gain in the look direction. The derivative constraint is applied by equating the first derivative of the square of magnitude of $p(\theta) = w^H v(\theta)$ to be zero. This method of imposing constraints is preferred as the problems associated with the selection of phase reference points can be avoided.

For avoiding ill-conditioning of the signal covariance matrix and to avoid large numerical values of weights, a properly chosen positive quantity is added to the matrix A (Lebret 1996).

$$A \leftarrow A + \sigma_n^2 I \tag{4.52}$$

where I is the identity matrix.

The initialisation procedure makes sure that the sidelobes are reduced and a peak is maintained in the desired look direction. However, this procedure does not ensure that the sidelobes are at a desired level. The level at which one wishes to limit the sidelobes is denoted by ε. To achieve this, an iterative procedure is employed, in which the weights are augmented with Δw at each step. This method assures that the algorithm converges uniformly to the desired pattern. The quantity added to the weights Δw is computed using constrained least square method. This restricts the pattern within the desired response.
The problem is defined as

$$\min_{\Delta w} \Delta w^H A \Delta w \tag{4.53}$$

The constraint for the unity gain towards the look direction is

$$v_s^H w = 0 \text{, and } \mathrm{Re}\{v_d^H w\} = 0$$

The constraint that restrict the sidelobes to be at the desired level is expressed as

$$v_i^H w = f_i \text{, for } i = 1, 2, \ldots, m \tag{4.54}$$

The choice of f_i determines the success of the algorithm. The constraints control the array gain at the peaks of sidelobes. The number of peaks that are to be controlled depends on m, and the peak values at these m locations are to be identified *a priori*.

The solution of this problem provides Δw, which is augmented to the weight vector, w for the generation of pattern. However, this process does not assure that the peaks of the new response will remain at the locations defined by the steering vectors v_i. Moreover, the algorithm does not converge in one single step. Hence, this process is repeated for certain number of iterations.

$$w \leftarrow w + \Delta w \tag{4.55}$$

There is a chance that a closed form solution for Δw may not be obtained if the value of m is too large. To make sure that a converged solution is obtained, the equations must not be over-determined. The value of m is, therefore, restricted at $M_{max} = N - 2$. This implies that only M_{max} sidelobes can be controlled at each iteration. As a rule, M_{max} is chosen corresponding to the largest sidelobe peaks (Tseng 1992). If the main peak levels are controlled, even though the algorithm might not find a solution with every sidelobe peak at the desired level, no peak is expected to be above the desired sidelobe level.

Since the choice of values of f_i is of prime importance for the success of the algorithm, care must be taken in the determination of f_i at θ_i. The array response is considered after the weight vector is being augmented.

$$\upsilon_i^H w = c_i + f_i \tag{4.56}$$

where c_i and f_i are the current and residual responses of the array at θ_i. These values are complex in nature. The sidelobe at the angle θ_i is desired to have response

$$|c_i + f_i| = \varepsilon \tag{4.57}$$

This defines a circle in complex plane, with centre c_i and radius ε. Any point lying on the circle f_i will satisfy (4.57). There can be infinite choices for f_i. There are two possible values of f_i, $\varepsilon - c_i$ and $-\varepsilon - c_i$ corresponding to the phase term, $(c_i + f_i)$ for which the output pattern is completely real, similar to the broadside radiation pattern of a uniform linear array.

The minimum norm value of f_i is zero when the sidelobe has a desired magnitude, i.e. when the current response is $|c_i| = \varepsilon$ in (4.57). If this value of f_i is reached accidentally by any other way, the algorithm will fail to converge. The minimum norm value of f_i is obtained by solving

$$f_i = \frac{c_i}{|c_i|}(\varepsilon - |c_i|) \tag{4.58}$$

Moreover, this solution ensures that the new response value has the same phase as that of the current response, and the difference between the values of responses in between the iterations is minimum. The algorithm is repeated till the peak locations remain invariant. It has been pointed out that this kind of weight adaptation might encounter numerical difficulties (Tseng 1992). The recommended approach is the QRD of C matrix so as to reduce the matrix to reasonable values.

Pattern generation: A 20-element linear array with half-wavelength uniform inter-element spacing is considered. The number of sidelobes considered is 18 (i.e. $N - 2$). It is assumed that look direction is 45° and the sidelobe boundaries are –90° to 35° and 55° to 90°. The desired sidelobe level is taken as –40 dB.

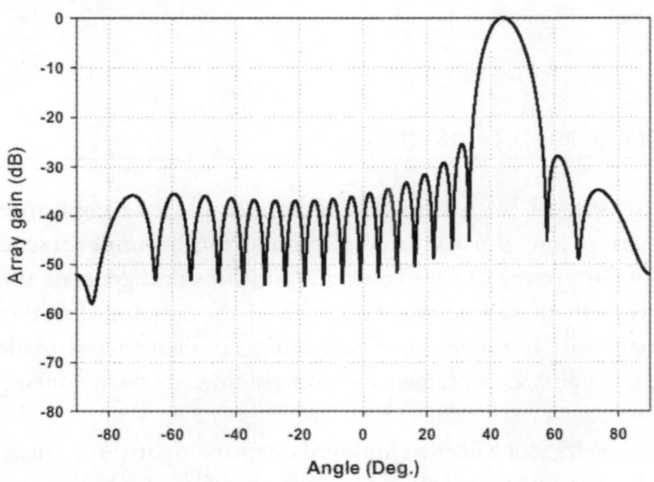

Figure 4.23 Pattern of 20-element linear array with half-wavelength inter-element spacing using linearly constrained least square algorithm (main lobe = 35° to 55°)

Figure 4.23 shows the computed pattern with a desired main beam in the region from 35° to 55°. The peaks are determined up to an accuracy of 0.5°. The convergence of the solution is achieved within three iterations. This proves the speed and efficiency of the linearly constrained least square algorithm in the pattern synthesis of phased array.

The look direction is next changed to 45°. Figure 4.24 shows the radiation pattern of a 32-element linear array with 0.4λ uniform spacing and main beam directed towards 45°. These simulations corroborate the capability of the algorithm in accurate pattern generation within a few iterations. This algorithm works for planar and non-planar arrays with uniform and non-uniform inter-element spacing.

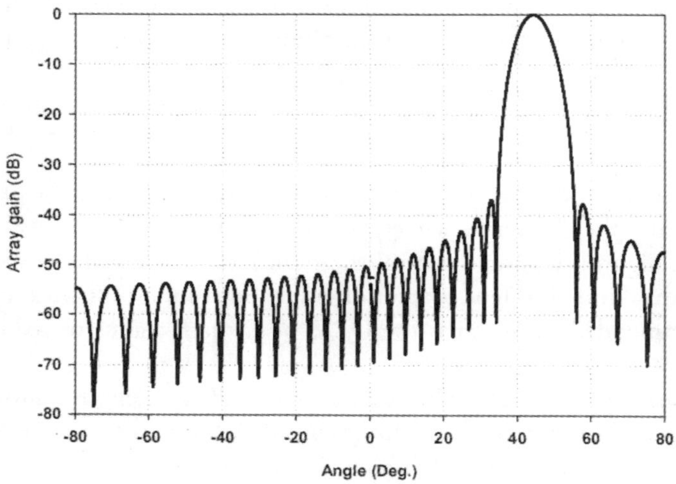

Figure 4.24 Radiation pattern of 32-element uniform array with 0.4λ inter-element spacing (main lobe = 35° to 55°)

4.3 Probe suppression in phased arrays

Adaptive algorithms such as LMS and SMI have been employed to optimise the sidelobe level in the radiation pattern of linear/planar array with uniform/non-uniform spacing. The modified improved LMS algorithm is one of the efficient algorithms that generate the radiation pattern of a given array with prescribed nulls towards each of the probing directions and main lobes towards desired directions. If the spacing between array elements is made non-uniform or random, the optimal weights are obtained accordingly towards the adapted pattern for a given signal scenario.

This depends on the efficiency of the employed adaptive algorithm. The signal environment may consist of multiple hostile sources along with multiple desired sources. The impinging signals can be either narrowband or wideband. Moreover, in actual scenario, the radar signals can be correlated owing to the multipath effect. The algorithm is expected to cater to such complex signal environment by maintaining sufficient gain towards each of the desired source and suppress each of the probing sources by placing accurate deep nulls towards them. These suppression capabilities of phased arrays when mounted on any aerospace platform or base station can be exploited towards active RCSR. If the transmitted energy from the antenna array can be minimised adaptively towards the hostile sources attempting to probe the antenna structure, it leads to active RCSR.

4.3.1 Theoretical background

The radiation behaviour of a phased array depends on the array geometry, aperture distribution, spacing between elements, array size, etc. For a linear array of N antenna elements with uniform spacing, the far-field is expressed as

$$E_i(\theta,\phi) = f(\theta,\phi)\sum_{i=1}^{N} I_i e^{-j(kd\cos\theta+\alpha)} \tag{4.59}$$

where $f(\theta,\phi)$ represents the far-field antenna element pattern, and I_i and α_i are the amplitude and phase of element excitations, respectively. The array pattern may be obtained by taking modulus of (4.59) as

$$|E_i(\theta,\phi)| = |f(\theta,\phi)| \cdot |S| \tag{4.60}$$

where $S = \sum_{i=1}^{N} I_i e^{-j(kd\cos\theta+\alpha)}$ is the space factor or array factor.

The array factor essentially refers to the spatial distribution of the antenna array elements, for a given aperture distribution. In other words, the element excitations determine the distribution of radiated energy in space (Balanis 2005). For phased array with planar geometric configuration, the far-field radiation pattern is determined by multiplying array factors for x- and y-axis.

$$\begin{aligned} E(\theta,\phi) &= f(\theta,\phi) \sum_{m=-N_x}^{N_x} \sum_{n=-N_y}^{N_y} I_{mn} \exp\left[j\left(mkd_x \sin\theta\cos\phi + \alpha_x\right)\right] \\ &\qquad \times \exp\left[j\left(mkd_y \sin\theta\sin\phi + \alpha_y\right)\right] \\ &= f(\theta,\phi) S_x S_y \end{aligned} \tag{4.61}$$

where $S_x = \sum_{m=-N_x}^{N_x} I_{mo} \exp\left[j(mkd_x \sin\theta\cos\phi + \alpha_x)\right]$

$$S_y = \sum_{n=-N_y}^{N_y} I_{on} \exp\left[j(nkd_y \sin\theta\sin\phi + \alpha_y)\right]$$

where $\alpha_x = -mkd_x \sin\theta_o \cos\phi_o$, $\alpha_y = -nkd_y \sin\theta_o \sin\phi_o$, (θ_o, ϕ_o) specifies the main lobe position, I_{mn} and $\alpha_x + \alpha_y$ are the amplitude and phase of element excitations, respectively.

It may be noted that for the array of rectangular cross section, the radiation pattern is determined from the product of the patterns of two linear arrays in orthogonal directions. If the spacing between the antenna elements is unequal, the radiation behaviour changes drastically. This is because of the variation in the elemental phases. As a result, the array manifold, i.e. a steering vector and, hence, the array pattern is modified.

In a phased array, the output from each sensor is multiplied with the weight vector so that the gain is maintained towards the desired signal and each probing signal is suppressed efficiently. If signal scenario consists of a single desired signal, the antenna weights are selected such that the main lobe is generated towards the desired signal. On the contrary, for signal environment consisting of multiple desired signals impinging simultaneously from different directions, the weights should be chosen in such a way that the multiple main lobes are generated towards each of the desired directions. SNR of any received signal would be the maximum only when the optimum weights are chosen for the generation of main lobes. The array adapts according to the signal environment depending upon the power level of impinging signals and the weights fed to the antenna elements. In fact the optimum weights maximise the output SINR by suppressing the probing sources, provided their impinging angle do not coincide with the main lobes.

If the incident signal is wideband, each spectral line of the signal may be treated as a monochromatic source. In other words, a wideband signal is equivalent to several narrowband sources depending upon the signal bandwidth. Thus, the signal covariance matrix will be different and, hence, the optimal weights for generation of adapted pattern. The optimality of antenna weights depends on the efficacy of algorithm used for weight adaptation for a given signal environment. The parameters that govern the weight adaptation are the geometrical configuration of array, impinging angles of signals, steering vector, eigenvalues/eigenvectors of covariance matrix, bandwidth of the impinging signals, etc.

Furthermore, if the two impinging signals (friendly or hostile) are correlated, the weight adaptation becomes critical. The correlation between the signals can be either due to multipath effect or when probing sources intentionally give rise to correlated interferences. The conventional adaptive array methods, which do not consider correlation among the impinging signals, suffer from self-signal cancellation effect. The array gets confused between the desired signal and the probing signal, thereby suppressing the desired signal. This is due to the fact that in the presence of correlated/coherent signals, the signal covariance matrix loses its Toeplitz structure. The eigenvalues and eigenvectors will be no more distinct and well-spread. This results in distortions in the adapted pattern with inaccurate nulls.

If there are multiple desired signals impinging simultaneously on the array, the weight updation changes to (Singh and Jha 2011)

$$W(t+1) = P[W(t) - \mu\, g(W(t))] + \frac{S_1}{S_1^H S_1} + \frac{S_2}{S_2^H S_2} + \ldots + \frac{S_m}{S_m^H S_m} \tag{4.62}$$

with the projection operator given by

$$P = I - \frac{S_1 S_1^H}{S_1^H S_1} - \frac{S_2 S_2^H}{S_2^H S_2} - \ldots - \frac{S_m S_m^H}{S_m^H S_m} \tag{4.63}$$

where S_1, S_2,..., S_m represent the steering vectors corresponding to each impinging desired signals (1, 2, ...m).

The Toeplitz–Hermitian correlation matrix of the signal received by an array is obtained as

$$R_x = SXS^H + \sigma_n^2 I \tag{4.64}$$

where $X = E\{x(t)x(t)^H\}$, $E\{\}$ is the expectation value and $S = \left[S_1(\theta_1),\ S_2(\theta_2)\ S_3(\theta_3) \ldots S_k(\theta_k)\right]$.

If the impinging signals are not correlated with each other, the source correlation matrix X is diagonal. However, the correlation between the impinging signals makes the matrix X non-conditional and non-singular. For coherent signals, i.e. when the impinging signals are fully correlated, the correlation matrix X becomes non-diagonal but singular (Lee and Hsu 2000). If in a given signal scenario, two of the impinging signals are correlated, i.e. $x_2(t) = \alpha\, x_1(t)$, α being the correlation coefficient. This coefficient signifies amplitude scaling as well as the signal delay due to multipath effect. Thus, the dimension of $x(t)$ becomes a $(k - 1) \times 1$, i.e. $x(t) = \left[(1+\alpha)x_1(t)\ \ x_3(t)\ \ \ldots x_k(t)\ \right]^T$ and A will be a $\{(k - 1) \times N\}$ matrix, i.e. $S = \left[S_1(\theta_1) + \alpha\, S_2(\theta_2)\ \ S_3(\theta_3) \ldots S_k(\theta_k)\right]$.

Due to correlation between the signals, the matrix R_x changes to a $\{(k - 1) \times (k - 1)\}$ non-singular matrix. The Toeplitz eigenstructure of the correlation matrix is disturbed, resulting in the degradation of array performance. The correlation matrix can be reconstructed to restore its eigenstructure using an iterative matrix reconstruction method (Lee and Hsu 2000). The algorithm helps in retrieving the Toeplitz form of correlation matrix with distinct eigenvalues and eigenvectors. The steps involved are as follows:

Step I: Estimate correlation matrix, R_x from the signal received by array

Step II: Reconstruct R_x to obtain a Toeplitz form

First, consider the N diagonals of the upper triangular section of the correlation matrix. For each diagonal, the average of matrix elements is determined. The mean value computed is stored in each diagonal element.

Next, consider $N - 1$ diagonals of the lower triangular section of the correlation matrix. Each diagonal element is the complex conjugate of the corresponding element in the upper triangular section of the matrix. This forms new Hermitian auto-correlation matrix (ACM), $\hat{R}_x$.

Step III: Compute matrix $\tilde{R}_{xs}$ using eigenvalues and eigenvectors of $\hat{R}_x$.

$$\tilde{R}_{xs} = \sum_{j=1}^{p} \lambda_j e_j e_j^H + \lambda_{av} \sum_{j=p+1}^{N} e_j e_j^H \tag{4.65}$$

where $\lambda_1 \geq \lambda_2 \geq \ldots \geq \lambda_N$ are eigenvalues, e_j, $j = 1,2,\ldots N$ are the eigenvectors and λ_{av} represents the average of λs.

Step IV: If $\left|\tilde{R}_{xs} - \hat{R}_x\right| > \varepsilon$, where ε is a scalar, store $\tilde{R}_{xs}$ in R_x and then repeat the reconstruction process starting from Step II. If the condition is not satisfied, then stop and use $\tilde{R}_{xs}$ as the final correlation matrix.

Thus, the final form of correlation matrix has the Toeplitz–Hermitian structure and, hence, the distinct eigenvalues and eigenvectors.

For correlated signals, the optimal weight (Lee and Hsu 2000) is given by

$$w_o = \tilde{R}_{xs}^{-1} C (C^H \tilde{R}_{xs}^{-1} C)^{-1} c \tag{4.66}$$

where C is the constraint vector expressed as $C = [S_1(\theta_1)\ S_2(\theta_2) \ldots S_p(\theta_p)]$ and $c = [c_1\ c_2 \ldots c_p]$ is the gain vector. The dimension of the constraint matrix C depends on the number of antenna elements. The optimum weight vector obtained is used to determine the array output $y(t) = w^H x(t)$.

4.3.2 Probe suppression with single desired source

The modified improved LMS algorithm is used to determine the optimal antenna weights. Figure 4.25 shows the suppression of two probing sources (20°, 120°) by non-uniform linear array of 10 identical antenna elements. The spacing is assumed to start from 0.484 λ, then increase with the step-size of 0.01 λ till the mid-point of the array, beyond which it decreases at the same rate till the end of the array. An individual antenna element is taken as an isotropic radiator. Mutual coupling effect is not considered. The desired signal (90°) power ratio is assumed to be 10. The probing sources are assumed to be incident at 20° and 120° with the power ratio of 10 and 100, respectively. The antenna elements are excited by uniform unit-amplitude distribution.

Both the quiescent and adapted patterns are included. The desired source is shown as a solid arrow, whereas probing sources are shown as dashed arrows. It may be observed that even for non-uniform spacing between antenna elements, the phased array is capable of tackling the multiple signal environments. Both the incident probing sources are actively cancelled.

The adapted pattern for planar array can be obtained using improved LMS algorithm similarly as the linear arrays. Figure 4.26 presents the adapted and quiescent pattern of a 16 × 10 planar array. The element spacing is considered as $d_x = 0.484\ \lambda$ and $d_y = 0.771\ \lambda$. The probing sources arrive at 20° and –20°. It is apparent that both the probing sources are suppressed efficiently.

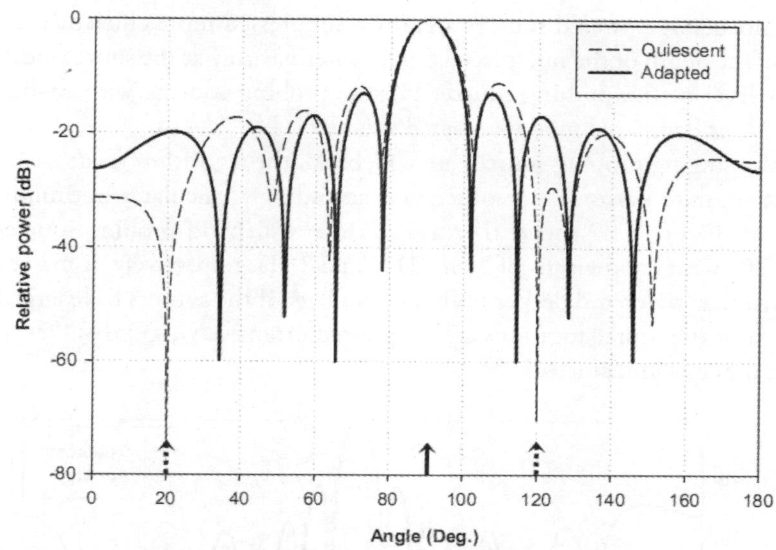

Figure 4.25 Probe suppression in non-uniform 10-element linear array. The probing sources impinge at 20° and 120°. The desired source is shown as a solid arrow, whereas probing sources are shown as dashed arrows along the *x*-axis

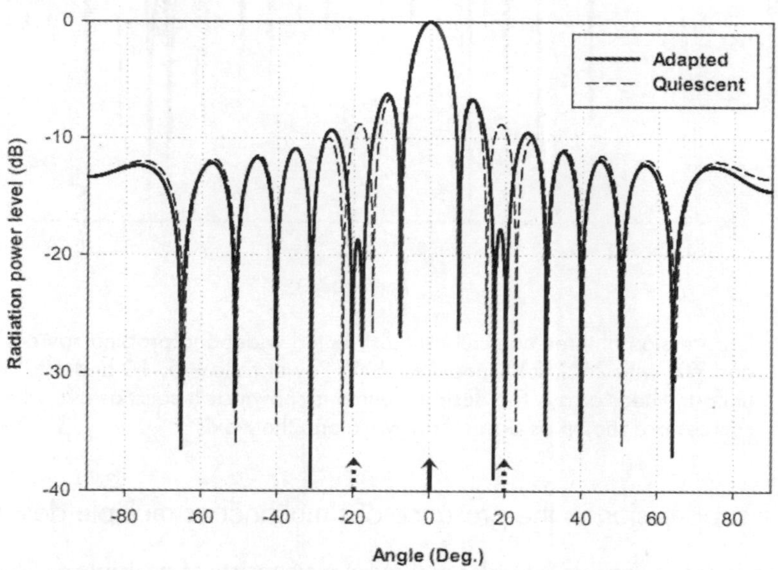

Figure 4.26 Probe suppression in 16 × 10 uniform planar array. The probing sources impinge at 20° and –20°. The desired source is shown as a solid arrow, whereas probing sources are shown as dashed arrows along the *x*-axis

The available degrees of freedom (DOF) for the suppression of probing signals by phased array depend on the number of hostile sources. Equal-powered sources consume more number of DOF. The unequally powered sources are, thus, suppressed more efficiently than the equal powered ones. The depth of the null placed by the antenna array at the source location depends on the power level of the probing source. Hostile probing sources with higher powers are suppressed more as compared to low-powered sources.

When the impinging probing sources are wideband, the algorithm treats each spectral line of source as a separate narrowband source and accordingly calculates optimum weights for adapted pattern. Figure 4.27 shows the case of three wideband probing sources arriving at 42°, 50° and 70° with bandwidths of 26%, 21% and 19%, respectively. It may be seen in the adapted pattern that wider and deeper nulls are produced if the sources have large bandwidths. The sidelobes near the source locations are suppressed extensively (–35 dB). Yet the main lobe in the pattern remains undistorted.

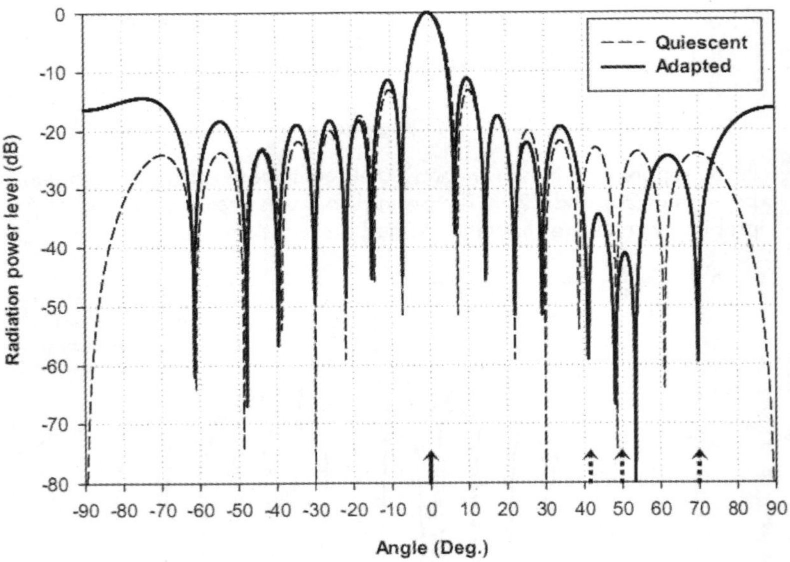

Figure 4.27 Suppression of three continuously distributed wideband probing sources (42°, 50° and 70° with 26%, 21% and 19% BW; Power ratio = 5, 10 and 100) in 16-element uniform linear array. The desired source is shown as a solid arrow, whereas probing sources are shown as dashed arrows along the *x*-axis

4.3.3 Probe suppression in the presence of simultaneous multiple desired signals

In this sub-section, the considered radar environment consists of multiple desired signals and uncorrelated signals trying to probe the aircraft with the on-board antenna array from different directions. Modified improved LMS algorithm is used for computing weights in such a scenario so as to minimise the output noise power and improve the reception of desired signals even in the presence of multiple uncorrelated probing signals.

Figure 4.28 presents the adapted pattern of 16 × 10 array for a signal environment consisting of two desired signals (–20°, 20°; 0 dB) and one probing signal (60°; 100). The array in conjunction with the efficient modified improved LMS algorithm (Singh and Jha 2011) is able to simultaneously generate main lobes towards each desired direction and place deep nulls in probing directions. This is apparent from the deep nulls up to –55 dB towards the probing sources in the adapted pattern.

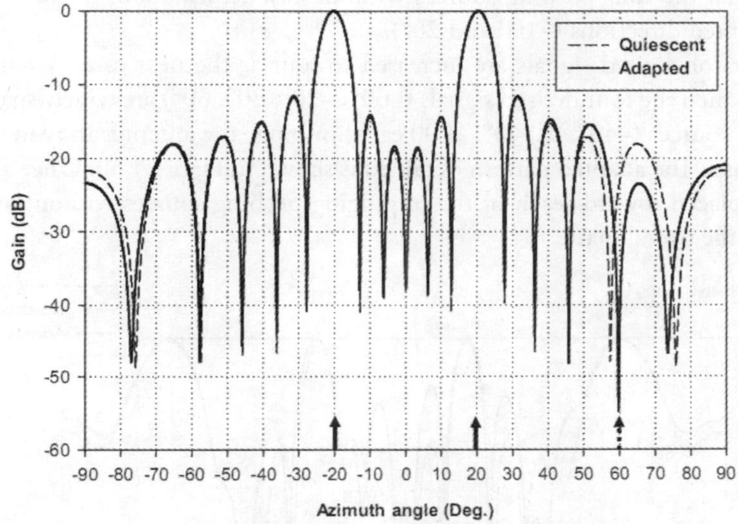

Figure 4.28 Adapted pattern of 16 × 10 antenna array. Two desired signals (20°, –20°; 1, 1) and one probing source (60°; 100). The desired source is shown as a solid arrow, while probing sources are shown as dashed arrows along the *x*-axis

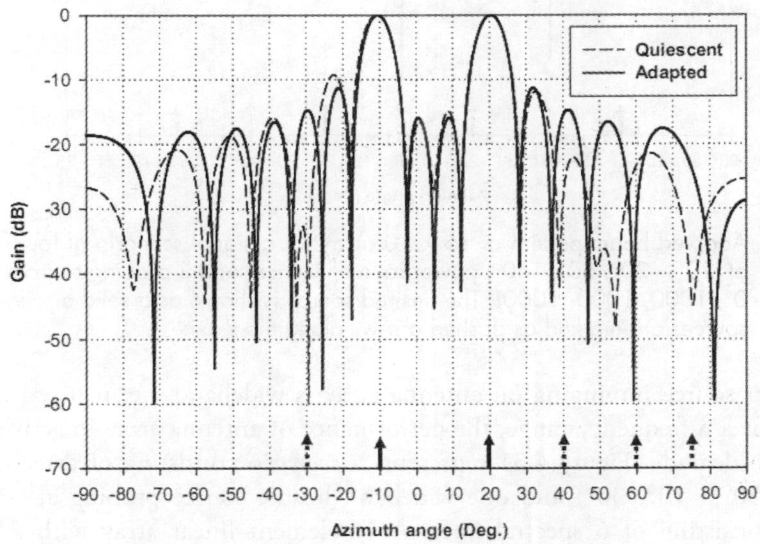

Figure 4.29 Adapted pattern of 16 × 10 array for two desired signals at (20°, –10°, power ratio of 1 each) and four probing sources at (–30°, 40°, 55°, 75°; 1000, 100, 100, 100). The desired source is shown as a solid arrow, while probing sources are marked as dashed arrows along the *x*-axis

The number of hostile probing sources is next taken as four (–30°, 40°, 55°, 75°) with input signal power ratio of 1000, 100, 100 and 100, respectively. The two desired signals are taken at –10° and 20°. In this case too, the results (Figure 4.29) show that the array is efficient in suppressing all the four probing sources without any deformation in the two main lobes towards the desired directions (–10° and 20°).

The numbers of desired signals are increased to four as the next case. A radar scenario is considered in which the four desired signals (–60°, –20°, 20°, 60°) are widely spaced and three hostile probing sources (–40°, 0°, 40°; 1000 each) impinge the antenna array in between these desired directions. The adapted pattern obtained is shown in Figure 4.30. Once again the nulls are accurately placed towards each of the impinging probing sources without any significant disturbance in the overall pattern.

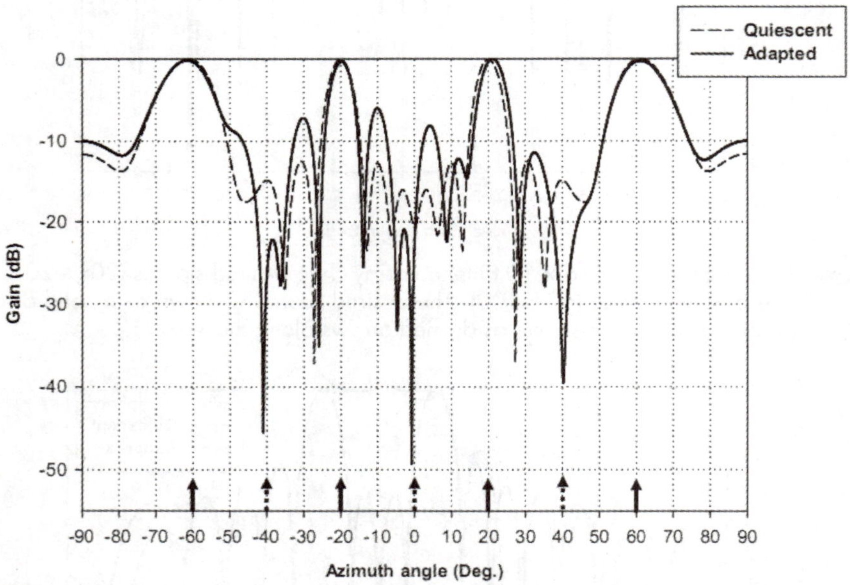

Figure 4.30 Adapted beam pattern of 16 × 10 array for a signal scenario of four desired signals at (20°, –20°, 60°, –60°; power ratio of 1 each) and 3 probing sources at (40°, –40°, 0°; 1000, 1000, 1000). The desired source is shown as a solid arrow, while probing sources are marked as dashed arrows along the x-axis

If the radar source impinging on antenna array is wideband, i.e. if it has finite spectral distribution over a frequency range, the performance of antenna array in active cancellation is expected to degrade. Figure 4.31 represents a scenario consisting of three desired signal directions (–50°, –10°, 30°) and one wideband hostile source probing at –30° with 5% bandwidth consisting of 6 spectral lines. A 16-element linear array with $\lambda/2$ spacing is considered. The adapted pattern for a narrowband/monochromatic probing source is included. On comparing with the quiescent pattern, it may be observed that the null corresponding to the wideband source is wider than that of the narrowband one. This is due to the fact that every spectral line acts as a narrowband source, and thus the eigenvalues and eigenvectors of the corresponding correlation matrix are distinct. This results in wider null towards the wideband probing source.

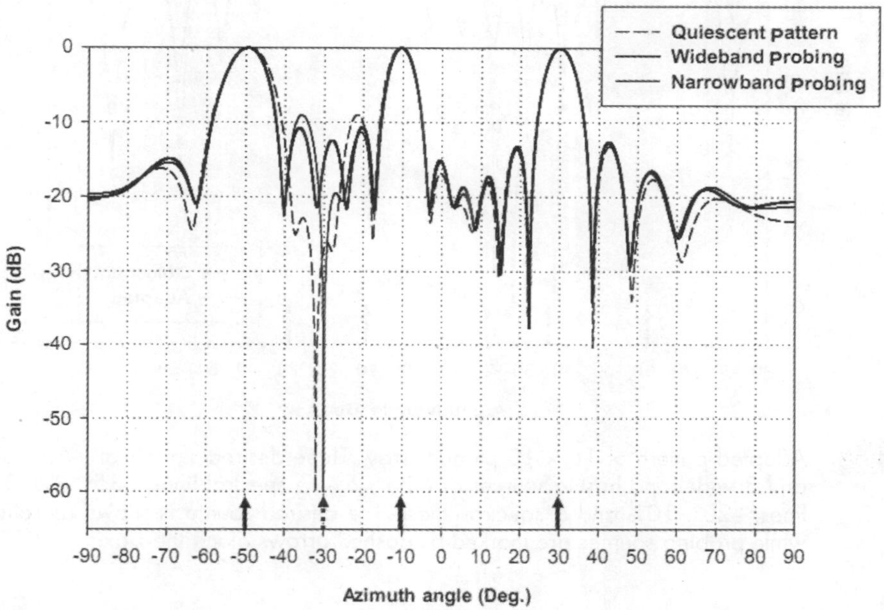

Figure 4.31 Suppression of one wideband source (–30°, 5%, 6 spectral lines) in the presence of three desired signals at (–50°, –10°, 30°) by a 16-element linear array. The desired source is shown as a solid arrow while probing sources are shown as dashed arrows along the *x*-axis

The important fact is that in either case, the main lobes of the pattern remain undistorted. In other words, there will not be any loss of relevant information related to the radar sources that the antenna array is tracking or transmitting to, and at the same time it is able to suppress the hostile radar probing sources.

Assume a slightly complex signal scenario with three desired signal directions (–60°, 10°, 30°) and three wideband hostile radar sources with different bandwidths and a small angular separation (–25°, 5%, 6 spectral lines; –35°, 2%, 3 spectral lines; –20°, 10%, 5 spectral lines). It may be noticed from Figure 4.32 that each wideband probing source is suppressed with deep and wide null placed in the adapted pattern. This demonstrates that the planar array with an efficient weight adaptation algorithm is capable of placing nulls in the pattern around the central frequency of each of the hostile radar probing source and attains significant level of suppression over the entire spectrum.

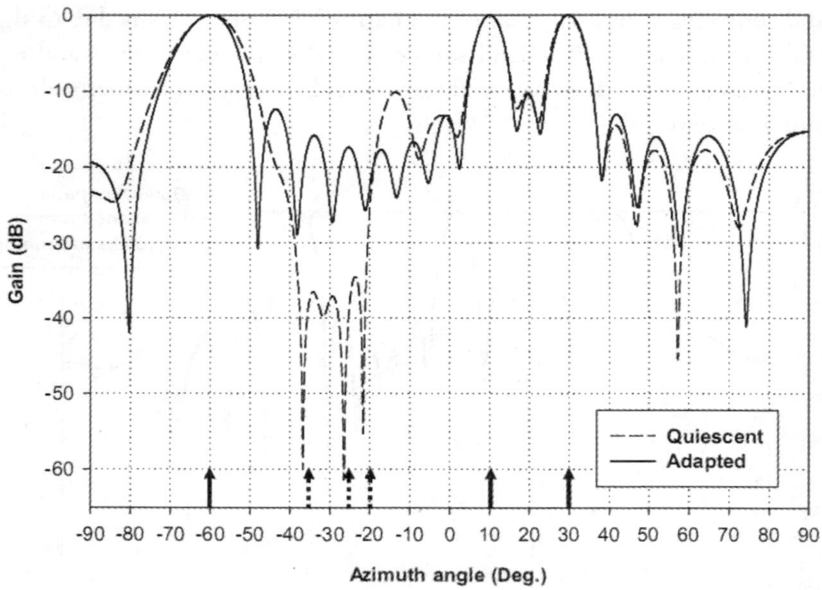

Figure 4.32 Adapted pattern of 16 × 10 planar array. Three desired signals at (–60°, 10°, 30°) and 3 wideband hostile sources at (–25°, 5%, 6 spectral lines; –35°, 2%, 3 spectral lines; –20°, 10% and 5 spectral lines). The desired source is shown as solid arrow, while probing sources are marked as dashed arrows along the *x*-axis

4.3.4 Probe suppression in the presence of correlated signals

In the practical scenario, a transmitted signal undergoes multiple reflections and diffractions due to several obstacles in its path. This multipath effect results in various correlated signals arising from the same source. To mitigate the multipath effect, the formulation for ACM, steering vector and complex weight coefficients has to be modified. This includes the correlation coefficient in the expressions, facilitating de-correlation of the radar signals. This correlation coefficient is basically a complex quantity, real part of which is the scaling factor and the imaginary part signifies the phase delay or lead among the radar signals.

Figure 4.33 illustrates such a case of hostile correlated probing source. Two desired signal directions (–30°, 0°) and two hostile probing sources (30°, 50°) are incident on a 10-element uniform linear array with $\lambda/2$ spacing. Probing signal at 30° is correlated {α = (–0.1, 0.09)} with the desired signal impinging at –30°.

The constraint vector is [1 1 0]. If the hostile source attempting to probe array at 50° is wideband, having 5% bandwidth with 6 spectral lines, the adapted pattern generated by the array shows an efficient active cancellation of probing with main lobe looking towards the desired sources. Each spectral line of the wideband probing source is suppressed without disturbing the rest of the pattern. This is evident that the array efficiently de-correlates the two impinging correlated signals and maintains the desired adapted pattern in accordance with the signal scenario. However, the depth of the null towards the wideband probing signal is less than the narrowband probing signal. This depends on the available degrees of freedom for the array in a given signal environment.

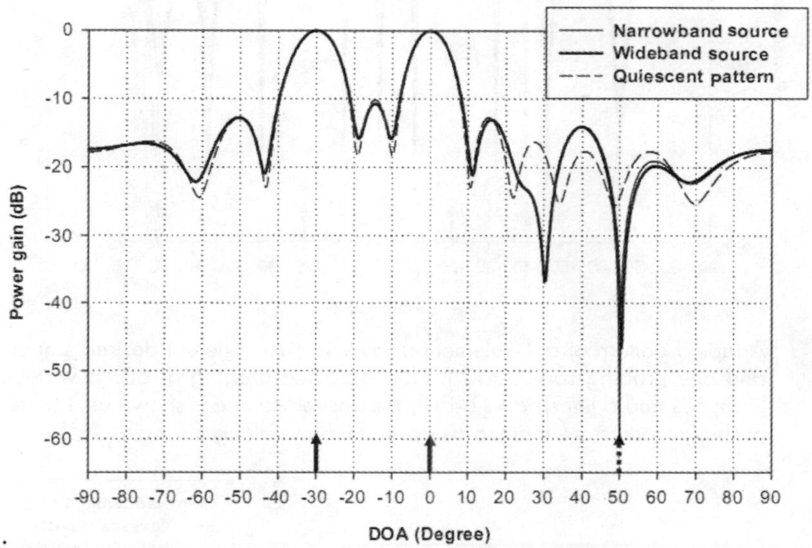

Figure 4.33 Adapted beam pattern of a 10-element phased array with two desired (–30°, 0°) and two probing sources (30°, 50°). Signal probing at 30° is correlated to the desired signal impinging at –30°. Probing signal at 50° is wideband (5%, 6 spectral lines). $c = [1\ 1\ 0]$; the desired source is shown as a solid arrow, while probing sources are marked as dashed arrows along the x-axis

To further analyse the role of bandwidth of the probing source in probe suppression, the bandwidth of the hostile source impinging at 60° is increased to 5% with 6 lines and, then, to 15% and 9 lines. The array performance in active cancellation of such wideband probing sources is compared in Figure 4.34. It may be noted that in both the scenarios, the null is placed accurately but the null-depth degrades as the bandwidth becomes wider.

The number of desired signals is next taken as four (–20°, 20°, –40°, 40°). The signal environment consisting of two hostile probing sources (–60°, 50°) is considered. The desired signals at (–20°, 20°) and (–40°, 40°) are coherent with $\alpha = (1, 0)$. Two probing sources at (–60° and 50°) are uncorrelated and wideband (2%, 3 spectral lines each). The constraint vector is taken as $c = [1\ 1\ 0\ 0]$.

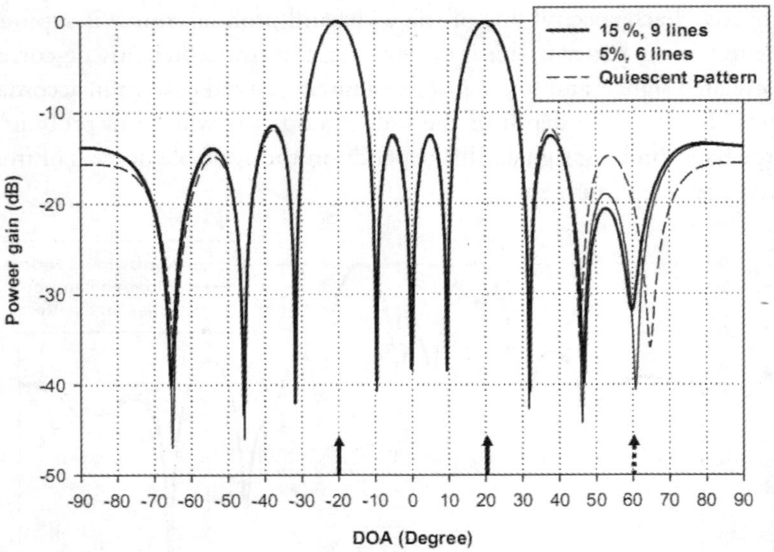

Figure 4.34 Adapted pattern of a 10-element array with two coherent desired signals (–20°, 20°) and one probing source (60°). Hostile source probing at 60° is wideband (15%, 9 lines; 5% and 6 lines), *c* = [1 1 0]; the desired source is shown as a solid arrow, while probing sources are marked as dashed arrows along the x-axis

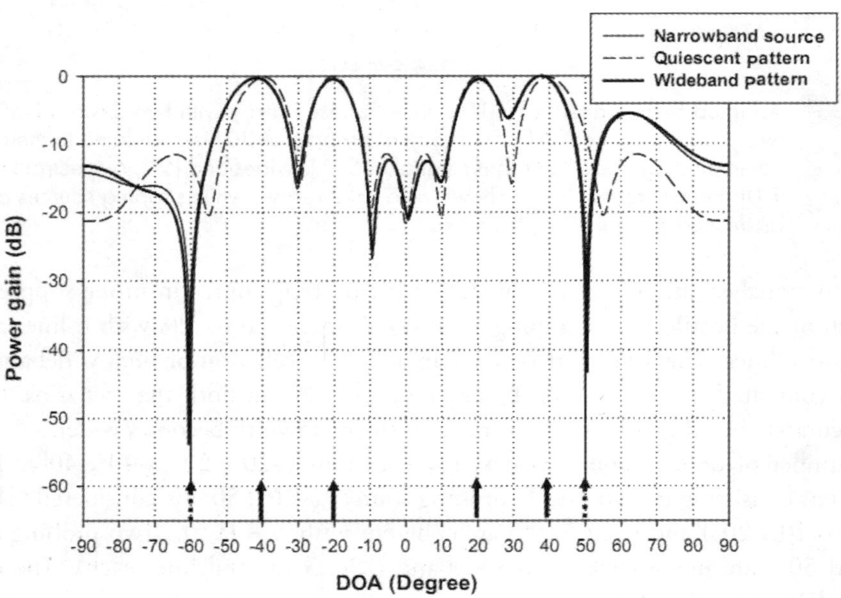

Figure 4.35 Adapted pattern of a 10-element array with four coherent desired signals (–20°, 20°, –40°, 40°) and two probing sources (–60°, 50°). Two probing sources (–60° and 50°) are uncorrelated and wideband (2%, 3 lines each), *c* = [1 1 0 0]. The desired source is shown as a solid arrow, while probing sources are marked as dashed arrows along the *x*-axis

In Figure 4.35, the adapted pattern for such a complex signal scenario is shown. It may be seen from the adapted pattern that the nulls are accurately placed even for coherent desired signals. One can notice very small shift (<0.1°) in the main lobes at 40° and –40°. The reason for this shift may be the wideband probing sources incident at –60° and 50° with multiple spectral lines. Moreover, these probing sources are also spatially close to the impinging angles of desired signals.

However, it cannot be ignored that the array is capable of de-correlating the incident coherent/correlated signals and subsequently generating the adapted pattern according to the signal environment.

In Figure 4.36, a signal scenario of two wideband probing sources with different bandwidths (–60°, 2%, 3 lines; 50°, 11%, 6 lines) is considered. Adapted beam pattern of an array with four desired (–20°, 20°, –40°, 40°) and two probing sources (60°, 50°) is shown. Four desired signals are assumed to be coherent, $\alpha = (1, 0)$. The constraint vector is c = [1 1 0 0]. In this case too, the array is able to maintain the main lobes towards each of the coherent desired signals. The nulls placed towards each probing source are sufficiently deep (–60 dB). These results demonstrate the capability of algorithm in tackling the correlated signals and, hence, the multipath effect.

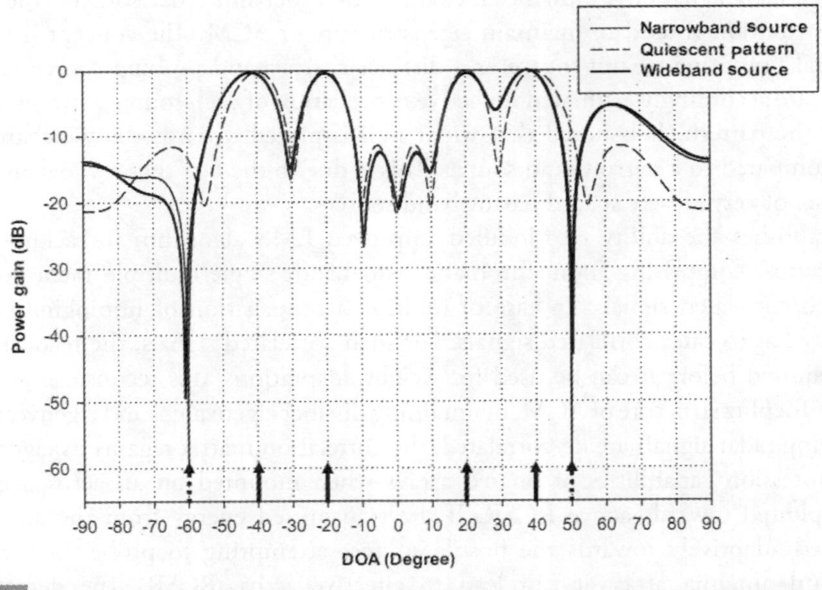

Figure 4.36 Adapted beam pattern of a 10-element array with four coherent desired signals (–20°, 20°, –40°, 40°) and two wideband probing signals (–60°, 2%, 3 lines; 50°, 11%, 6 lines), c = [1 1 0 0]; the desired source is shown as a solid arrow, while probing sources are marked as dashed arrows along the x-axis

4.4 Conclusion

The phased array in a given radar environment performs well depending upon the weight adaptation towards the optimum value and, hence, the correct adapted pattern. The correctness of the array pattern is judged by the gain towards each of the desired sources and suppression of the probing signals. This capability is based on the efficacy of the weight adaptation algorithm. The estimation of optimum weights has great control over the generation of antenna pattern, the main lobe width sidelobe distribution and the null placement. The algorithm should work for arbitrary geometric configuration of phased array, whether linear, planar, uniform or non-uniform element spacing.

The efficiency of adaptive algorithm is determined from the convergence rate, output SINR, flexibility and ease of implementation. The robustness of the algorithm can be further enhanced by including constraints in the weight adaptation process. The efficiency of algorithm can be improved by using optimisation techniques. An appropriate combination of constraints, steering vector and its derivative ensures the resultant pattern to have a desired gain towards the look angle and deep nulls towards the undesired probing sources.

If the phased array is involved in tracking multiple simultaneous radar sources, its radiation pattern is expected to have main lobes directed towards each source. For such scenario, the steering vectors towards each source, the corresponding projection vectors and the optimal weights are computed accordingly. The Modified Improved LMS algorithm is shown to be able to cater such signal environment efficiently. For wideband radar sources, the associated spectral distribution is used to maintain eigenstructure of ACM. The concept is to consider each spectral line of wideband source as a distinct narrowband probing source. The weight coefficients are accordingly estimated for active cancellation of each source, without disturbing the rest of the pattern. It is noted that wider null is placed towards the wideband probing source as compared to a narrowband source. This is due to the fact that a wideband source is taken as a set of very closely spaced narrowband sources.

This establishes the ability of Modified Improved LMS algorithm in tackling arbitrary signal scenarios, comprising monochromatic, wideband, single/multiple radar sources and uncorrelated/correlated signals. In case of multipath propagation of impinging signals, the phased array has to cater correlated signals. For such correlated signals, the reconstruction of ACM is required before it can be used for weight adaptation. This reconstruction of ACM restores the Toeplitz structure of ACM, maintaining distinct eigenvalues and eigenvectors. Once the impinging radar signals are de-correlated, the correlation matrix regains its eigenstructure.

The suppression capabilities of phased arrays when mounted on an aerospace platform may be exploited towards active RCSR. If the transmitted energy from the antenna array is minimised adaptively towards the hostile sources attempting to probe the aircraft with the on-board antenna array, it can lead to effective active RCSR. The detectability of the array is highly reduced as the pattern of phased array has main lobes only towards the desired directions.

References

Abdolee, R., V. Vakilin, and T. A. Rahaman. 2008. 'Element space and amplitude perturbation using genetic algorithm for antenna sidelobe cancellation.' *Signal Processing: An International Journal* 2: 10–16.

Alexander, S. T. and A. L. Ghirnikar. 1993. 'A method for recursive least squares filtering based upon an inverse QR decomposition.' *IEEE Transactions on Signal Processing* 41: 20–30.

Andreasen, M. G. 1962. 'Linear arrays with variable inter-element spacing.' *IRE Transactions on Antennas and Propagation* 10: 137–43.

Applebaum, S. P. 1976. 'Adaptive arrays.' *IEEE Transactions on Antennas and Propagation* 24: 585–99.

Balanis, C. A. 2005. *Antenna Theory, Analysis and Design*. Third edition. New Jersey: John Wiley and Sons, 1117.

Bershad, N. J. 1986. 'Analysis of the normalised LMS algorithm with Gaussian inputs.' *IEEE Transactions on Acoustics, Speech, Signal Processing* 34: 793–806.

Brennan, L. E. and I. S. Reed. 1982. 'An adaptive array signal processing algorithm for communications.' *IEEE Transactions on Aerospace and Electronics Systems* 18: 124–30.

Buehring, W. 1978. 'Adaptive orthogonal projection for rapid converging interference suppression.' *Electronics Letters* 14(16): 515–16.

Carlson, B. D. and D. Willner. 1992. 'Antenna pattern synthesis using weighted least squares.' *IEE Proceedings, Pt. H.* 139(1): 11–16.

Carlson, B. D. 1988. 'Covariance matrix estimation errors and diagonal loading in adaptive arrays.' *IEEE Transactions on Aerospace and Electronic System* 24: 397–401.

Chandran, S. 2004. *Adaptive Antenna Arrays: Trends and Applications.* New York: Springer, 660.

Chen, K., X. Yun, Z. He, and C. Han. 2007. 'Synthesis of sparse planar arrays using modified real genetic algorithm.' *IEEE Transactions on Antenna and Propagation* 55: 1067–73.

Chern, S. -J. and C. -Y. Chang. 2002. 'Adaptive linearly constrained inverse QRD–RLS beamforming algorithm for moving jammers suppression.' *IEEE Transactions on Antennas and Propagation* 50: 1138–50.

Cioffi, J. M. and T. Kailath. 1984. 'Fast recursive-least square, transversal filters for adaptive filtering.' *IEEE Transactions on Acoustics, Speech, Signal Processing* 32: 998–1005.

Elliot, R. S. 2005. *Antenna Theory and Design.* New Jersey: John Wiley & Sons, 624.

Evans, J. B., P. Xue, and B. Liu. 1993. 'Analysis and implementation of variable step size adaptive algorithms.' *IEEE Transactions on Signal Processing* 41: 2517–35.

Fabre, P. and C. Gueguen. 1986. 'Improvement of the fast recursive least-square algorithms via normalisation: A comparative study.' *IEEE Transactions on Acoustics, Speech, Signal Processing* 34: 296–308.

Fenn, A. J. 2008. *Adaptive Antennas and Phased Arrays for Radar and Communications*. Boston: Artech House, 394.

Frost III, O. L. 1972. 'An algorithm for linearly constrained adaptive array processing.' *Proceedings of IEEE*. 60: 926–35.

Fuchs, B. and J. J. Fuchs. 2010. 'Optimal narrow beam low sidelobe synthesis for arbitrary arrays.' *IEEE Transactions on Antenna and Propagation* 58: 2130–35.

Gallaudet, T. C. and C. P. D. Moustier. 2000. 'On optimal shading for arrays of irregularly-spaced or noncoplanar elements.' *IEEE Journal of Oceanic Engineering* 25(4): 553–67.

Ganz, M. W., R. L. Moses, and S. L. Wilson. 1990. 'Convergence of the SMI and the diagonally loaded SMI algorithms with weak interference.' *IEEE Transactions on Antennas and Propagation* 38: 394–99.

Givens, M. 2009. 'Enhanced-convergence normalised LMS algorithm.' *IEEE Signal Processing Magazine* 26(3): 81–95.

Godara, L. C. and A. Cantoni. 1986. 'Analysis of constrained LMS algorithm with application to adaptive beamforming using perturbation sequences.' *IEEE Transactions on Antennas and Propagation* 34: 368–79.

Godara, L. C. and D. A. Gray. 1989. 'A structured gradient algorithm for adaptive beamforming.' *Journal of the Acoustical Society of America* 86(3): 1040–46.

Godara, L. C. 1986. 'Error analysis of the optimal antenna array processors.' *IEEE Transactions on Aerospace and Electronics Systems* 22: 395–409.

Godara, L. C. 1990. 'Improved LMS algorithm for adaptive beamforming.' *IEEE Transactions on Antennas and Propagation* 38: 1631–35.

Godara, L. C. 2004. *Smart Antennas.* Washington DC: CRC Press, 448.

Griffiths, L. J. 1969. 'A simple adaptive algorithm for real-time processing in antenna arrays.' *Proceedings of IEEE* 57: 1696–1704.

Griffiths, L. J. and K. M. Buckley. 1987. 'Quiescent pattern control in linearly constrained adaptive array.' *IEEE Transactions on Acoustic, Speech, Signal Processing* 35: 917–26.

Harrington, R. F. 1961. 'Sidelobe reduction by non-uniform element spacing.' *IRE Transactions on Antennas and Propagation* 9(2): 187–92.

Haykin, S. 1996. *Adaptive Filter Theory.* Englewood Cliffs, NJ: Prentice-Hall, 936.

Hirasawa, K. and B. J. Strait. 1971. 'On a method for array design by matrix inversion.' *IEEE Transactions on Antennas and Propagation* 19: 446–47.

Homer, J., P. J. Kootsookos, and V. Selvaraju. 2007. 'Enhanced NLMS adaptive array via DOA detection.' *IET Communications* 1(1): 19–26.

Huarng, K. -C. and C. -C. Yeh. 1992. 'Performance analysis of derivative constraint adaptive arrays with pointing errors.' *IEEE Transactions on Antennas and Propagation* 40: 975–81.

Hudson, J. E. 1981. *Adaptive Array Principles.* New York: Peter Peregrinus, 253.

Hung, H. and M. Kaveh. 1990. 'Coherent wide-band ESPRIT method for directions-of-arrival estimation of multiple wide-band sources.' *IEEE Transactions on Acoustics, Speech, and Signal Processing* 38: 354–60.

Isernia, T., P. D. Iorio, and F. Soldovieri. 2000. 'An effective approach for the optimal focusing of array fields subject to arbitrary upper bounds.' *IEEE Transactions on Antenna and Propagation* 48: 1837–47.

Ishimaru, A. 1962. 'Theory of unequally spaced arrays.' *IRE Transactions on Antennas and Propagation* 8: 691–702.

Khatib, H. H. -A., and R. T. Compton, Jr. 1978. 'A gain optimising algorithm for adaptive array.' *IEEE Transactions on Antennas and Propagation* 26: 228–35.

King, D. D., R. F. Packard, and R. K. Thomas. 1961. 'Unequally spaced, broad-band antenna arrays.' *IRE Transactions on Antennas and Propagation* 9: 496–500.

Krim, H. and M. Viberg. 1996. 'Two decades of array signal processing research: The parametric approach.' *IEEE Signal Processing Magazine* 13(4): 67–94.

Kumar, B. P. and G. R. Branner. 2005. 'Generalised analytical technique for the synthesis of unequally spaced arrays with linear, planar, cylindrical or spherical geometry.' *IEEE Transactions on Antennas and Propagation* 53: 621–34.

Lebret, H. and S. Boyd. 1997. 'Antenna array pattern synthesis via convex optimisation.' *IEEE Transactions on Signal Processing* 45: 526–31.

Lebret, H. 1996. 'Optimal beamforming via interior point methods.' *Journal of VLSI Signal Processing* 14(1) 29–41.

Lee, J. -H. and C. -L. Cho. 2004. 'GSC-based adaptive beamforming with multiple-beam constraints under random array position errors.' *Signal Processing* 84: 341–50.

Lee, J. -H. and T. F. Hsu. 2000. 'Adaptive beamforming with multiple-beam constraints in the presence of coherent jammers.' *Signal Processing* 80: 2475–80.

Lee, J. -H., K. -P. Cheng, and C. -C. Wang. 2006. 'Robust adaptive array beamforming under steering angle mismatch.' *Signal Processing* 86: 296–309.

Lee, Y. and W. -R. Wu. 2005. 'A robust adaptive generalised sidelobe canceller with decision feedback.' *IEEE Transactions on Antennas and Propagation* 53: 3822–32.

Lin, H. -C. 1982. 'Spatial correlations in adaptive arrays.' *IEEE Transactions on Antennas and Propagation* 30: 128–38.

Lueng, H. and S. Haykin. 1989. 'Stability of recursive QRD–RLS algorithm using finite precision systolic array implementation.' *IEEE Transactions on Acoustics, Speech, and Signal Processing* 37: 760–63.

Mailloux, R. J. 1994. *Phased Array Antenna Handbook*. Second edition. Norwood, MA: Artech House, 496.

Najm, W. G. 1990. 'Constrained least squares in adaptive imperfect arrays.' *IEEE Transactions on Antennas and Propagation* 38: 1874–82.

Nitzberg, R. 1985. 'Application of the normalised LMS algorithm to MSLC.' *IEEE Transactions on Aerospace and Electronics Systems* 21: 79–91.

Ogawa, Y., M. Ohmiya, and K. Itoh. 1985. 'An LMS adaptive array using a pilot signal.' *IEEE Transactions on Aerospace and Electronics Systems* 21: 777–82.

Podilchak, S. K., A. P. Freundrofer, and Y. M. M. Antar. 2009. 'Planar antenna for directive beam steering at end-fire using an array of surface-wave launchers.' *Electronics Letters* 45(9): 444–45.

Reed, I. S., J. D. Mallett, and L. E. Brennan. 1974. 'Rapid convergence rate in adaptive arrays.' *IEEE Transactions on Aerospace and Electronics Systems* 10: 853–63.

Shor, S. W. W. 1966. 'Adaptive technique to discriminate against coherent noise in a narrow-band system.' *Journal of Acoustical Society of America* 39(1): 74–78.

Singh, H. and R. M. Jha. 2011. 'A novel algorithm for suppression of wideband probing in adaptive array with multiple desired signals.' *Defense Science Journal* 61(4): 325–30.

Singh, H. and R. M. Jha. 2013. 'Efficacy of modified improved LMS algorithm in active cancellation of probing in phased array.' *Journal of Information Assurance and Security* 8: 10.

Thng, I., A. Cantoni, and Y. H. Leung. 1993. 'Derivative constrained optimum broadband antenna arrays.' *IEEE Transactions on Signal Processing* 41: 2376–88.

Trees, H. L. V. 2002. *Detection, Estimation, and Modulation Theory: Optimum Array Processing*. New York: Wiley, 1443.

Tseng, C. Y. and L. J. Griffiths. 1992. 'A simple algorithm to achieve desired patterns for arbitrary arrays.' *IEEE Transactions on Signal Processing* 40: 2737–46.

Tseng, C. Y. 1992. 'Minimum variance beam-forming with phase independent derivative constraints.' *IEEE Transactions on Antenna and Propagation* 40: 285–94.

Unz, H. 1960. 'Linear arrays with arbitrarily distributed elements.' *IRE Transactions on Antennas and Propagation* 8(2): 222–23.

Veen, B. D. V. and K. M. Buckley. 1988. 'Beamforming: A versatile approach to spatial filtering.' *IEEE Acoustics, Speech, and Signal Processing Magazine* 5(2): 4–24.

Vinoy, K. J. and R. M. Jha. 1996. *Radar Absorbing Materials: From Theory to Design and Characterisation.* Norwell, Boston: Kluwer Academic Publishers, 209.

Ward, C., P. Hargrave, and J. McWhirter. 1986. 'A novel algorithm and architecture for adaptive digital beamforming.' *IEEE Transactions on Antennas and Propagation* 34: 338–46.

Widrow, B. and J. M. McCool. 1976. 'Comparison of adaptive algorithms based on the methods of steepest descent and random search.' *IEEE Transactions on Antennas and Propagation* 24: 615–37.

Widrow, B. and S. D. Stearns. 1985. *Adaptive Signal Processing.* Englewood Cliffs, NJ: Prentice-Hall, 474.

Widrow, B., P. E. Mantey, L. J. Griffiths, and B. B. Goode. 1967. 'Adaptive antenna systems.' *Proceedings of IEEE* 55: 2143–58.

Yasin, M., P. Akhtar, J. M. Khan, and N. S. H. Zaheer. 2010. 'Enhanced sample matrix inversion is a better beamformer for a smart antenna system.' *World Applied Sciences Journal* 10: 1167–75.

5 Mutual Coupling Effects in Phased Arrays

5.1 Introduction

When two antennas are placed close to each other, phenomena like reflection, diffraction, scattering and radiation cause interchange of energy between them. This is true irrespective of their being in transmitting or receiving mode. This energy exchange between the antenna elements is identified as mutual coupling effect. The coupling effect in phased array affects the antenna gain and beam width and is dependent on (i) radiation behaviour of each antenna element, (ii) inter-element spacing, (iii) feed network and (iv) the orientation of antenna element respect to the neighbouring element. The coupling effect becomes important especially when the spacing between the antennas is reduced below half-wavelength. The presence of coupling in an array changes the terminal impedances, the reflection coefficients and the array gain of the antennas. These fundamental properties of the array influence its radiation characteristics, output signal-to-interference-noise ratio (SINR) and radar cross section (RCS). Furthermore, it affects the steady-state response, transient response, speed of response, resolution capability, interference rejection ability and direction of arrival (DOA) estimation competence of the array.

Several techniques have been proposed for the estimation and mitigation of mutual coupling effect in different types of adaptive arrays. These include Yagi array (Leviatan et al. 1983), least mean square (LMS) and Applebaum arrays (Gupta and Ksienski 1983), power inversion array (Compton 1979), circular array of isotropic elements and semicircular array of printed dipoles (Fletcher and Darwood 1998), microstrip patch antenna arrays, linear arrays of dipole, sleeve dipole and spiral antennas (Pasala and Friel 1994), conformal dipole arrays (Guo and Li 2009), helical arrays (Hui et al. 2003) and arrays of arbitrary geometry (Strait and Hirasawa 1969).

The parameters governing the array performance are obtained using various techniques like the method of moments (MoM), multiple signal classification (MUSIC), estimation of signal parameters via rotational invariance techniques (ESPRIT), scheme for spatial multiplexing of local elements (SMILE) and direct data domain (DDD) algorithms. The algorithms used to estimate the coupling can be extended towards its compensation. Moreover, optimisation techniques like the genetic algorithm (GA), particle swarm optimisation (PSO) and linear programming (LP) can be used in conjuncture with these techniques towards the enhancement of array efficiency.

Most of the proposed methods focus on the compensation of mutual coupling in receiving arrays (Li and Nie 2004), (Goossens and Rogier 2007) and the array feed network (Jenn and Lee 1995), (Jenn and Flokas 1996), (Lau et al. 2006), (Coetzee and Yu 2008). The behaviour of antenna arrays in receiving and transmitting modes need not be the same. The factors like difference in signal paths within the transmitting and receiving array, antenna excitations, etc. make the mutual coupling effect different in the two modes (Lui et al. 2009). The techniques of estimating the mutual impedance for transmitting arrays like the conventional impedance matrix method (Gupta and Ksienski 1983) differ from those for the receiving array. The techniques to determine the receive mode impedance include full-wave methods (Pasala and Friel 1994), (Adve and Sarkar 2000), the calibration methods (Friedlander and Weiss 1991), (Dandekar et al. 2002) and the receiving-mutual impedance method (RMIM) (Hui 2004a, b), (Lui and Hui 2010).

The efficiency and accuracy of receiving mutual impedances is further improved by preserving the original boundary conditions of the compact array during the analysis (Lui and Hui 2010). Niow et al. (2011) suggested the method for the compensation of mutual coupling in transmitting arrays. The compensation network designed in terms of mutual impedances is employed to predict the radiation pattern using the principle of pattern multiplication. This approach facilitates the port decoupling and design of matching circuits towards the maximum power transfer. Wang and Hui (2011) proposed a system identification method for the wideband mutual coupling compensation of receiving arrays. A multiport compensation network is obtained on the basis of receiving mutual impedances of an array over the frequency band.

This chapter presents the mutual coupling effect in adaptive arrays. The mutual resistance, reactance, impedance and output SINR for dipole arrays in different configurations are discussed. The role of the factors such as the array size, spacing between elements, dipole length and dipole height and the aspect angle in determining the output noise power and SINR of the array is analysed. Once the mutual impedance in dipole array is known, it can be incorporated into the performance analysis of arrays.

5.2 Theoretical background for mutual impedance

In an adaptive array of N antenna elements (Figure 5.1), the signal received at each element is weighted and, then, summed to produce the array output. The output SINR is used as the performance index of adaptive array. The weights used to gain antenna elements need to be adjusted appropriately to maximise the output SINR for a given signal environment.

Using the network theory, an array of N antenna elements can be considered as $N + 1$ terminal linear bilateral network (Figure 5.2), driven by a generator characterised by open-circuit voltage V_g and impedance Z_g. According to Kirchoff's approximation, the voltage at each element is given by (Gupta and Ksienski 1983)

$$\begin{aligned} v_1 &= i_1 Z_{11} + i_2 Z_{12} + \ldots + i_N Z_{1N} + i_s Z_{1s} \\ v_2 &= i_1 Z_{21} + i_2 Z_{22} + \ldots + i_N Z_{2N} + i_s Z_{2s} \\ &\vdots \\ v_N &= i_N Z_{N1} + i_2 Z_{N2} + \ldots + i_N Z_{NN} + i_s Z_{Ns} \end{aligned} \tag{5.1}$$

where the impedance matrix is given by

$$Z = \begin{bmatrix} Z_{11} & Z_{12} & \cdots & Z_{1N} \\ Z_{21} & Z_{22} & \cdots & Z_{2N} \\ \vdots & \vdots & \ddots & \vdots \\ Z_{N1} & Z_{N2} & \cdots & Z_{NN} \end{bmatrix} \tag{5.1a}$$

Z_{ii} is the self or input impedance of the i-th port. It is the parameter that determines the antenna efficiency. Z_{ij} represents the mutual impedance between i^{th} port and j^{th} port (antenna elements).

The self impedance is a complex quantity and is expressed as $Z_{self} = R_{self} + jX_{self}$ where R_{self} is the self resistance and X_{self} is self reactance of the antenna. Similarly, the mutual impedance is given by $Z_{mutual} = R_{mutual} + jX_{mutual}$, where R_{mutual} is the mutual resistance and X_{mutual} is the mutual reactance between the antenna elements.

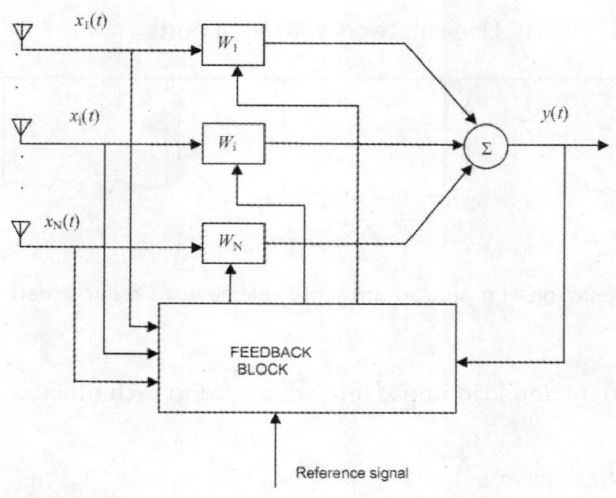

Figure 5.1 An *N*-element adaptive array

Since each port of an adaptive array has load impedance Z_L, it is added to each diagonal element of the impedance matrix. Thus, the impedance matrix takes the form

$$Z = \begin{bmatrix} Z_{11}+Z_L & Z_{12} & \cdots & Z_{1N} \\ Z_{21} & Z_{22}+Z_L & \cdots & Z_{2N} \\ \vdots & \vdots & \ddots & \vdots \\ Z_{N1} & Z_{N2} & \cdots & Z_{NN}+Z_L \end{bmatrix} \tag{5.1b}$$

The normalised impedance matrix with respect to Z_L is given by

$$Z_o = \begin{bmatrix} 1+\frac{Z_{11}}{Z_L} & \frac{Z_{12}}{Z_L} & \cdots & \frac{Z_{1N}}{Z_L} \\ \frac{Z_{21}}{Z_L} & 1+\frac{Z_{22}}{Z_L} & \cdots & \frac{Z_{2N}}{Z_L} \\ \vdots & \vdots & \ddots & \vdots \\ \frac{Z_{N1}}{Z_L} & \frac{Z_{N2}}{Z_L} & \cdots & 1+\frac{Z_{NN}}{Z_L} \end{bmatrix} \tag{5.1c}$$

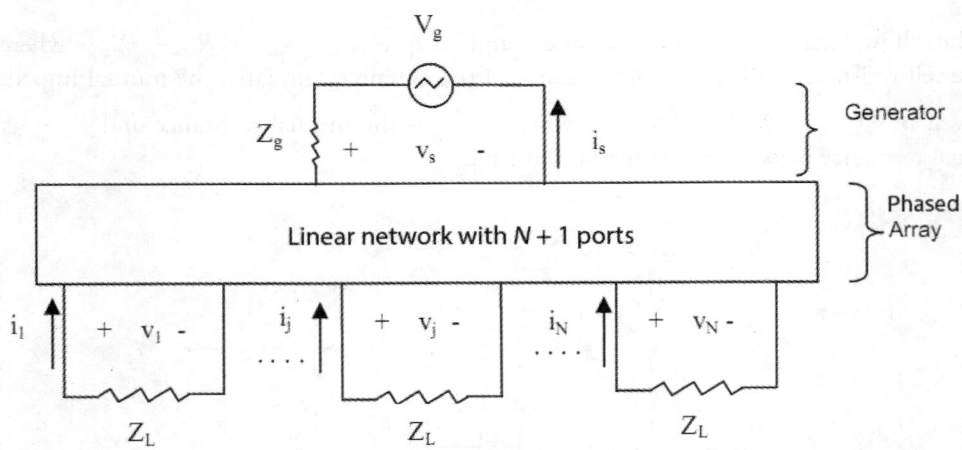

Figure 5.2 Representation of a phased array of *N*-elements as a linear network with *N* +1 ports

The terminal current and load impedance are related to each other as

$$i_j = -\frac{v_j}{Z_L}, \qquad j = 1, 2, \ldots, N \tag{5.2}$$

For open circuit, $i_j = 0$. Thus,

$$v_j = v_{oj} = Z_{js} i_s \tag{5.3}$$

Substituting (5.2) and (5.3) in (5.1), one gets

$$\begin{bmatrix} 1+\frac{Z_{11}}{Z_L} & \frac{Z_{12}}{Z_L} & \cdots & \frac{Z_{1N}}{Z_L} \\ \frac{Z_{21}}{Z_L} & 1+\frac{Z_{22}}{Z_L} & \cdots & \frac{Z_{2N}}{Z_L} \\ \vdots & \vdots & \ddots & \vdots \\ \frac{Z_{N1}}{Z_L} & \frac{Z_{N2}}{Z_L} & \cdots & 1+\frac{Z_{NN}}{Z_L} \end{bmatrix} \begin{bmatrix} v_1 \\ v_2 \\ \vdots \\ v_N \end{bmatrix} = \begin{bmatrix} v_{o1} \\ v_{o2} \\ \vdots \\ v_{oN} \end{bmatrix} \tag{5.4}$$

$$\text{or } Z_o V = V_o \tag{5.5}$$

where Z_o represents the normalised impedance matrix, and V_o is the matrix of open circuit voltage at the terminals of antenna array. The output voltages at the terminals of antenna array are given by

$$V = Z_o^{-1} V_o \tag{5.6}$$

Hence, the matrix Z_o behaves like a transformation matrix. It transforms the open circuit voltages of the antenna elements into the corresponding antenna terminal voltages. In the absence of coupling effect, the impedance matrix becomes a diagonal and is given by

$$Z_o = \begin{bmatrix} 1+\frac{Z_{11}}{Z_L} & 0 & \cdots & 0 \\ 0 & 1+\frac{Z_{22}}{Z_L} & \cdots & 0 \\ \vdots & \vdots & \ddots & \vdots \\ 0 & 0 & \cdots & 1+\frac{Z_{NN}}{Z_L} \end{bmatrix} \tag{5.7}$$

The expression for self and mutual impedances is dependent on the array geometry, the inter-element spacing and the radiation behaviour of each antenna element. For an array consisting of $\lambda/2$ dipoles with side-by-side geometric configuration, the self and mutual impedances are derived in Appendix A.

5.3 Steady-state performance of dipole array with mutual coupling

Let us consider the case of one desired signal and m probing signals incident on an LMS adaptive array of $\lambda/2$ dipoles (Figure 5.3).

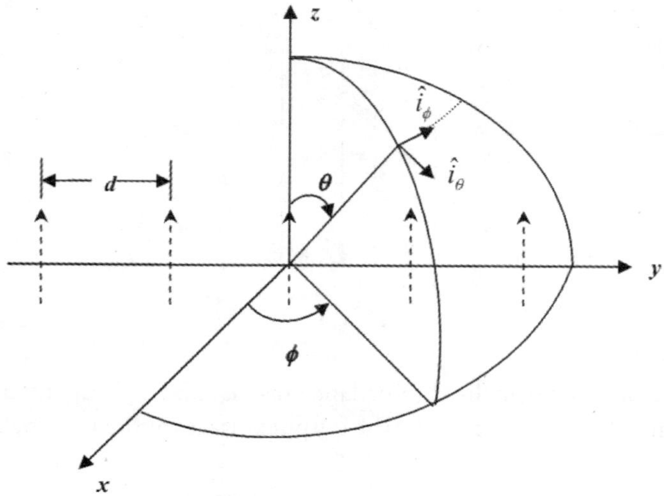

Figure 5.3 An array of N half-wavelength, centre-fed dipoles

The desired and probing signals are taken to be of the same frequency. Then, at each antenna element, the open circuit voltages (Gupta and Ksienski 1983) are given by

$$V_o = X_d + \sum_{l=1}^{m} X_{il} \tag{5.8}$$

where $X_d = A_d e^{j(\omega_o t+\psi_d)} U_d$ and $X_{il} = A_{il} e^{j(\omega_o t+\psi_{il})} U_{il}$ (5.8a)

A_d and A_{il} are the amplitudes of the desired and l^{th} probing signals, respectively. The symbol ω_o denotes the carrier frequency, whereas ψ_d and ψ_{il} are the phases of the desired and l^{th} probing signals at the origin of the coordinate space. U_d and U_{il} are the signal vectors for the desired and l^{th} probing signals.

The desired signal vector may be defined as

$$U_d = \begin{bmatrix} a_1(\theta_d,\phi_d,p_d)e^{j\rho_{d1}} \\ a_2(\theta_d,\phi_d,p_d)e^{j\rho_{d2}} \\ \vdots \\ a_N(\theta_d,\phi_d,p_d)e^{j\rho_{dN}} \end{bmatrix} \tag{5.9}$$

where (θ_d, ϕ_d) are the elevation and azimuth angles of the signal impinging from the desired direction, p_d is the corresponding polarisation, $a(\theta,\phi,p)$ is the amplitude distribution used and ρ_{dj} is the desired signal phase at the jth element with respect to the coordinate origin. The phase of the signal is given by $\rho = k(N-1)d\sin\theta\sin\phi$.

Similarly, the signal vector of the lth probing signal is given by

$$U_{il} = \begin{bmatrix} a_1(\theta_{il},\phi_{il},p_{il})e^{j\rho_{il1}} \\ a_2(\theta_{il},\phi_{il},p_{il})e^{j\rho_{il2}} \\ \vdots \\ a_N(\theta_{il},\phi_{il},p_{il})e^{j\rho_{ilN}} \end{bmatrix} \tag{5.10}$$

From (5.6) and (5.8), the input signal to the adaptive array is given by

$$V = Z_o^{-1}\left(X_d + \sum_{l=1}^{m} X_{il} \right) \tag{5.11}$$

If thermal additive noise is included, then the total input signal of the adaptive array becomes

$$X = V + X_n \tag{5.12}$$

From (5.11) and (5.12), one gets

$$X = Z_o^{-1}\left(X_d + \sum_{l=1}^{m} X_{il} \right) + X_n \tag{5.13}$$

where X_n is the vector representing the received noise, expressed as

$X_n = \left[n_1(t), n_2(t), \ldots, n_N(t) \right]^T$, and T denotes the transpose.

The signal covariance matrix is expressed as

$$R = E\left\{ X^* X^T \right\} \tag{5.14}$$

The asterisk denotes the complex conjugate and $E\{\ \}$ denotes the expectation. Substituting (5.13) in (5.14), one gets

$$R = \frac{\sigma^2}{Z_o^* Z^T}\left[Z_o^* Z_o^T + \sum_{l=1}^{m} \xi_{il} U_{il}^* U_{il}^T + \xi_d U_d^* U_d^T \right] \tag{5.15}$$

where σ^2 is the power of thermal noise, ξ_d is the ratio of the power of desired signal to σ^2 and ξ_{il} is the ratio of the power of lth probing signal to σ^2.

The normalised signal covariance matrix due to probing signals and thermal noise is given by

$$R_n = Z_o^* Z_o^T + \sum_{l=1}^{m} \xi_{il} U_{il}^* U_{il}^T \tag{5.16}$$

Thus, (5.15) takes the form

$$R = \frac{\sigma^2}{Z_o^* Z^T}\left[R_n + \xi_d U_d^* U_d^T\right] \tag{5.17}$$

With its inverse given by

$$R^{-1} = \frac{Z_o}{\sigma^2 Z_o^T}\left\{\left[R_n + \xi_d U_d^* U_d^T\right]\right\}^{-1} \tag{5.18}$$

Let us consider a matrix inversion lemma (Householder 2006)

$$(A - \alpha U^* U^T)^{-1} = A^{-1} - \beta A^{-1} U^* U^T A^{-1}$$

where the scalars α and β are related as $\frac{1}{\alpha} + \frac{1}{\beta} = U^T A^{-1} U^*$ Thus, from (5.18), one has

$$(R_n - (-\xi_d U_d^* U_d^T))^{-1} = R_n^{-1} - \chi R_n^{-1} U_d^* U_d^T R_n^{-1}$$

where $\frac{1}{\chi} = \frac{1}{\xi_d} + U_d^T R_n^{-1} U_d^*$

$$\text{or } \chi = \frac{\xi_d}{1 + \xi_d U_d^T R_n^{-1} U_d^*} \tag{5.19}$$

Hence, (5.18) can be expressed as

$$R^{-1} = \frac{Z_o Z_o^*}{\sigma^2}\left\{R_n^{-1} - \tau R_n^{-1} U_d^* U_d^T R_n^{-1}\right\} \tag{5.20}$$

Using LMS algorithm (Godara 2004), the weight vector W in steady-state may be expressed as

$$W = \frac{S}{R} \tag{5.21}$$

where the reference correlation vector, $S = E\left\{X^* r(t)\right\}$, in which, $r(t) = A_r e^{j(\omega_o t + \psi_d)}$ is a complex reference signal (Riegler and Compton 1973). Thus, the reference correlation vector is given by $S = A_r A_d (Z_o^{-1})^* U_d^*$.

The steady-state weight vector will be

$$W = KZ_o^T R_n^{-1} U_d^* \tag{5.21a}$$

where $K = \dfrac{A_r A_d}{\sigma^2}(1 - \tau U_d^T R_n^{-1} U_d^*)$

The output SINR of the adaptive array is expressed as

$$SINR = \frac{P_d}{\sum_{l=1}^{m} P_{il} + P_n} \tag{5.22}$$

where the output power due to desired signal P_d is given by

$$P_d = \frac{A_d^2}{2}\left|\frac{U_d^T W}{Z_o^T}\right|^2 \tag{5.23}$$

P_{il} is the output power due to l^{th} probing signal and is given by

$$P_{il} = \frac{A_{il}^2}{2}\left|\frac{U_{il}^T W}{Z_o^T}\right|^2 \tag{5.24}$$

and P_n is the output power due to thermal noise, expressed as

$$P_n = \frac{\sigma^2}{2}|W|^2 \tag{5.25}$$

Thus, (5.22) yields the output SINR of an N-element array in steady-state given by

$$SINR = \xi_d U_d^T R_n^{-1} U_d^* \tag{5.26}$$

An adaptive phased array of N centre-fed, half-wavelength dipoles is considered. All the elements are taken to be identical with an inter-element spacing d. The matrix elements of dipole impedance matrix, including both self and mutual components are obtained using expressions (A.17) through (A.20) listed in Appendix A.

5.3.1 Side-by-side dipole array

Figure 5.4 presents the dependence of mutual impedance between two centre-fed $\lambda/2$ dipoles on the inter-element spacing. Both the amplitude and phase of the impedance are shown. It is apparent that the mutual impedance is large for small inter-element spacing. The coupling effect degrades the array performance significantly.

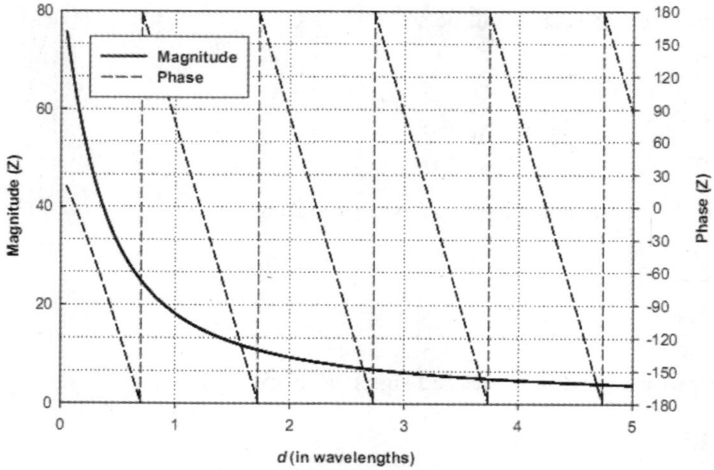

Figure 5.4 Dependence of mutual impedance between two centre-fed $\lambda/2$ dipoles on the dipole spacing

Steady-state performance: For a 6-element array of half-wavelength, centre-fed dipoles, the variation of the magnitude of mutual impedances between the antenna elements with inter-element spacing is shown in Figure 5.5a. It may be observed that for a given spacing, the mutual impedance decreases as one moves away from the first antenna element. The corresponding phases of the mutual impedances are presented in Figure 5.5b. It is apparent that as one moves away from the antenna element, the change in the phase of the mutual impedance tends to increase.

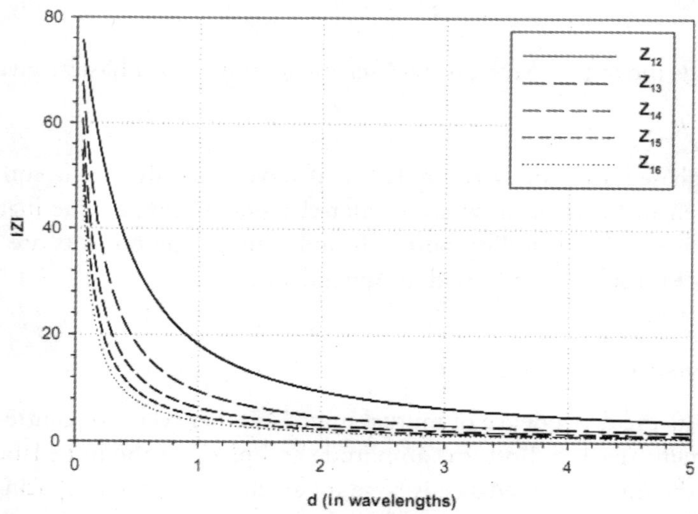

Figure 5.5a Magnitude of mutual impedance between half-wavelength, centre-fed dipoles in 6-element array

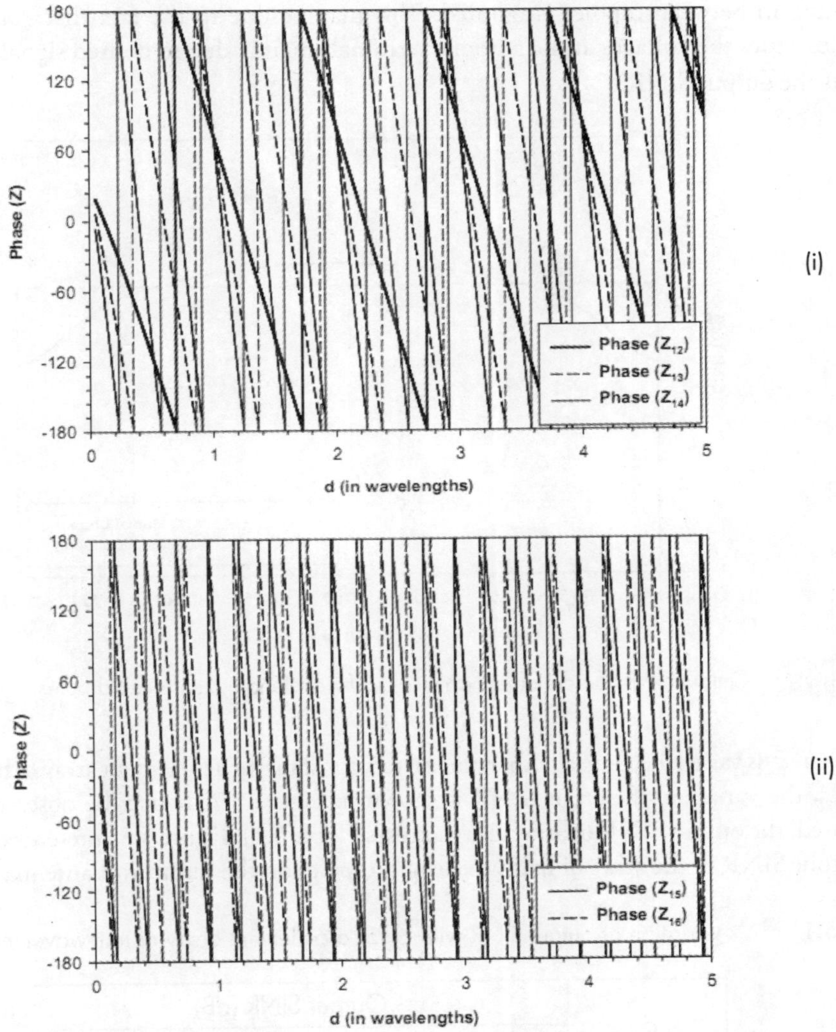

Figure 5.5b Phase of mutual impedance between half-wavelength, centre-fed dipoles in 6-element array. (i) Phase of Z_{12}, Z_{13} and Z_{14} (ii) Phase of Z_{15} and Z_{16}

Next, the dependence of output SINR of a 6-element dipole array on the impinging angle of desired signal is presented in Figure 5.6. The spacing between the dipoles is kept half-wavelength. It is assumed that each dipole is terminated with the load impedance ($Z_L = Z_{ii}^*$). The ratio of powers of desired signal and that of thermal noise, ξ_d is taken as 5 dB.

An assumption of no probing signals incident on the dipole array is made. The output SINR of an array is computed for no mutual coupling as well as with mutual coupling case. The array performance, i.e. the output SINR degrades with the impinging angle of the desired signal, when coupling effect is considered. On the other hand, this is not true when mutual coupling in between dipoles is ignored. This may be due to the fact that coupling between dipoles varies the voltages at the antenna terminals mainly due to desired signal, which in turn affects the output SINR.

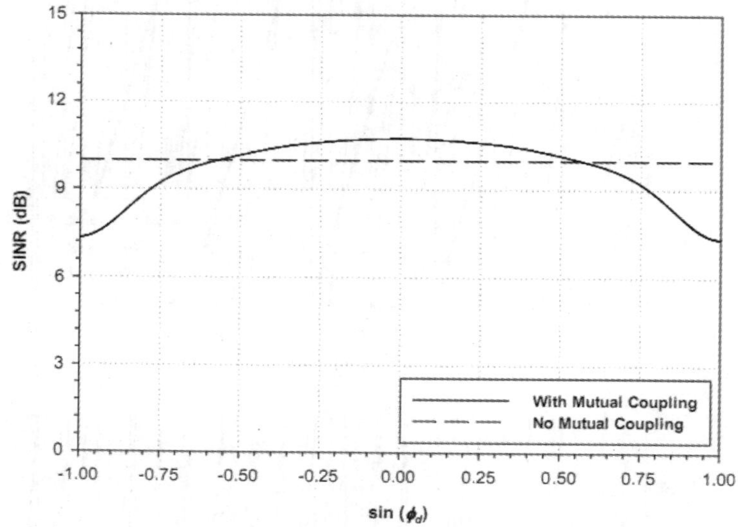

Figure 5.6 Output SINR of a 6-element array of half-wavelength, centre-fed dipoles. ξ_d= 5 dB, $d = 0.5\lambda$

When ξ_d is varied, keeping all other parameters constant for a 6-element array of half-wavelength dipoles, the variation of output SINR is shown in Figure 5.7. It may be observed that as ξ_d is increased, the output SINR of array also increases (Table 5.1). Figure 5.8a presents the dependence of output SINR of the array of half-wavelength dipoles on the number of antenna elements.

Table 5.1 Variation of output SINR with ξ_d for a 6-element array of half-wavelength dipoles

ξ_d	Output SINR (dB)	
	With mutual coupling	No mutual coupling
5 dB	10.75	10
10 dB	15.75	15
20 dB	25.75	25

The ratio of desired signal power to thermal noise power is 10 dB. It may be observed that the output SINR of array increases with the number of antenna elements in the array (Table 5.2). This may be due to the increase in the total contribution of antenna array towards the output SINR. Figure 5.8b shows the effect of variation of ξ_d on the output SINR. As ξ_d is increased, the output SINR also increases.

Table 5.2 Variation of output SINR for *N*-element array of half-wavelength dipoles; ξ_d = 10 dB

N	Output SINR (dB)	
	With mutual coupling	No mutual coupling
6	15.75	15
16	20.07	19.26
32	23.10	22.27
64	26.13	25.28
256	32.15	31.30
512	35.17	34.31
1024	38.18	34.72

Table 5.3 Variation of output SINR with inter-element spacing for a 16-element array of half-wavelength dipoles; ξ_d = 10 dB

Interelement spacing, *d*	Output SINR (dB)	
	With mutual coupling	No mutual coupling
0.25 λ	16.70	19.26
0.5 λ	20.07	19.26
λ	19.51	19.26

Next, the spacing between the antenna elements in 16-element dipole array is varied, and its effect on the performance of array in terms of output SINR is analysed (Table 5.3).

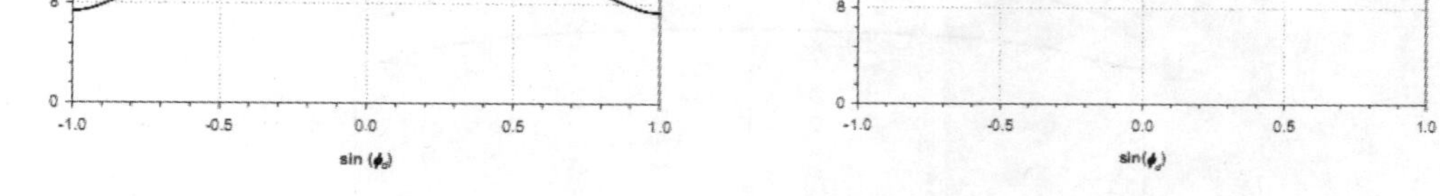

(a) (b)

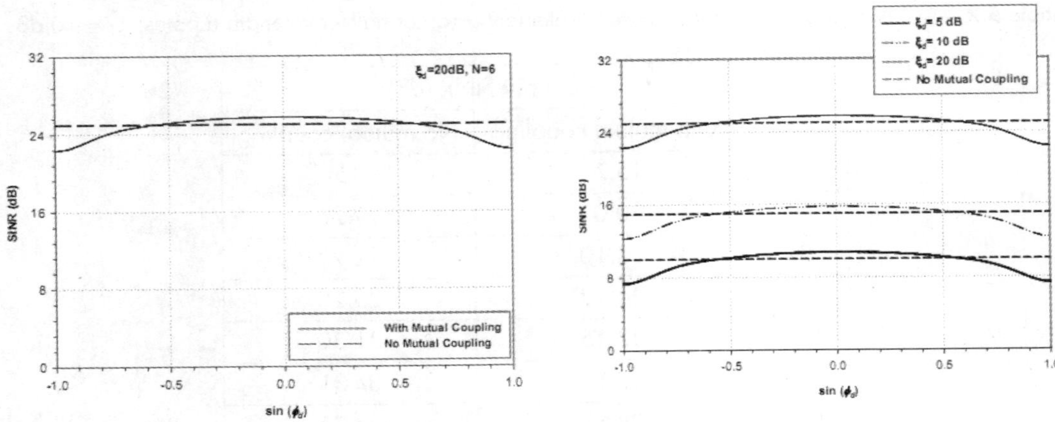

(c) (d)

Figure 5.7 Effect of ratio of desired signal power to thermal noise power, ξ_d on output SINR of a 6-element array of half-wavelength, centre-fed dipoles. (a) ξ_d= 5 dB, (b) ξ_d= 10 dB, (c) ξ_d= 20 dB (d) Characteristic curves

Figure 5.9 presents the array performance in terms of output SINR of 16-element dipole array with ξ_d= 10 dB for three different inter-element spacing. There is no change in the level of output SINR for no mutual coupling case. However, the output SINR in the presence of mutual coupling changes with the change in inter-element spacing. When the inter-element spacing between the dipoles is more than 0.5 λ, the level of output SINR is almost the same as for no mutual coupling case. In other words, as expected, the coupling effect is negligible if the dipoles are placed far apart.

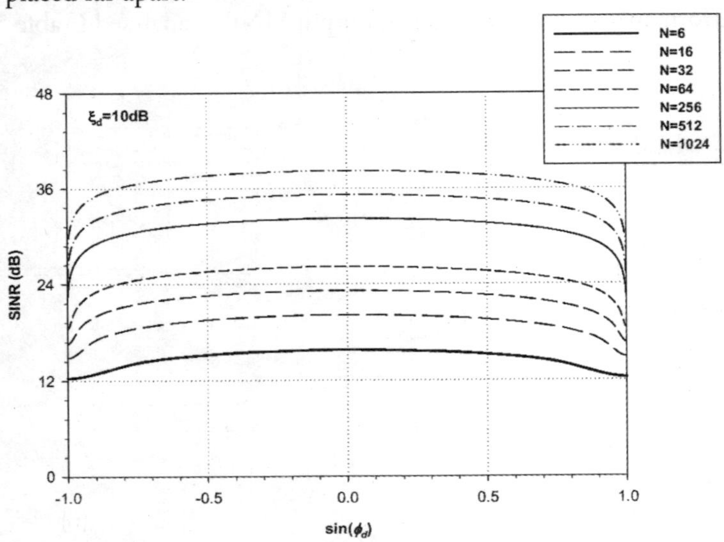

Figure 5.8a Effect of number of antenna elements on output SINR of an array of $\lambda/2$, centre-fed dipoles; ξ_d = 10 dB, d = 0.5 λ, θ_d = 90°

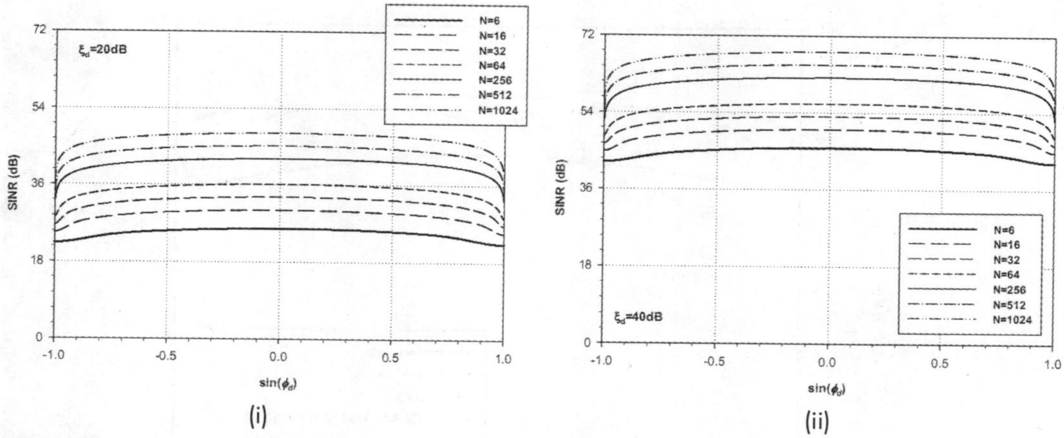

Figure 5.8b Effect of the ratio of desired signal power to thermal noise power on output SINR of centre-fed $\lambda/2$ dipole array; $d = 0.5\ \lambda$, $\theta_d = 90°$ (i) $\xi_d = 20$ dB (ii) $\xi_d = 40$ dB

The degradation of output SINR and, hence, the performance of antenna array may be explained on the basis of total incident power due to desired signal. If the inter-element separation is reduced, the total size of the array aperture also reduces. This in turn decreases the total incident power due to desired signal. However, the thermal noise level remains the same as the array size is unaltered. This results in the drop in signal-to noise ratio.

Table 5.4 Variation of output SINR with ξ_d and number of elements for half-wavelength dipoles

ξ_d	Output SINR (dB)					
	$N = 6$		$N = 16$		$N = 32$	
	No mutual coupling	With mutual coupling	No mutual coupling	With mutual coupling	No mutual coupling	With mutual coupling
5 dB	10.0	12.4	14.3	16.9	17.3	20.0
10 dB	15.0	17.4	19.3	21.9	22.3	25.0
20 dB	25.0	27.4	29.3	31.9	32.3	35.0

Similarly, keeping the aperture size fixed, if the inter-element spacing is reduced by adding more number of antenna elements, the output SINR degrades. This is due to the increase in the thermal noise power generated by more antenna elements, without rise in desired signal power. This drops the performance of the array (Compton 1982). Table 5.4 presents the effect of the number of dipole elements and the ratio of desired signal power to thermal noise power on the output SINR.

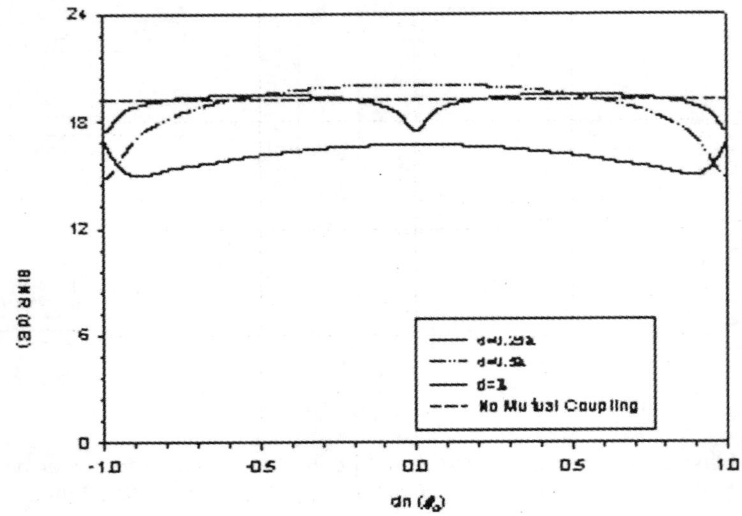

Figure 5.9 Effect of inter-element spacing on output SINR of a 16-element array of half-wavelength, centre-fed dipoles; ξ_d = 10 dB, θ_d = 90°

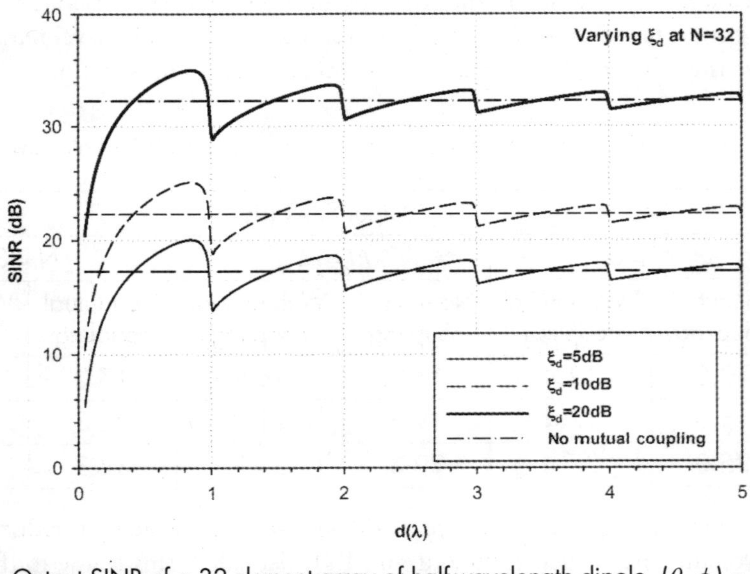

Figure 5.10 Output SINR of a 32-element array of half-wavelength dipole. (θ_d, ϕ_d) = (90°,0°)

Keeping the aperture size fixed at twice the wavelength (Gupta and Ksienski 1983), if the output SINR is computed for an array of half-wavelength dipoles for different number of antenna elements, the trend will be as in Figure 5.11. The ratio of desired signal power to thermal noise power, ξ_d, is taken as 5 dB. For no mutual coupling case, the output SINR is directly proportional to the number of dipole elements. However, this is true when coupling effect is considered. The output SINR degrades with the increase in the number of array elements for a given array aperture size.

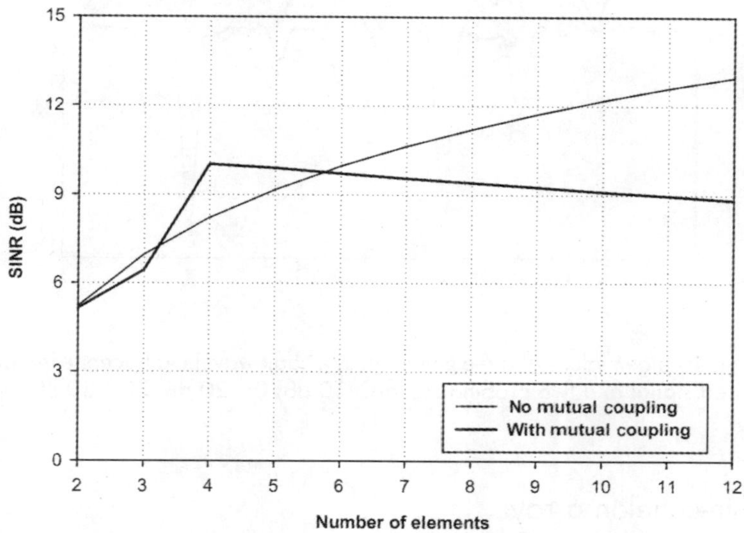

Figure 5.11 Output SINR of a 6-element array of half-wavelength, centre-fed dipoles of fixed aperture

Transient response: The eigenvalues of a signal covariance matrix and, hence, the performance of adaptive array will be different for no mutual coupling and with mutual coupling case (Gupta and Ksienski 1983). The covariance matrix can be written as

$$\Phi = \sigma^2 \left[I + \sum_{l=1}^{m} \xi_{il} \left(\frac{U_{il}}{Z_0} \right)^* \left(\frac{U_{il}}{Z_0} \right)^T + \xi_d \left(\frac{U_d}{Z_0} \right)^* \left(\frac{U_d}{Z_0} \right)^T \right] \tag{5.27}$$

If one desired signal and m probing signals are incident on N-element antenna array, the received signal covariance matrix will at least have $(N–m–1)$ eigenvectors having unity eigenvalues. The other eigenvectors will have non-unity eigenvalues.

The mutual coupling effect tends to change these non-unity eigenvalues. The coupling between the elements affects the transient response of array towards the desired signal and probing signals, for smaller inter-element spacing. Figure 5.12 shows the non-unity eigenvalues of the covariance matrix when one desired signal (0°;10 dB) and two probing signals (30°;20 dB), (–45°;30 dB) are incident on dipole array.

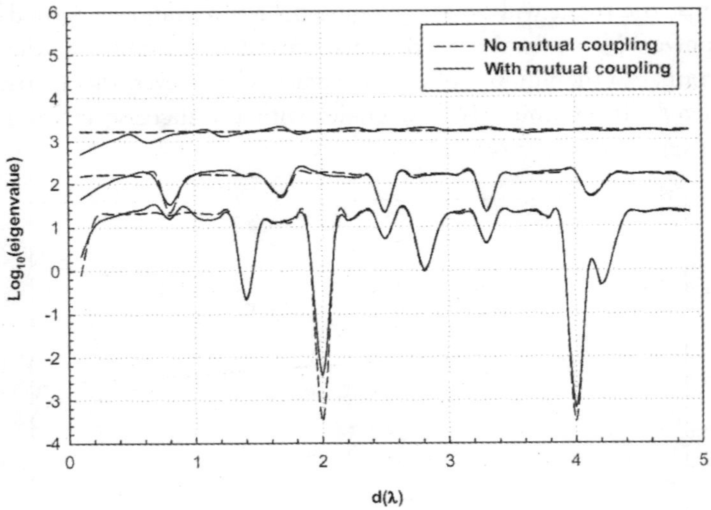

Figure 5.12 Non-unity eigenvalues of a 6-element array of half-wavelength, centre-fed dipoles. One desired signal and two probing signals. (10 dB, 0°; 20 dB, 30°; 30 dB, –45°)

5.3.2 Parallel-in-echelon array

The schematic of parallel-in-echelon dipole array is shown in Figure 5.13 in which the height h of one of the elements of a pair of an echelon dipole array is zero with respect to the reference base, while the other element is considered at some height h. The length of an N-element dipole array is l, and d is the spatial distance between the two adjacent dipoles.

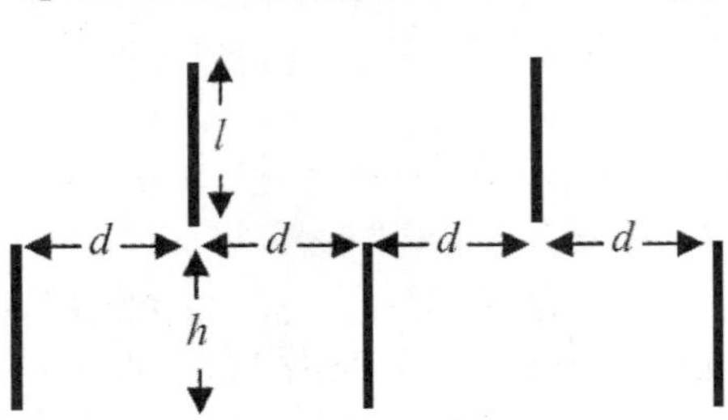

Figure 5.13 Schematic of parallel-in-echelon configuration of a dipole array

For a pair of $\lambda/2$ dipoles placed in parallel-in-echelon configuration (Figure 5.13), spaced at distance d, the self and mutual impedances are expressed in terms of cosine, $C_i(x)$, and sine, $S_i(x)$, integrals. The resistance and reactance of self impedance of dipoles in parallel-in-echelon configuration (King 1957) may be expressed as

$$R_{self} = \frac{\eta}{2\pi}\left\{C + \ln(kl) - C_i(kl) + \frac{1}{2}\sin(kl)\left[S_i(2kl) - 2S_i(kl)\right]\right.$$

$$+\frac{1}{2}\cos(kl)\left[C+\ln(kl/2)+C_i(2kl)-2C_i(kl)\right]\Big\} \tag{5.28}$$

$$X_{self}=\frac{\eta}{4\pi}\Big\{2S_i(kl)+\cos(kl)\left[2S_i(kl)-S_i(2kl)\right]$$
$$-\sin(kl)\left[2C_i(kl)-C_i(2kl)-C_i\left(\frac{2ka^2}{l}\right)\right]\Big\} \tag{5.29}$$

On the other hand, the resistance and reactance of mutual impedance of parallel-in-echelon are given by

$$R_{mutual}=-\frac{\eta}{8\pi}\cos(kh)\Big[-2C_i\left(k\left(\sqrt{d^2+h^2}+h\right)\right)-2C_i\left(k\left(\sqrt{d^2+h^2}-h\right)\right)+$$
$$C_i\left(k\left[\sqrt{d^2+(h-l)^2}+(h-l)\right]\right)+C_i\left(k\left[\sqrt{d^2+(h-l)^2}-(h-l)\right]\right)+$$
$$C_i\left(k\left[\sqrt{d^2+(h+l)^2}+(h+l)\right]\right)+C_i\left(k\left[\sqrt{d^2+(h+l)^2}-(h+l)\right]\right)\Big]$$
$$+\frac{\eta}{8\pi}\sin(kh)\Big[-2S_i\left(k\left(\sqrt{d^2+h^2}+h\right)\right)-2S_i\left(k\left(\sqrt{d^2+h^2}-h\right)\right)+$$
$$S_i\left(k\left[\sqrt{d^2+(h-l)^2}+(h-l)\right]\right)+S_i\left(k\left[\sqrt{d^2+(h-l)^2}-(h-l)\right]\right)+$$
$$S_i\left(k\left[\sqrt{d^2+(h+l)^2}+(h+l)\right]\right)+S_i\left(k\left[\sqrt{d^2+(h+l)^2}-(h+l)\right]\right)\Big] \tag{5.30}$$

$$X_{mutual}=\frac{\eta}{8\pi}\sin(kh)\Big[-2C_i\left(k\left(\sqrt{d^2+h^2}+h\right)\right)-2C_i\left(k\left(\sqrt{d^2+h^2}-h\right)\right)-$$
$$C_i\left(k\left[\sqrt{d^2+(h-l)^2}+(h-l)\right]\right)+C_i\left(k\left[\sqrt{d^2+(h-l)^2}-(h-l)\right]\right)-$$
$$C_i\left(k\left[\sqrt{d^2+(h+l)^2}+(h+l)\right]\right)+C_i\left(k\left[\sqrt{d^2+(h+l)^2}-(h+l)\right]\right)\Big]$$
$$-\frac{\eta}{8\pi}\cos(kh)\Big[-2S_i\left(k\left(\sqrt{d^2+h^2}+h\right)\right)+2S_i\left(k\left(\sqrt{d^2+h^2}-h\right)\right)-$$
$$S_i\left(k\left[\sqrt{d^2+(h-l)^2}+(h-l)\right]\right)-S_i\left(k\left[\sqrt{d^2+(h-l)^2}-(h-l)\right]\right)-$$
$$S_i\left(k\left[\sqrt{d^2+(h+l)^2}+(h+l)\right]\right)-S_i\left(k\left[\sqrt{d^2+(h+l)^2}-(h+l)\right]\right)\Big] \tag{5.31}$$

where the sine and cosine integrals are given by (Abramowitz and Stegun 1965)

$$S_i(x) = \sum_{n=0}^{\infty} \frac{(-1)^n x^{2n+1}}{(2n+1)(2n+1)!} \quad (5.32)$$

$$C_i(x) = C + \ln(x) + \sum_{n=1}^{\infty} (-1)^n \frac{x^{2n}}{2n(2n)!} \quad (5.33)$$

where $C = 0.5772$ is the Euler's constant.

The details of the formulation for self and mutual impedance are given in Appendix A.

Equal length dipoles: A uniform array of N half-wavelength centre-fed dipoles in parallel-in-echelon configuration is considered. Three different cases, *viz.* (a) non-staggered ($h = -\lambda/4$), (b) staggered by $h = 0$ and (c) staggered by $h = \lambda/4$ are analysed (Figure 5.14). In the first case, i.e. dipole array with non-staggered elements, the height of centre point for both the dipoles is same from the reference ground (or $x = 0$; x is the height of element from the reference ground).

In the second case of staggered dipole array, the height h of the base of the second element from the centre is zero. In the third case of staggered dipole array, the antenna elements are staggered by a height of half-wavelength ($x = \lambda/2$). It is assumed that all the elements are identical with d inter-element spacing and with an alternate height of $h = 0$ and $h = x$ from the reference ground. In other words, if the base of the first element of an echelon dipole array is at a height $x = 0$ from the reference ground, then the height of the antenna dipole elements at an even position should be at x, while the height of the antenna elements at an odd position should be at $x = 0$.

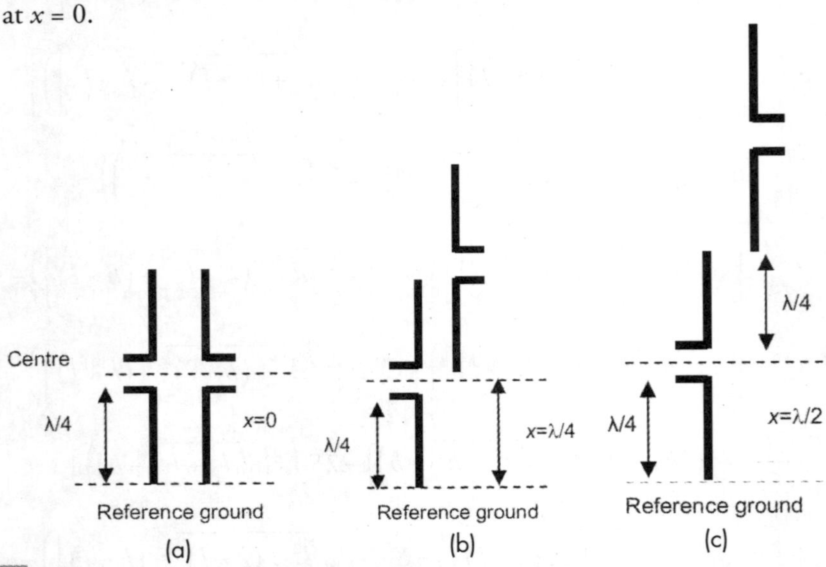

Figure 5.14 Schematic of parallel-in-echelon configuration for three different cases. (a) non-staggered ($h = -\lambda/4$), (b) staggered by $h = 0$, (c) staggered by $h = \lambda/4$

The mutual resistance, mutual reactance and magnitude/phase of mutual impedance for the three cases are computed varying the inter-element spacing *d*. Figure 5.15 shows that the trend of the three parameters remains same for each case. However, on closer scrutiny, it is observed that the value of these parameters decreases when the height of the base of the second element from the reference ground (*x*) increases. This is due to the increase in the separation between the dipoles and, thus, reduction in the coupling effect.

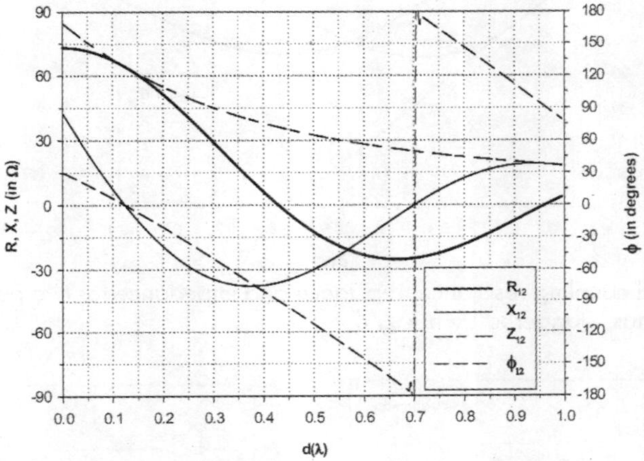

Figure 5.15 Mutual coupling resistance, reactance and impedance for two parallel half-wavelength dipole antennas, non-staggered

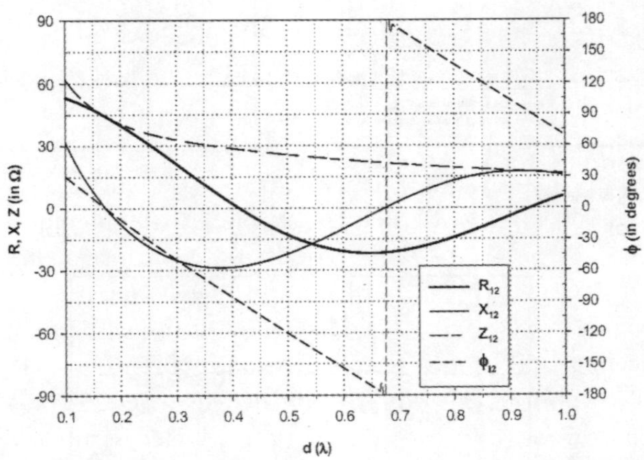

Figure 5.16 Mutual coupling resistance, reactance and impedance for two parallel half-wavelength antennas, staggered by $h = 0$

Further, it is clear from Figure 5.15 that the discontinuity point in phase (ϕ_{12}) shifts towards left with the increase in the height from the reference ground. Thus, among three cases, one gets minimum mutual impedance for the third case when the two dipole elements are staggered by a height of $\lambda/4$, i.e. height of the base of the second element from the reference ground is $\lambda/2$.

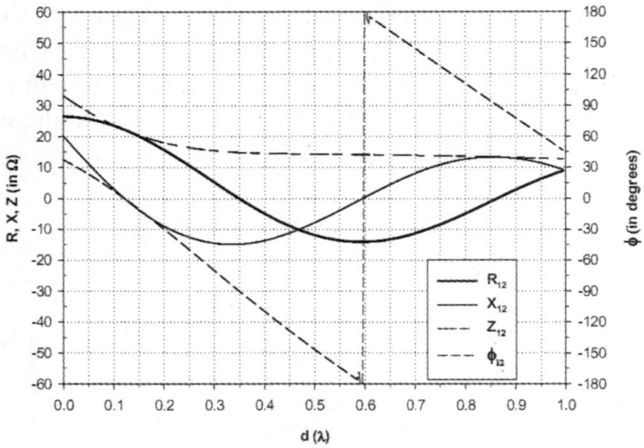

Figure 5.17 Mutual coupling resistance, reactance and impedance for two parallel half-wavelength antennas, staggered by $h = \lambda/4$

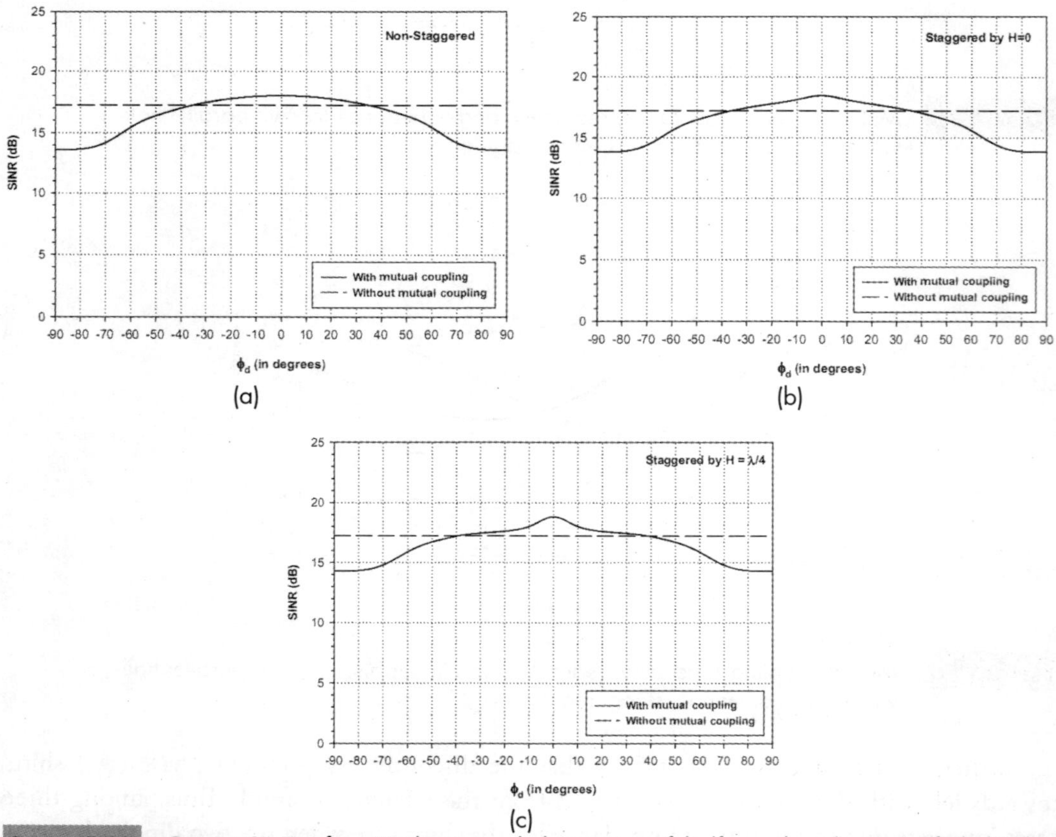

Figure 5.18 Output SINR of a 10-element echelon array of half-wavelength, centre-fed dipoles. σ_d = 10 dB, θ_d = 90°, $d = 0.5\lambda$ (a) non-staggered, (b) staggered by $h = 0$, (c) staggered by $h = \lambda/4$

Assuming the signal environment with no probing signals, output SINR is calculated for all the three types of dipole arrays. Figure 5.18 shows the variation of output SINR with the angle of incidence. It can be observed that the maximum output SINR is obtained for aspect angle, 0° and as one moves away from this aspect angle, the value of output SINR decreases. The output SINR value is found to be the maximum for case 3, i.e. when the height of the base of the second antenna element is of half-wavelength from the reference ground.

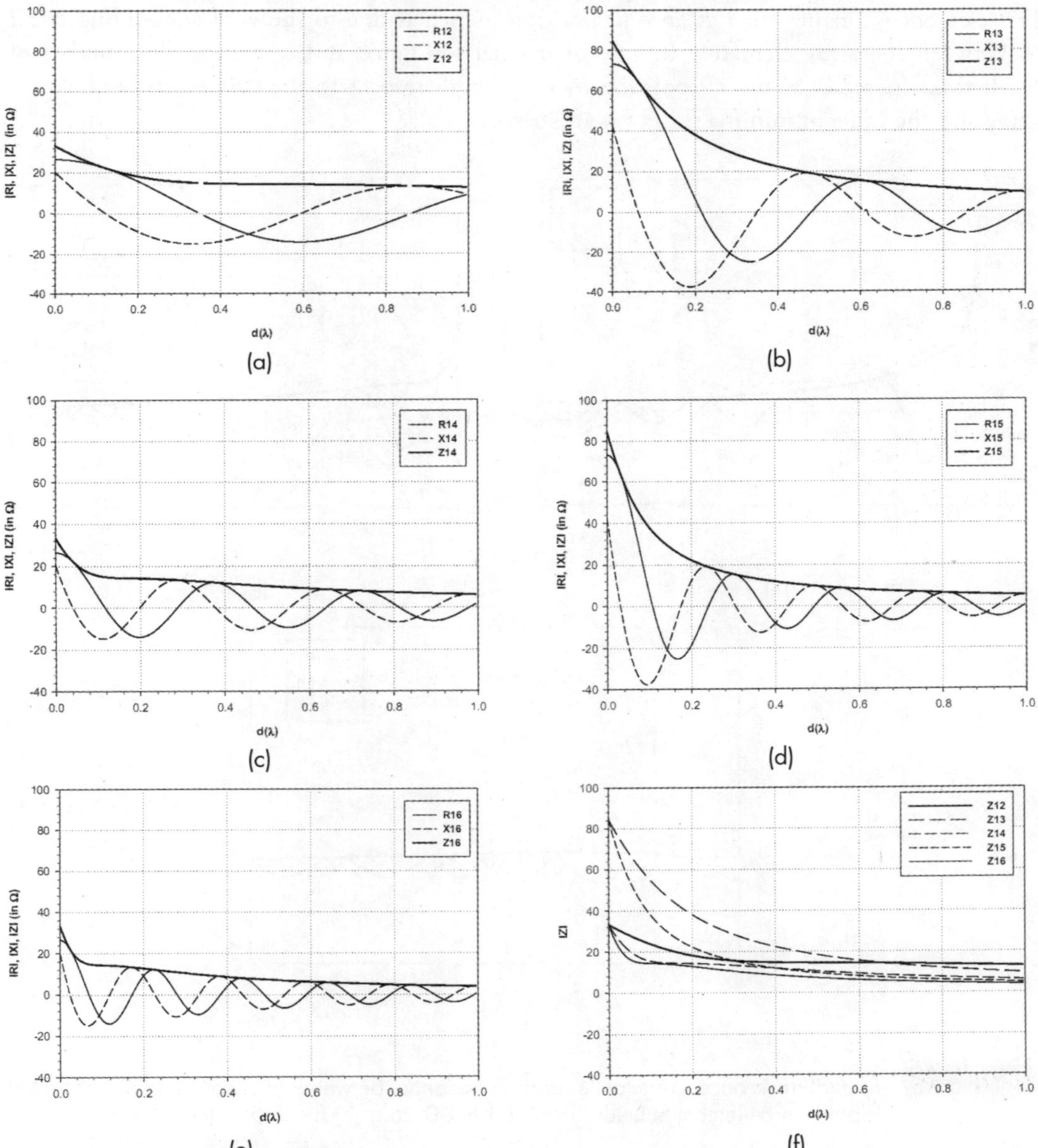

Figure 5.19 Mutual resistance, reactance and impedance between half-wavelength, centre-fed dipoles in 6-element echelon array. (a) Z_{12}, (b) Z_{13}, (c) Z_{14}, (d) Z_{15}, (e) Z_{16}, (f) $|Z|$

Figure 5.19 shows the variation of the mutual resistance, mutual reactance and mutual impedance with inter-element spacing *d* for different elements with respect to the first element of a 6-element echelon array of half-wavelength centre-fed dipoles. For a given inter-element spacing, the value of the mutual resistance and reactance and, hence, the mutual impedance between the antenna elements decrease as one moves away from the first element. It can also be observed that the magnitude of mutual impedance at an even position with respect to the first element is smaller than at an odd position, which is due to the weaker coupling effect in between the array elements. In case of mutual resistance and reactance, the number of oscillations increases as the nth position increases with respect to the first position of dipole array and the value of minima shifts towards zero.

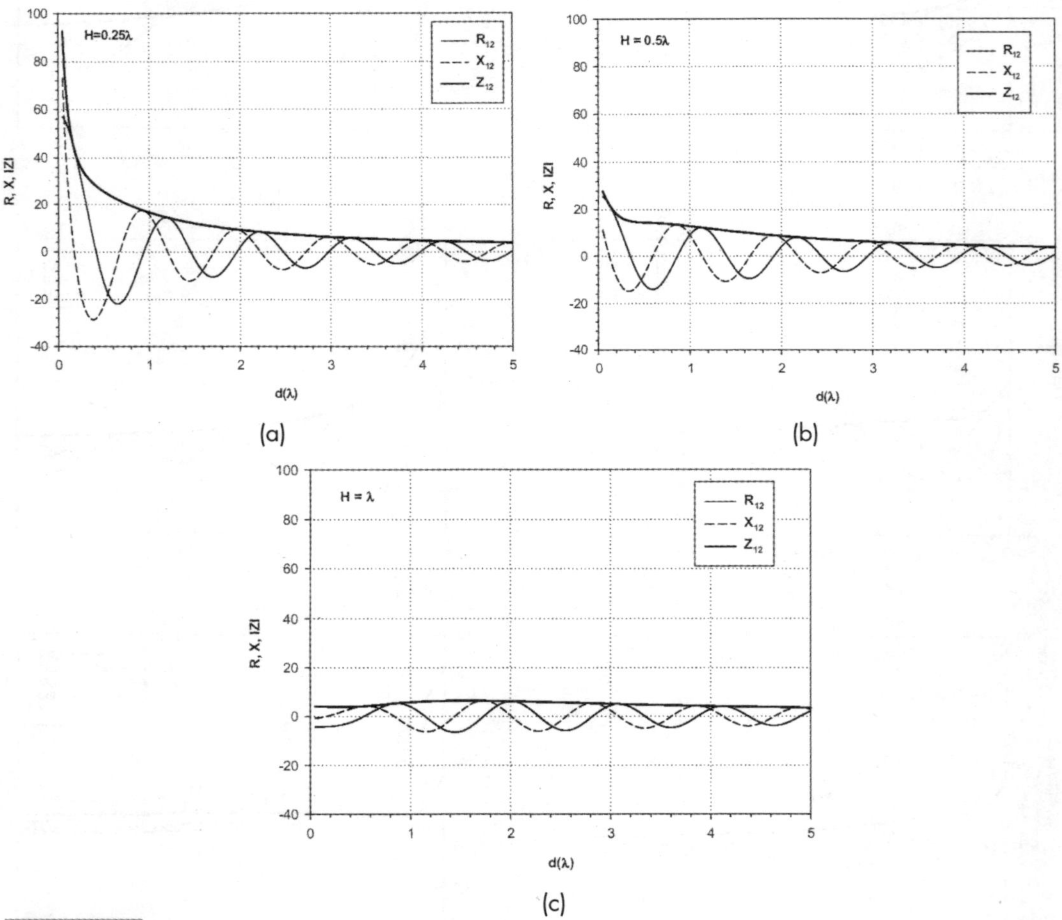

Figure 5.20 Mutual resistance, reactance and impedance between half-wavelength, centre-fed dipoles in 6-element echelon array. (a) $h = 0.25\lambda$, (b) $h = 0.5\ \lambda$, (c) $h = \lambda$

For a 6-element array of half-wavelength, centre-fed echelon dipoles, the variation of mutual resistance, reactance and impedance with inter-element spacing for different heights is shown in Figure 5.20. It is noted that with small separation between the elements, the magnitude of mutual impedance is large for $h = 0.25\ \lambda$, while it is very small for $h = \lambda$. This is due to the weaker mutual coupling effect at a larger height of dipole antenna element.

Figure 5.21 shows the dependence of mutual resistance, reactance and impedance of two-element echelon dipole array ($d = \lambda/2$) on height h. It may be observed that the mutual impedance is high for very small heights, and its magnitude decreases as the height of the second dipole from the reference ground is increased. Beyond $h = \lambda$, the mutual impedance between the two dipole elements approaches zero.

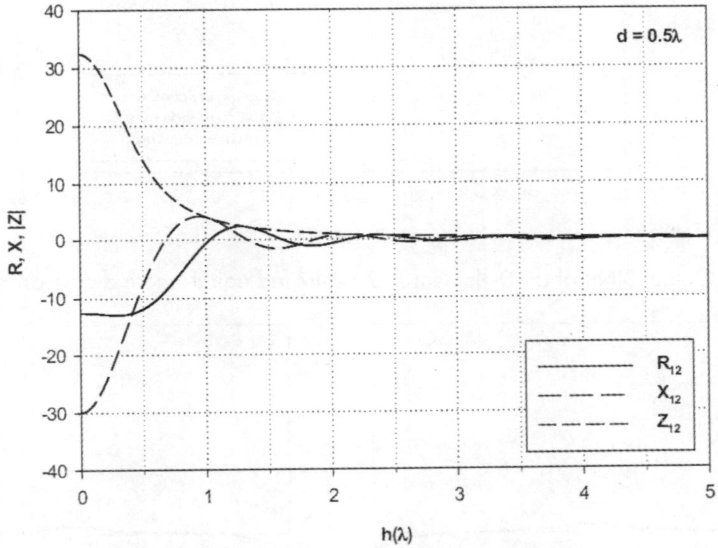

Figure 5.21 Mutual resistance, mutual reactance and mutual impedance between half-wavelength, centre-fed dipoles in 2-element echelon array for $d = 0.5\ \lambda$

Next, the performance of dipole array is analysed for two configurations, *viz.* side-by-side and parallel-in-echelon 10-element half-wavelength, centre-fed equal-length dipole arrays. Figure 5.22 shows the variation of the output SINR with DOA of desired signal for both the configurations. The load impedance of each dipole antenna element is equal to the complex conjugate of its self impedance. The ratio of powers of the desired signal and the thermal noise is taken as 10 dB. An assumption of no probing signals being incident on the antenna array is considered. It may be observed that in both the configurations, the output SINR (in the presence of coupling effect) reduces with the impinging angle of the desired signal. Moreover, the output SINR of the array is higher for the broadside aspect (±10°) in parallel-in-echelon dipole array as compared to that of the side-by-side dipole array.

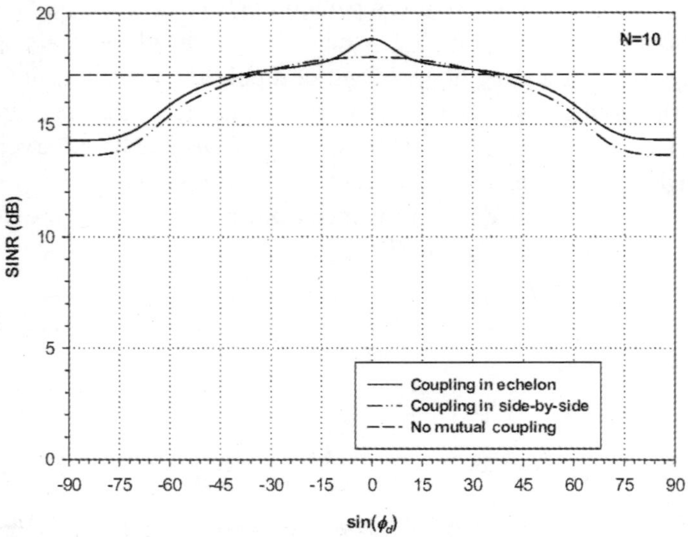

Figure 5.22 Output SINR of a 10-element $\lambda/2$ centre-fed equal-length dipole array

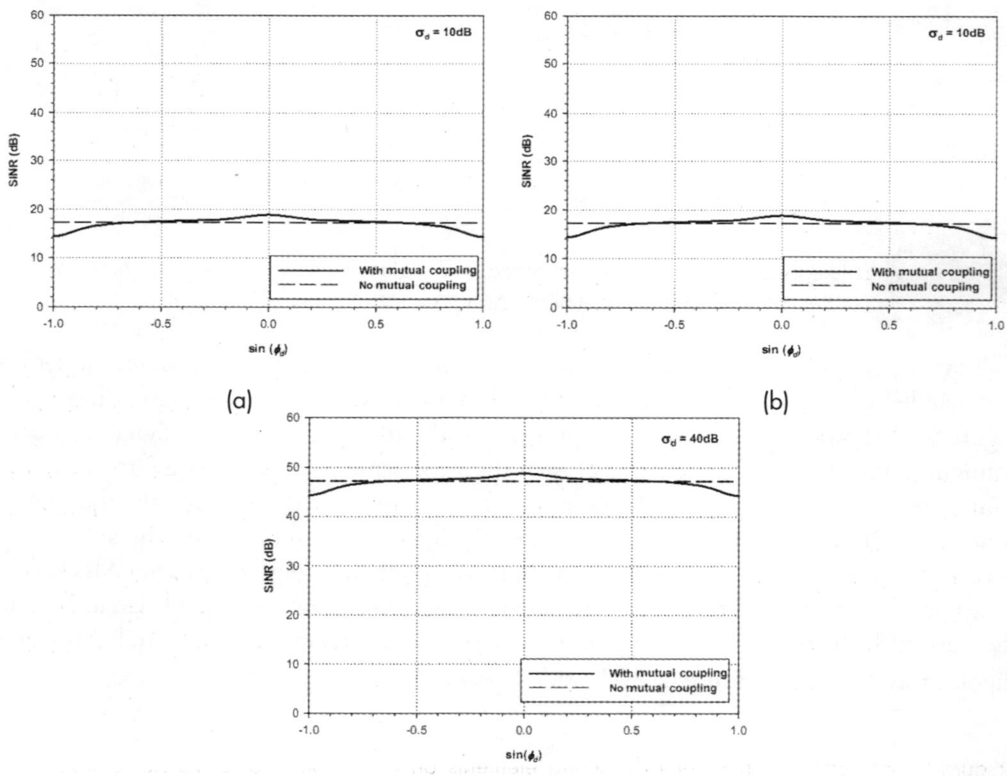

Figure 5.23 Effect of σ_d on output SINR of a 10-element echelon array of half-wavelength, centre-fed dipoles. (a) σ_d = 10 dB, (b) σ_d = 20 dB, (c) σ_d = 40 dB

Figure 5.23 presents the variation in output SINR with r espect to the desired signal direction for a different σ_d. Here, the output SINR of a dipole array is seen to increase with σ_d. Figure 5.24 presents the dependence of output SINR of an echelon array on the number of dipole antenna elements and, hence, the array size. Once again the number of dipoles in the array is directly proportional to the output SINR. This is due to the increase in the total contribution of antenna array towards the output SINR.

The role of inter-element spacing in 10-element parallel-in-echelon array in controlling the array performance is analysed next. Figure 5.25 shows the output SINR of 10-element echelon dipole array with σ_d = 10 dB and an alternate height of h = 0 and h = $\lambda/2$ for three different element spacing (d = 0.25 λ, 0.5 λ, λ). The output SINR of array is found to be the lowest for d = 0.25 λ. Further, the output SINR is maximum at 0° for d = 0.5 λ, while it is minimum for d = λ. Hence, it can be inferred that the performance of parallel-in-echelon array in terms of output SINR is better for d = 0.5 λ.

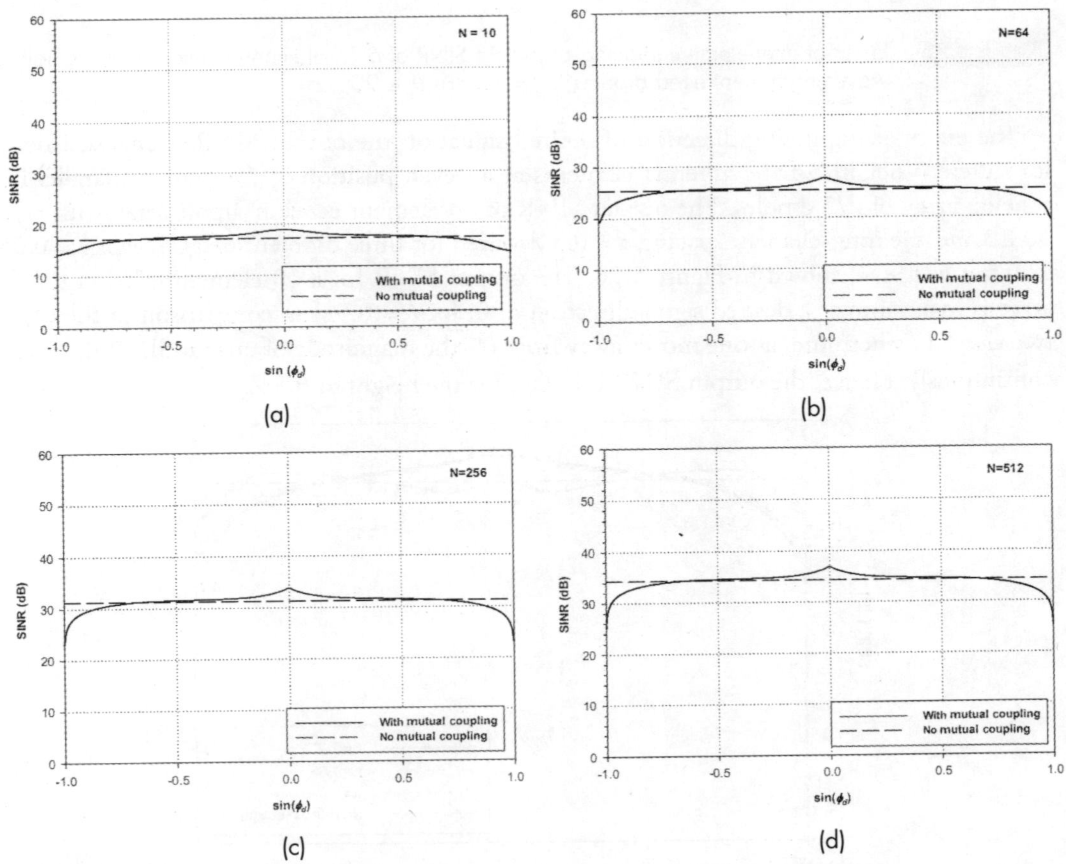

Figure 5.24 Effect of number of antenna elements on output SINR of a 10-element equal-length echelon array of half-wavelength, centre-fed dipoles; σ_d = 10 dB, d = 0.5 λ, θ_d = 90°. (a) N = 10, (b) N = 64, (c) N = 256, (d) N = 512

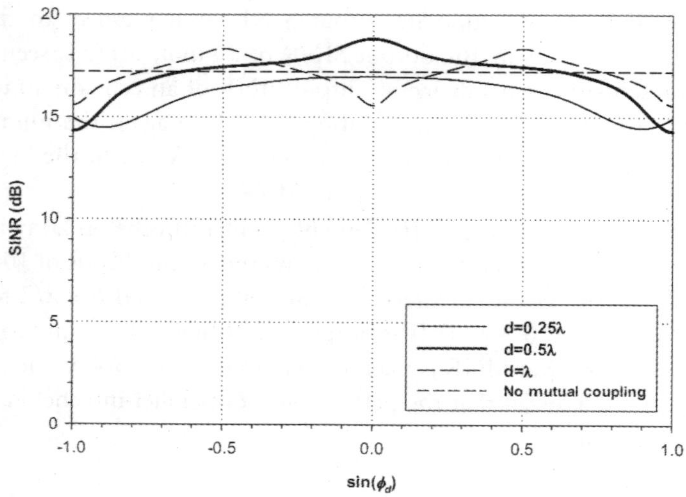

Figure 5.25 Effect of inter-element spacing on output SINR of a 10-element echelon array of half-wavelength, centre-fed dipoles; σ_d = 10 dB, θ_d = 90°

The effect of impinging direction of desired signal on the output SINR is analysed next, for different heights of the antenna elements at an even position in 10-element parallel-in-echelon array of $\lambda/2$ dipoles. The output SINR of 10-element echelon dipole array with σ_d = 10 dB and the inter-element spacing $d = 0.5\ \lambda$ varied for three different heights, *viz.* $h = 0.25\ \lambda$, $0.5\ \lambda$ and λ, is shown in Figure 5.26. The output SINR for a 10-element echelon dipole array is maximum at a desired signal direction of 0° for $h = 0.5\ \lambda$ in comparison to the other two cases. Furthermore, as one moves away from 0°, the magnitude of output SINR decreases continuously. Hence, the output SINR is higher for the height of $0.5\ \lambda$.

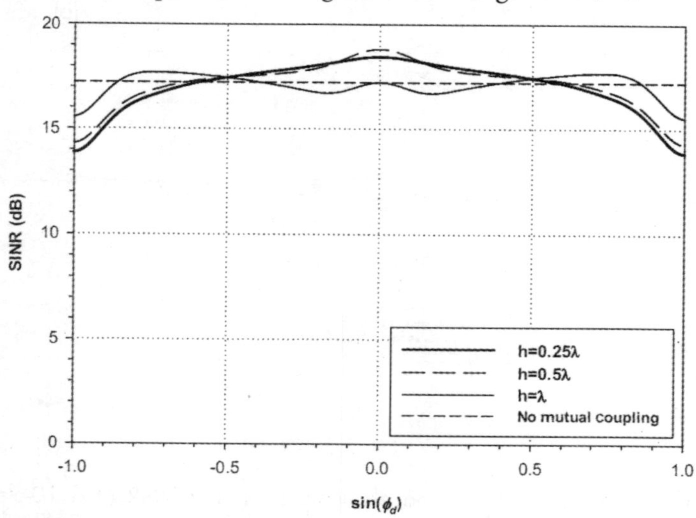

Figure 5.26 Effect of height h on output SINR of a 10-element echelon array of half-wavelength, centre-fed dipoles; σ_d = 10 dB, θ_d = 90°

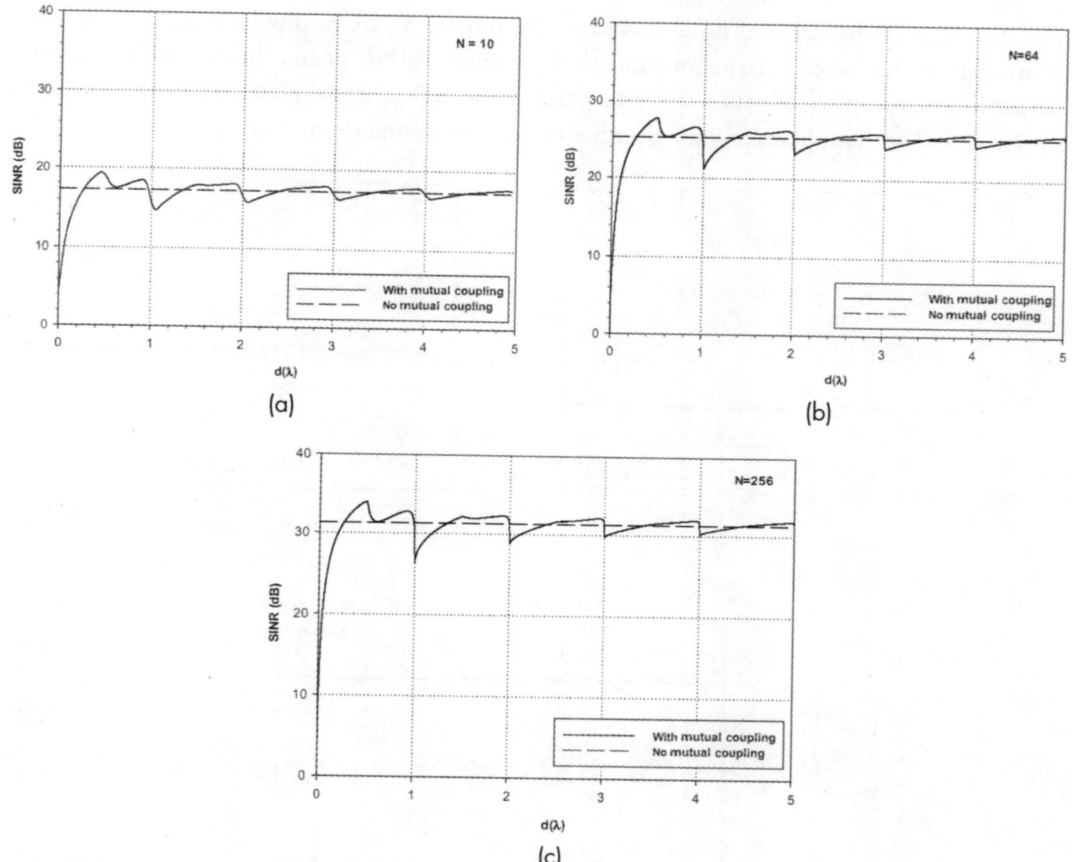

Figure 5.27 Effect of number of elements on output SINR for different number of elements in echelon array of half-wavelength, centre-fed dipoles; σ_d = 10 dB, (θ_d, Φ_d) = (90°,0°). (a) N = 10, (b) N = 64, (c) N = 256

The effect of inter-element spacing on the output SINR for the different number of elements in half-wavelength echelon dipole array is shown in Figure 5.27. The inter-element spacing is varied up to 5 λ.

The desired signal is assumed to impinge from the broadside (90°,0°) with σ_d = 10 dB. The height of the elements at even positions is taken to be $\lambda/2$ from the reference ground. It can be seen that the coupling between the elements affects the output SINR of echelon array even when the elements are placed far apart. Further, the output SINR rises with the number of dipoles in parallel-in-echelon dipole array.

The role of inter-element spacing in the output SINR of a 10-element half-wavelength echelon array is analysed next, for different values of the ratio of desired signal power to the thermal noise power σ_d, while keeping the other parameters constant (Figure 5.28). It is assumed that the height of the elements at even position is taken to be $\lambda/2$ from the reference ground. From Figure 5.28, it is apparent that as the magnitude of the ratio of power associated with the desired signal to that of the thermal noise increases, the output SINR of dipole array also increases. Thus, Figures 5.27 and 5.28 demonstrate that the output SINR of an echelon dipole array may be significantly improved by increasing the ratio of the desired signal power to the thermal noise power σ_d rather than increasing the number of dipole antenna elements in array.

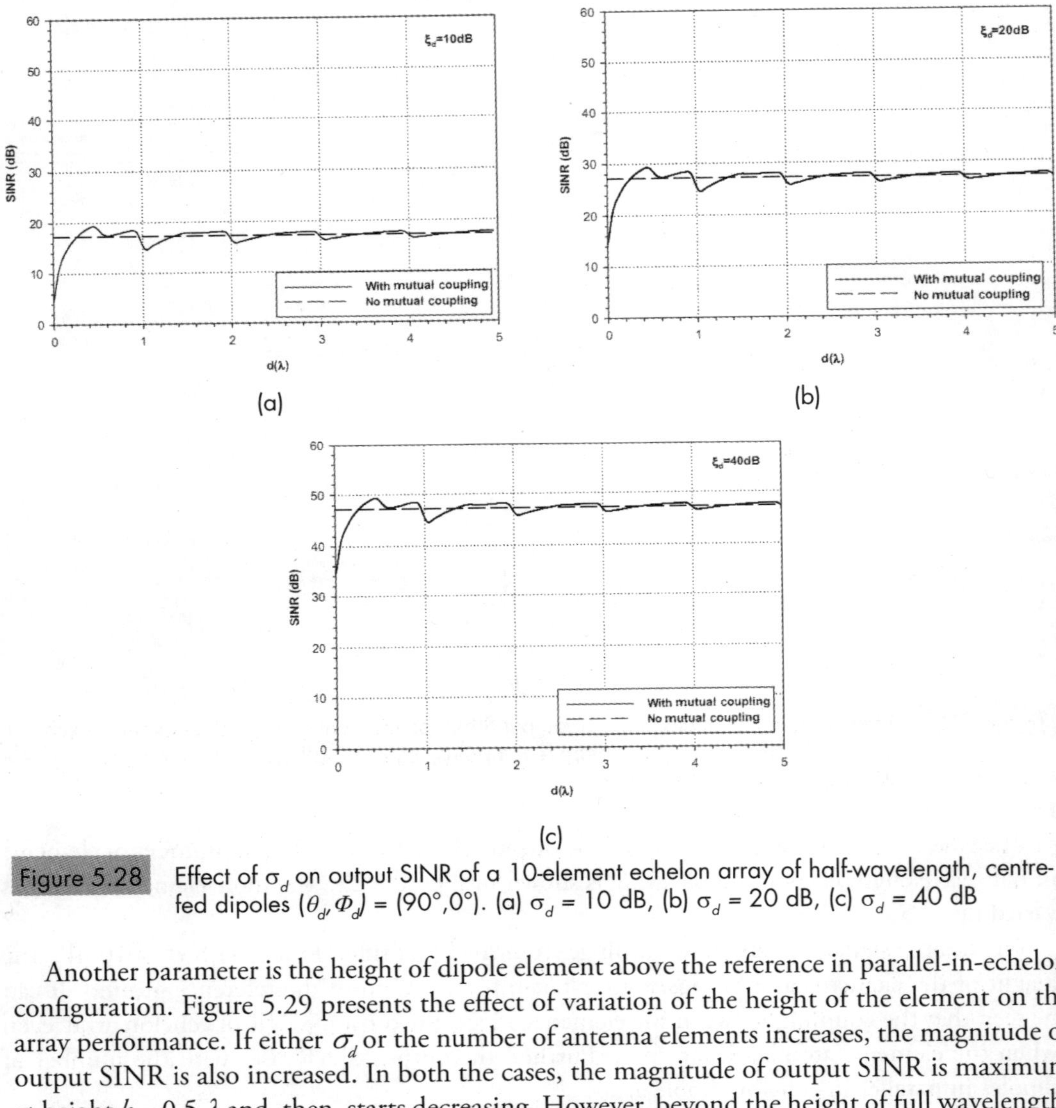

Figure 5.28 Effect of σ_d on output SINR of a 10-element echelon array of half-wavelength, centre-fed dipoles $(\theta_d, \Phi_d) = (90°, 0°)$. (a) σ_d = 10 dB, (b) σ_d = 20 dB, (c) σ_d = 40 dB

Another parameter is the height of dipole element above the reference in parallel-in-echelon configuration. Figure 5.29 presents the effect of variation of the height of the element on the array performance. If either σ_d or the number of antenna elements increases, the magnitude of output SINR is also increased. In both the cases, the magnitude of output SINR is maximum at height $h = 0.5\ \lambda$ and, then, starts decreasing. However, beyond the height of full wavelength, the value of output SINR decreases by a small amount and, then, becomes constant. Hence, the mutual coupling also affects the output SINR even for the larger heights of the elements at even position of the half-wavelength echelon dipole array. Moreover, if the number of elements in an echelon array increases then so does the output SINR difference between with and without mutual coupling cases.

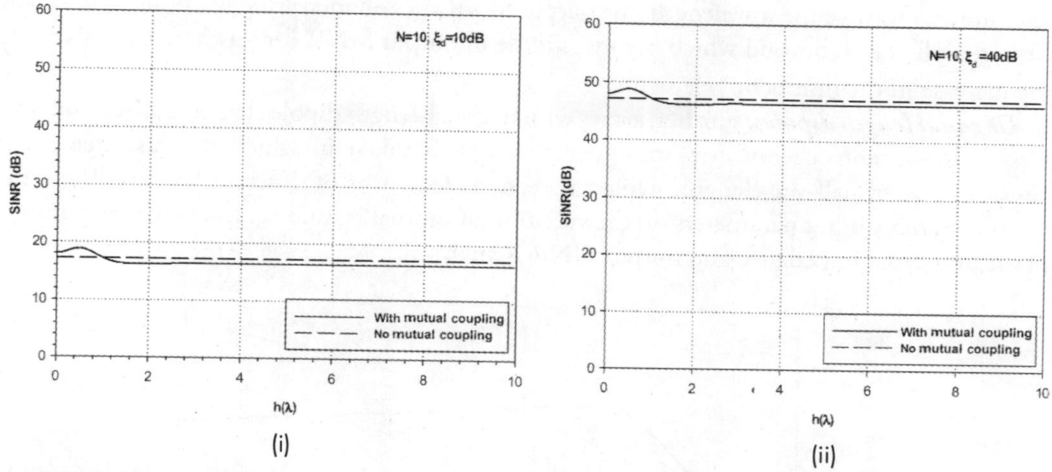

Figure 5.29a Effect of σ_d on output SINR of a 10-element echelon array of half-wavelength, centre-fed dipoles; σ_d = 10 dB, (θ_d, Φ_d) = (90°,0°). (i) σ_d = 10 dB, (ii) σ_d = 40 dB

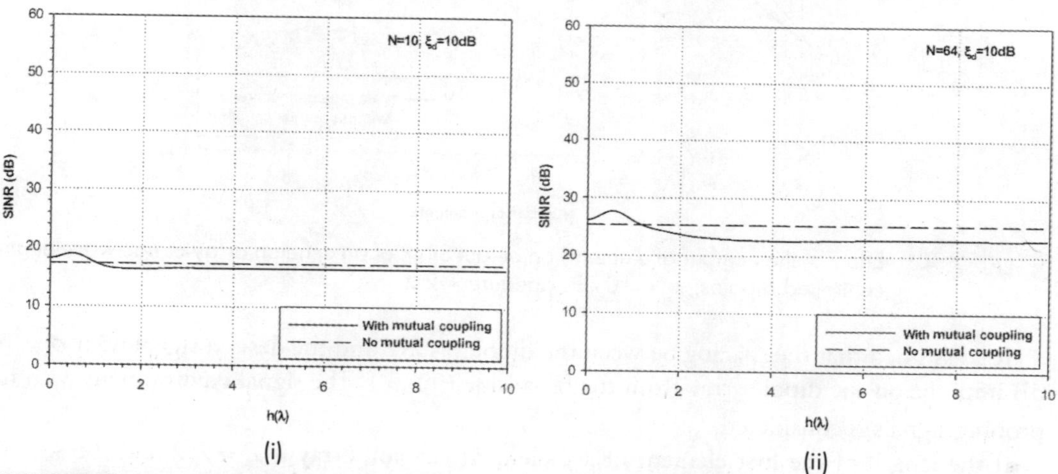

Figure 5.29b Effect of the number of elements on output SINR in echelon array of half-wavelength, centre-fed dipoles; σ_d = 10 dB, (θ_d, Φ_d) = (90°,0°). (i) N = 10, (ii) N = 64

Keeping the aperture size fixed at twice the wavelength, the output SINR is determined next by increasing the number of dipole antenna elements of a half-wavelength echelon array (Figure 5.30). It is assumed that only desired signal power with σ_d = 10 dB impinge on the dipole array from the broadside direction, and the height of the elements at an even position of an echelon array is taken to be $\lambda/2$. It may be observed from Figure 5.30 that the magnitude of output SINR is directly proportional to the number of dipole elements for no coupling case. However, this is not the case when coupling effect is included. It is apparent that for a fixed aperture size at twice the wavelength, the output SINR rises on increasing the number of dipole elements till $N = 7$, beyond which the magnitude of output SINR decreases, notwithstanding the increase in the number of elements.

Unequal length dipoles: Further, arrays with unequal-length dipole elements are considered. First, a 2-element echelon array of unequal length is taken in which the first element is considered as of half-wavelength, while the second element is of different lengths. Then, the effect of various input parameters on the variation of mutual resistance, reactance, magnitude/phase of mutual impedance and output SINR is analysed.

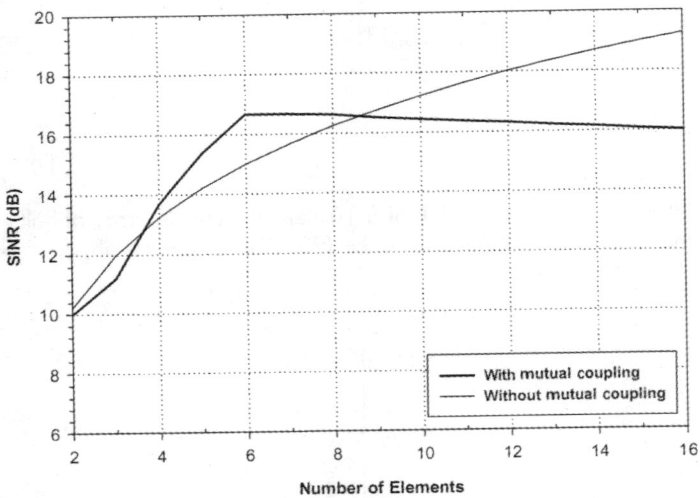

Figure 5.30 Effect of the number of elements on output SINR of an echelon array of half-wavelength, centre-fed dipoles; σ_d = 10 dB, aperture = 2 λ

It is assumed that the spacing between the dipoles is $\lambda/2$ and the desired signal with σ_d= 10 dB impinge on the dipole array from the broadside (90°,0°). The signal environment with no probing signals is considered.

If the length of the first element of a 2-element echelon array is $L = \lambda/2$ and the second element is $L = \lambda/3$, the variation of mutual resistance, reactance and magnitude/phase of mutual impedance for three different cases, *viz.* non-staggered, staggered elements with $h = 0$ and $h = \lambda/4$, is as shown in Figure 5.31. When the length of the second element of a 2-element echelon array is $L = \lambda/3$ (Figure 5.31a) instead of $L = \lambda/2$, the magnitude of mutual resistance, mutual reactance and, hence, mutual impedance falls significantly.

However, when this echelon array having the length of $\lambda/2$ and $\lambda/3$ is compared for the three cases, i.e. (i) non-staggered, (ii) staggered by $h = 0$ and (iii) staggered by $h = \lambda/4$, it is apparent that the minimum mutual impedance is obtained for the third case, i.e. staggered by $h = \lambda/4$. This is because of the weaker coupling effect owing to the large effective distance between the antenna elements.

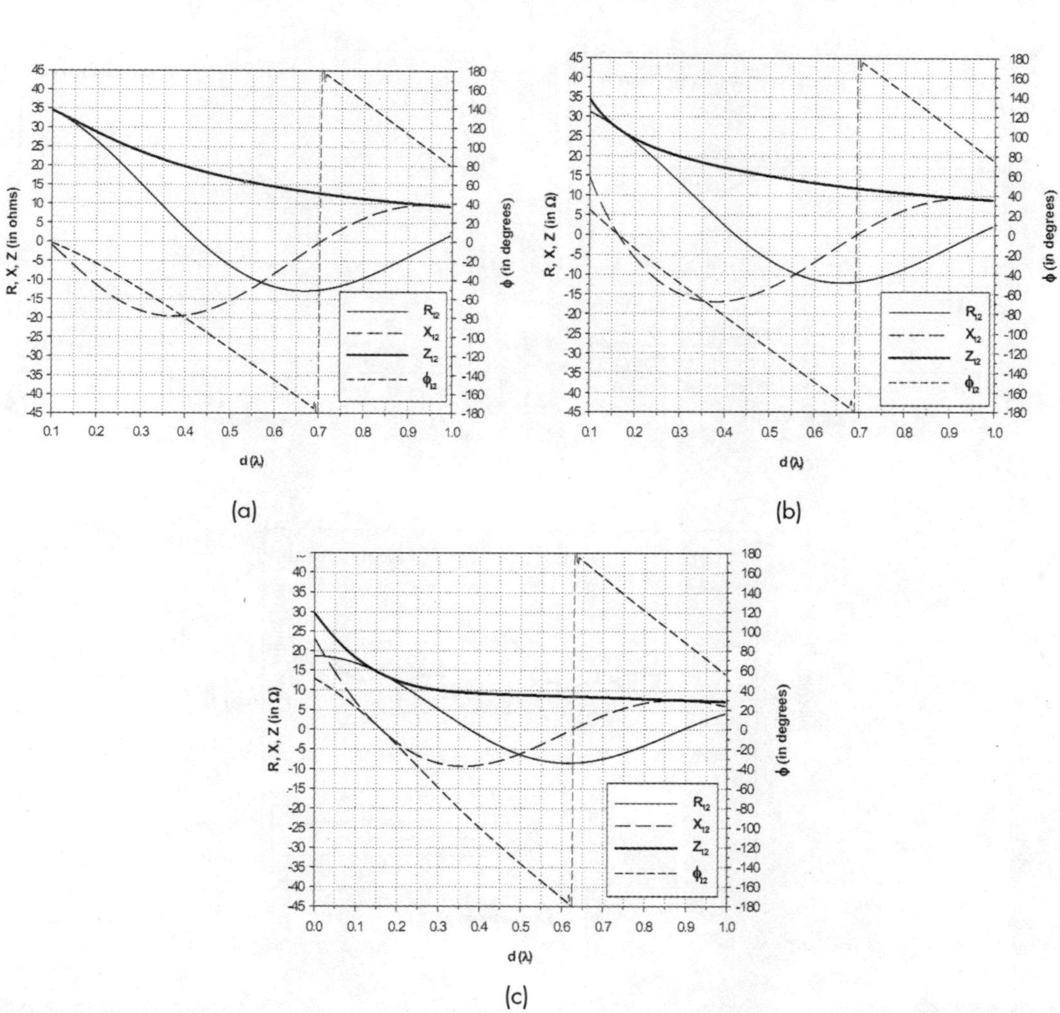

Figure 5.31 Mutual coupling resistance, reactance and impedance for two parallel unequal-length antennas of length $\lambda/2$ and $\lambda/3$. (a) non-staggered (b) staggered by $h = 0$ (c) staggered by $h = \lambda/4$

Next, a 10-element echelon array having an alternate length of $\lambda/2$ and $\lambda/3$ with spacing $d = \lambda/2$ is considered. It is assumed that only desired signal with $\sigma_d = 10$ dB impinges on the dipole array from the broadside direction. The effect of desired signal direction on the array performance in terms of output SINR is analysed for the three cases taking mutual coupling into account (Figure 5.32). It may be observed that the maximum output SINR ($\approx$ 40 dB) is obtained at 0° as compared to the magnitude of 18.8 dB in case of equal-length half-wavelength dipole array (Figure 5.18). The output SINR, however, decreases as one moves away from 0°. Moreover, the output SINR is maximum for the third case, when the antenna elements are staggered by $h = \lambda/4$.

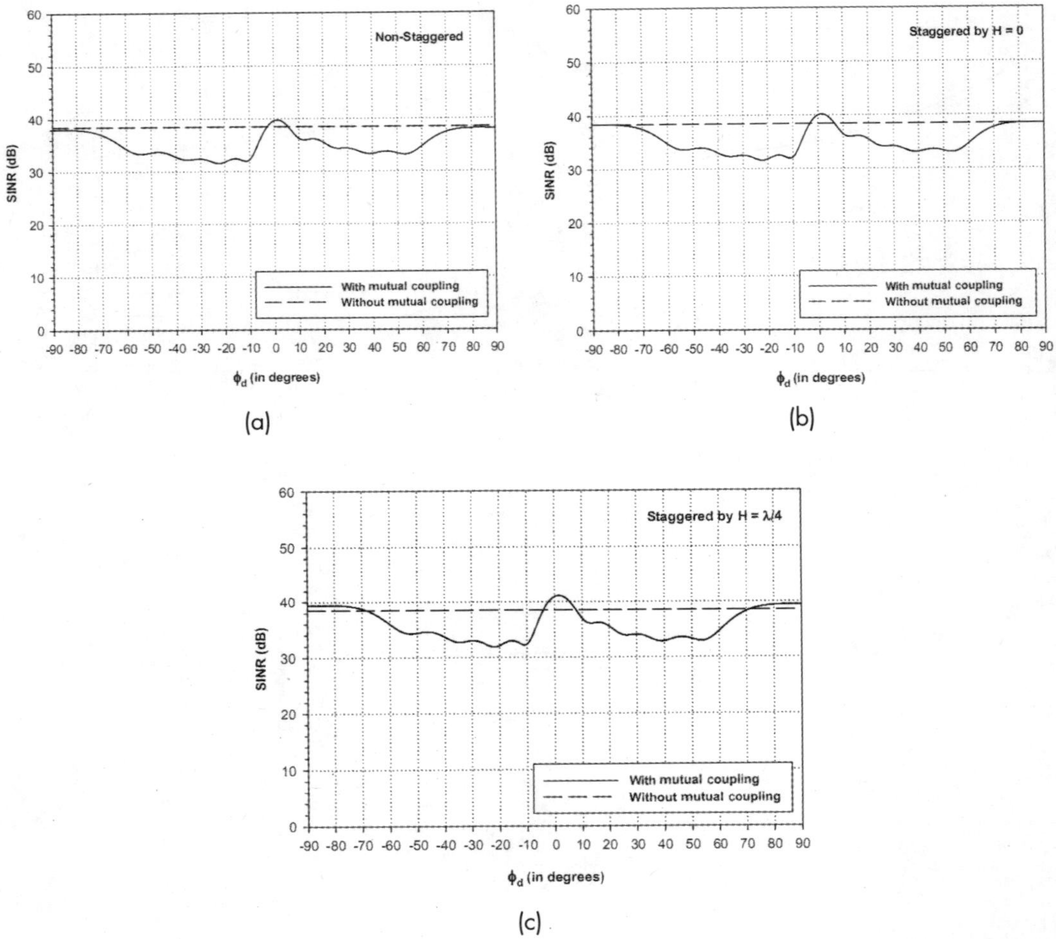

Figure 5.32 Output SINR for a 10-element echelon array having an alternate length of $\lambda/2$ and $\lambda/3$ (a) non staggered (b) staggered by $h = 0$, (c) staggered by $h = \lambda/4$

The number of elements is varied keeping all other parameters constant for an echelon array, with an alternate unequal length of $\lambda/2$ and $\lambda/3$. It is assumed that only the desired power signal with $\sigma_d = 10$ dB is incident on the array from the broadside (90°). Figure 5.33 presents the variation in the output SINR with respect to the desired signal direction with the number of elements for the above three cases. It may be observed that if the number of dipoles in an echelon array is increased, the magnitude of output SINR and the sharpness of the magnitude around 0° also increases. Moreover, as one moves away from 0°, the magnitude of output SINR goes on decreasing significantly. Again, the output SINR is best for the third case, when the antenna elements are staggered by $h = \lambda/4$.

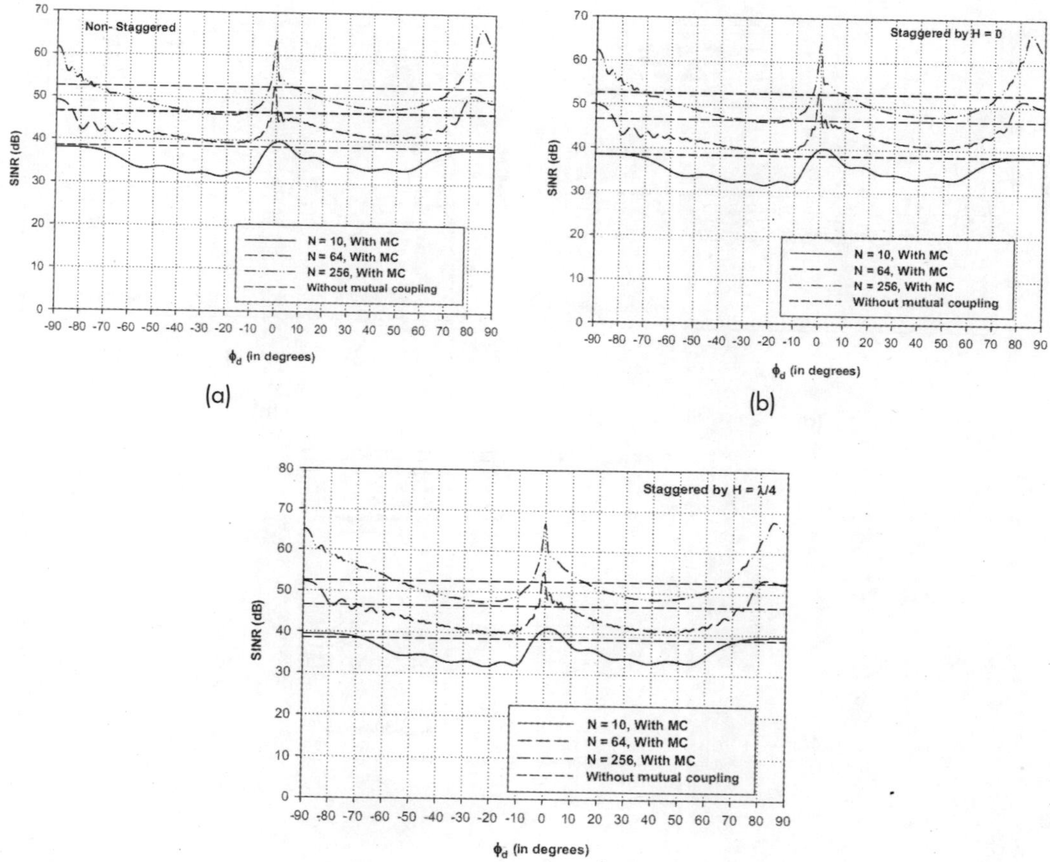

Figure 5.33 Effect of desired signal direction on output SINR for different number of elements in echelon array having an alternate length of $\lambda/2$ and $\lambda/3$ (a) non-staggered (b) staggered by h = 0, (c) staggered by $h = \lambda/4$

The effect of inter-element spacing d on the output SINR is then analysed for the different number of elements in an echelon array having an alternate unequal lengths of $\lambda/2$ and $\lambda/3$. In all the three cases, simulations are done for N = 10, 64 and 256. Figure 5.34 presents the variation of output SINR with respect to spacing d for the different number of dipoles in parallel-in-echelon configuration. It is observed that the output SINR rises with the number of dipole elements in an echelon array. This is true for all the three cases. For small spacing between dipole elements, the magnitude of output SINR rises rapidly. However, beyond $d = \lambda$, the variation in magnitude of output SINR becomes negligible.

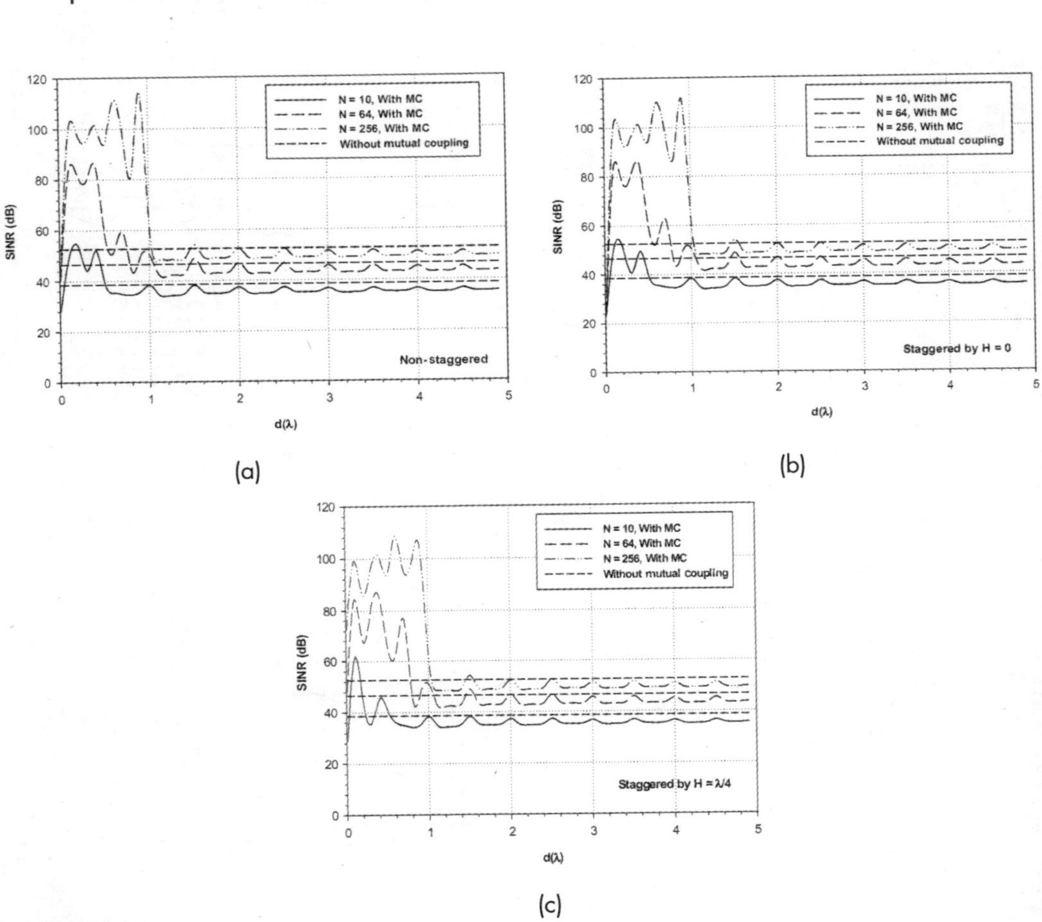

Figure 5.34 Effect of the number of elements on output SINR for an echelon array having an alternate length of $\lambda/2$ and $\lambda/3$ (a) non-staggered (b) staggered by $h = 0$, (c) staggered by $h = \lambda/4$

Finally, taking all the parameters same, the output SINR of a 10-element quarter-wavelength dipole array is computed in Figure 5.35. The effect of impinging direction of desired signal on the output SINR is analysed for two cases, when the height of elements at an even position of a 10-element quarter wavelength echelon array is (i) $\lambda/4$ and (ii) $\lambda/2$. It may be seen that the output SINR is maximum, ≈50db, at 0°, for both the cases. However, the trend of variation of output SINR about 0° is symmetrical, which is not the case when unequal length antennas were used. Further, it may be inferred that better SINR can be achieved if the height of the elements at even position is taken as $\lambda/2$. Thus, a quarter-wavelength echelon dipole array may be a preferred choice for achieving high output SINR.

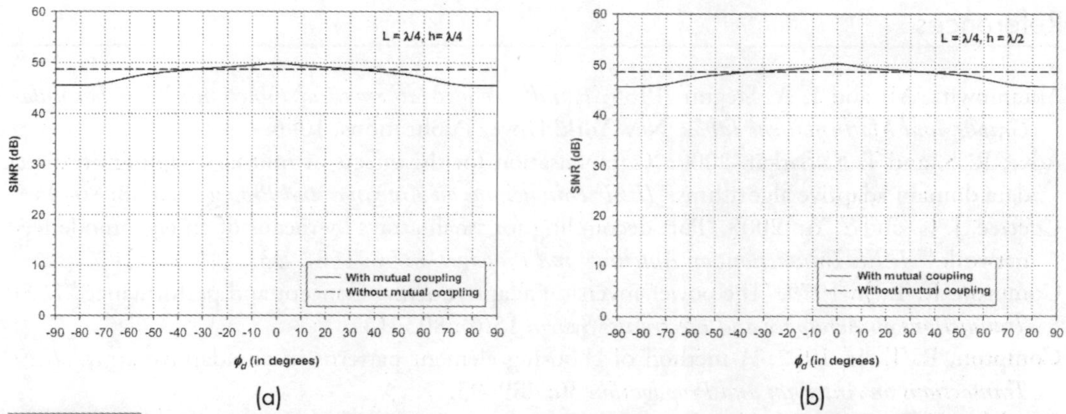

Figure 5.35 Effect of desired signal direction on the output SINR for a 10-element quarter-wavelength echelon array; σ_d = 10 dB, (θ_d, Φ_d) = (90°,0°), $d = \lambda/2$, (a) $h = \lambda/4$, (b) $h = \lambda/2$

5.4 Conclusion

In an antenna array, the elements are placed sufficiently close so that the coupling between the antenna elements cannot be ignored. The mutual coupling effect is known to degrade the array performance in terms of antenna gain and beam width. This effect is predominant for small inter-element spacing. Moreover, due to mutual coupling in between antenna elements, the signal weight vector is altered resulting in the output SINR degradation, which in turn affects the suppression capabilities of adaptive array. The mutual coupling, especially for smaller inter-element spacing drops the eigenvalues of signal covariance matrix. This results in slow response of antenna array. In other words, due to coupling between the elements, the adaptive array takes longer time to respond towards the impinging signals and placing nulls towards the probing signals.

In this chapter, the performance of dipole array including mutual coupling effect is discussed. The array design parameters are varied to observe the effect of the performance of dipole array. Both the equal-length and unequal-length dipole arrays are considered. Different configurations such as side-by-side dipoles, collinear dipoles and parallel-in-echelon dipole array are analysed. For a 10-element half-wavelength echelon dipole, the performance of a dipole array in terms of output SINR is found to be better in case of staggered echelon array with $h = \lambda/4$. The echelon array with unequal length dipole antenna having an alternative length of $\lambda/2$ and $\lambda/3$ is studied. It is shown that as one decreases the length of the elements at even positions of an echelon array, the performance of an antenna array improves in terms of gain and beam width. However, when $\lambda/3$ or $\lambda/4$ length antennas are placed at even positions of an echelon array, the output SINR decreases rapidly as one moves away from 0°. The performance can be improved if the parallel-in-echelon array with equal-length of quarter-wavelength is considered. This establishes the fact that intelligent choice of design parameters can contribute to the mitigation of coupling effect in the antenna arrays.

References

Abramowitz, M. and I. A. Stegun. 1965. *Handbook of Mathematical Functions: With Formulas, Graphs, and Mathematical Tables.* New York: Dover Publications, 1046.

Adve, R. S. and T. K. Sarkar. 2000. 'Compensation for the effects of mutual coupling on direct data domain adaptive algorithms.' *IEEE Transactions on Antennas and Propagation* 48: 86–94.

Coetzee, J. C. and Y. Yu. 2008. 'Port decoupling for small arrays by means of an eigenmode feed network.'" *IEEE Transactions on Antennas and Propagation* 56: 1587–93.

Compton, R. T., Jr. 1979. 'The power-inversion adaptive array: Concept and performance.' *IEEE Transactions on Aerospace and Electronics System* 15(6): 803–14.

Compton, R. T., Jr. 1982. 'A method of choosing element patterns in an adaptive array.' *IEEE Transactions on Antennas and Propagation* 30: 489–93.

Dandekar, K. R., H. Ling, and G. Xu. 2002. 'Experimental study of mutual coupling compensation in smart antenna applications.' *IEEE Transactions on Wireless Communications* 1(3): 480–87.

Fletcher, P. N. and P. Darwood. 1998. 'Beam forming for circular and semicircular array antennas for low cost wireless LAN data communication systems.' *IEE Proceedings on Microwave Antennas and Propagation* 145(2): 153–58.

Friedlander, B. and A. J. Weiss. 1991. 'Direction finding in the presence of mutual coupling.' *IEEE Transactions on Antennas and Propagation* 39: 273–84.

Godara, L. C. 2004. *Smart Antennas.* Washington DC: CRC Press, 448.

Goossens, R. and H. Rogier. 2007. 'A hybrid UCA-RARE/ Root-MUSIC approach for 2-D direction of arrival estimation in uniform circular arrays in the presence of mutual coupling.' *IEEE Transactions on Antennas and Propagation* 55: 841–49.

Guo, J. L. and J. Y. Li. 2009. 'Pattern synthesis of conformal array antenna in the presence of platform using differential evolution algorithm.' *IEEE Transactions on Antennas and Propagation* 57: 2615–21.

Gupta, I. J. and A. A. Ksienski. 1983. 'Effect of mutual coupling on the performance of adaptive arrays.' *IEEE Transactions on Antennas and Propagation* 31: 785–91.

Householder, A. S. 2006. *The Theory of Matrices in Numerical Analysis.* New York: Dover Publications, 272.

Hui, H. T. 2004. 'A new definition of mutual impedance for application in dipole receiving antenna arrays.' *IEEE Antennas and Wireless Propagation Letters* 3(1): 364–67.

Hui, H. T. 2004. 'A practical approach to compensate for the mutual coupling effect in an adaptive dipole array.' *IEEE Transactions on Antennas and Propagation* 52: 1262–69.

Hui, H. T., K. Y. Chan, and E. K. N. Yung. 2003. 'Compensating for the mutual coupling effect in a normal-mode helical antenna array for adaptive nulling.' *IEEE Transactions on Vehicular Technology* 52: 743–51.

Jenn, D. C. and S. Lee. 1995. 'In-band scattering from arrays with series feed networks.' *IEEE Transactions on Antennas and Propagation* 43: 867–73.

Jenn, D. C. and V. Flokas. 1996. 'In-band scattering from arrays with parallel feed networks.' *IEEE Transactions on Antennas and Propagation* 44: 172–78.

King, H. 1957. 'Mutual impedance of unequal length antennas in echelon.' *IRE Transactions on Antennas and Propagation* 5: 306–13.

Lau, B. K., J. B. Andersen, G. Kristensson, and A. F. Molisch. 2006. 'Impact of matching network on bandwidth of compact antenna arrays.' *IEEE Transactions on Antennas and Propagation* 54: 3225–38.

Leviatan, Y., A. T. Adams, P. H. Stockmann, and D. R. Miedaner. 1983. 'Cancellation performance degradation of a fully adaptive Yagi array due to inner-element coupling.' *Electronics Letters* 19(5): 428–32.

Li, X. and Z. P. Nie. 2004. 'Mutual coupling effects on the performance of MIMO wireless channels.' *IEEE Antennas Wireless Propagation Letters* 3(1): 344–47.

Lui, H. S. and H. T. Hui. 2010. 'Improved mutual coupling compensation in compact antenna arrays.' *IET Microwaves, Antennas and Propagation* 4: 1506–16.

Lui, H. S., H. T. Hui, and M. S. Leong. 2009. 'A note on the mutual coupling problems in transmitting and receiving antenna arrays.' *IEEE Antennas and Propagation Magazine* 51(5): 171–76.

Niow, C. H., Y. T. Yu, and H. T. Hui. 2011. 'Compensate for the coupled radiation patterns of compact transmitting antenna arrays.' *IET Microwaves, Antennas and Propagation* 5: 699–704.

Pasala, K. M. and E. M. Friel. 1994. 'Mutual coupling effects and their reduction in wideband direction of arrival estimation.' *IEEE Transactions on Aerospace and Electronic Systems* 30: 1116–22.

Riegler, R. L. and R. T. Compton, Jr. 1973. 'An adaptive array for interference rejection.' *Proceedings of IEEE* 61: 748–58.

Strait, B. and K. Hirasawa. 1969. 'Array design for a specified pattern by matrix methods.' *IEEE Transactions on Antennas and Propagation* 17: 237–39.

Wang, B. H. and H. T. Hui. 2011. 'Wideband mutual coupling compensation for receiving antennas using the system identification method.' *IET Microwaves, Antennas and Propagation* 5: 184–91.

6

RCS of Dipole Array Including Mutual Coupling Effects

6.1 Introduction

Radar cross section (RCS) of an aerospace vehicle significantly depends on the scattering of the on-board mounted antenna array. This antenna mode scattering is dependent on (i) the geometry of the phased array, (ii) the nature of the feed network employed to excite it, and (iii) the Mutual Coupling effect. A phased array, in general, requires a feed network to function as an effective transmitter or receiver. This feed network comprises radiating elements, phase-shifters, couplers and terminating resistors. Depending on their mode of arrangement, the feed network may be categorised as series feed, corporate-feed and space-feed (Volakis 2007).

To design a complete phased array system with specific radiation characteristics, several factors like the number of elements, their inter-element spacing, beam scan angle, arrangement of phase shifters, couplers, and loads have to be considered. Further, if the spacing between the antenna elements is less than half-wavelength, the effects of mutual coupling become predominant. These parameters of the array design are found to affect the RCS of the array considerably.

Mutual coupling affects the impedance of the array elements, reflection coefficients, radiation pattern and, thus, the RCS of phased array. Every element in the antenna array is of a definite physical size and shape. The incident wave on each of these elements will reradiate due to various reasons, such as the mismatches in the network, manufacturing defects, etc. The fields radiated from one antenna element interact with those of the surrounding antennas, giving rise to mutual coupling. These reradiated fields when scattered along different directions affect the RCS of the target significantly.

Further, the effect of mutual coupling on the array performance is influenced by (i) the feed network, (ii) array scan angle, and (iii) the geometry of the antenna array, *viz.* side-by-side, collinear and parallel-in-echelon configuration. The literature available in the open domain presents several methods to estimate the antenna RCS. These include finite difference time domain (FDTD) method (Ma et al. 2005), Shooting and Bouncing Ray (SBR) approach (Dikmen et al. 2010), (Tao et al. 2010), Physical Optics (PO)-based model (Zdunek and Rachowicz 2008), Scattering Matrix approach (Fan and Jin 1997), Method of Moments (MoM) (Pozar 1989), (Wang et al. 2010), Analytical Formulation approach (Zhang et al. 2010) and Method of Moments–Particle Swarm Optimisation (MoM–PSO) method (Wang et al. 2009).

Over the years, efforts have been put to reduce or compensate the Mutual Coupling effect (Hui 2007). Gupta and Ksienski (1983) presented an open-circuit voltage approach for determining the mutual coupling in transmitting and receiving antenna array. Another approach based on the compensation of receiving mutual impedance was proposed, especially for small and compact receiving monopole array (Hui 2004), (Yu and Hui 2011). Lee and Chu (2005) studied the Mutual Coupling effect for a finite non-linearly loaded antenna array using power series expansion technique. However, the effect of the feed network was neglected. Lu et al. (2009) proposed a method to estimate the RCS of antenna array as a product of array factor and element factor without inclusion of Mutual Coupling effect. Jenn and Lee (1995) estimated the RCS of a series-fed phased array by considering the feed components but ignoring the Mutual Coupling effect. Lee and Chu (1988) included the Mutual Coupling factor in RCS estimation of a parallel-fed phased array using iterative scattering matrix approach. The edge effects and multiple reflections within the feed network were included. Li et al. (2004) incorporated the effect of mutual coupling and feed mismatches in the analysis of phased array with Butler feed network.

In this chapter, the RCS estimation of a series-fed dipole array (Figure 6.1) including Mutual Coupling effect is presented. The formulation for the RCS of series-fed phased array is presented by tracing the path of the signal moving from the antenna aperture towards the feed network. The scattered field at each stage of the feed network is determined in terms of impedances. These individual level contributions, when coherently superimposed, provide the total RCS of phased array. The simulation results compare the RCS pattern with and without mutual coupling. The authenticity of the results is established by comparing them with the trends, inferences and results reported in the open domain.

6.2 Formulation for the RCS of series-fed dipole array

The RCS of an object for a plane wave incidence can be expressed as the ratio of the scattered field to the incident field (Knott et al. 1985). The mathematical expression for the antenna mode scattered field is given by

$$\vec{E}_n^s(\theta,\phi) = \left[\frac{j\eta}{4\lambda Z_a}\vec{h}\left\{\vec{h}.\vec{E}^i(\theta,\phi)\right\}\frac{e^{-j\vec{k}\vec{R}}}{R}\right]\Gamma_n^r(\theta,\phi) \tag{6.1}$$

where $\vec{E}_n^s$ is the scattered field of the nth element, $\vec{E}^i$ is the incident field, (θ,ϕ) is the direction of the signal, $Z_a = R_a + jX_a$ is the radiation impedance with $R_a = R_r + R_d$, R_a is the antenna resistance, X_a is the antenna reactance, R_r is the radiation resistance, R_d is the loss resistance of antenna (including conduction and dielectric losses), λ is the wavelength, η is the impedance of medium surrounding the antenna, $k = 2\pi/\lambda$ is the free space wave number, R is the distance between the target and the observation point, Γ_n^r is the total reflected signal returning to aperture element n and $\vec{h}$ is the effective height of the antenna element.

For a lossless antenna placed in free space, $\eta = \eta_o = 120\,\pi$ ohms with $R_d = 0$. Thus, $R_a = R_r$. The expression (6.1) can be, thus, rewritten as

$$\vec{E}_n^s(\theta,\phi) = \left[\frac{j\eta_o}{4\lambda Z_a}\vec{h}\left\{\vec{h}.\vec{E}^i(\theta,\phi)\right\}\frac{e^{-jk\bar{R}}}{R}\right]\Gamma_n^r(\theta,\phi) \tag{6.2}$$

For a unit amplitude incident plane wave (Aumann et al. 1989)

$$\vec{E}^i(\theta,\phi) = e^{-j\vec{k}.\vec{d}_n}\hat{\theta} \tag{6.3}$$

where $\vec{d}_n = \hat{x}(n-1)d$ represents the distance vector, with inter-element spacing, d. The wave vector $\vec{k}$ can be expressed in terms of direction cosines and the unit vectors (Chu 1991). It is given by

$$\vec{k} = k\left(\hat{x}\sin\theta\cos\phi + \hat{y}\sin\theta\sin\phi + \hat{z}\cos\theta\right) \tag{6.4}$$

The dot product in the phase factor of the incident signal can, thus, be simplified as

$$\vec{k}.\vec{d}_n = k\left(\hat{x}\sin\theta\cos\phi + \hat{y}\sin\theta\sin\phi + \hat{z}\cos\theta\right).\hat{x}(n-1)d$$
$$= (n-1)kd\sin\theta\cos\phi$$
$$\vec{k}.\vec{d}_n = (n-1)\alpha \tag{6.5}$$

For a closed body, with the coordinate system's origin located within it, the scalar or dot product of the outward normal and the incident wave propagation vector must be negative.

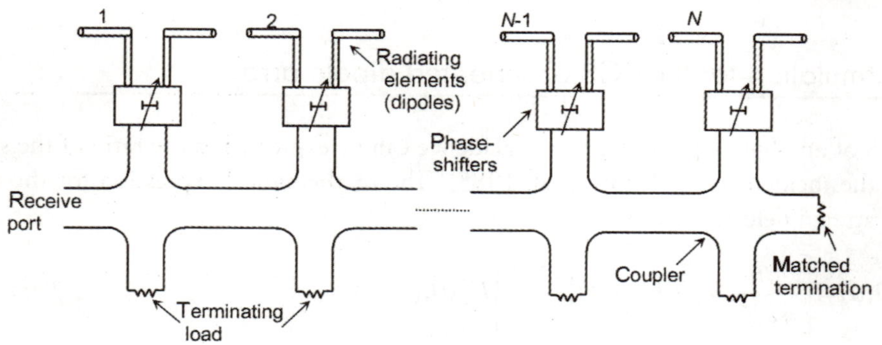

Figure 6.1 Typical series-fed network of phased array

Mathematically, $\vec{k}.\vec{d}_n < 0$. Hence, from (6.3) and (6.5), one gets

$$e^{-j\vec{k}.\vec{d}_n} = e^{j(n-1)\alpha} \tag{6.6}$$

For an *x*-polarised antenna, substitution of (6.6) in (6.2) yields

$$\vec{E}_n^s(\theta,\phi) = \left[\frac{j\eta_o}{4\lambda Z_a} h\hat{x}\left\{h\hat{x}.e^{j(n-1)\alpha}\hat{\theta}\right\}\frac{e^{-j\vec{k}\vec{R}}}{R}\right]\Gamma_n^r(\theta,\phi)$$

$$= \left[\frac{j\eta_o}{4\lambda Z_a} h^2\left(\hat{x}.\hat{\theta}\right)\left(\Gamma_n^r(\theta,\phi)e^{j(n-1)\alpha}\right)\frac{e^{-j\vec{k}\vec{R}}}{R}\right]\hat{x}$$

$$= \left[\frac{j\eta_o}{4\lambda Z_a} h^2\left(\cos\theta\right)\vec{E}_n^r(\theta,\phi)\right]\frac{e^{-j\vec{k}\vec{R}}}{R}\hat{x} \tag{6.7}$$

This is the scattered field due to a single dipole element in the array. The overall scattered field is obtained by summing up (6.7) over all the array elements.

$$\vec{E}^s(\theta,\phi) = \sum_{n=1}^{N}\vec{E}_n^s(\theta,\phi) = \sum_{n=1}^{N}\left[\left\{\frac{j\eta_o}{4\lambda Z_a} h^2\left(\cos\theta\right)\vec{E}_n^r(\theta,\phi)\right\}\frac{e^{-j\vec{k}\vec{R}}}{R}\hat{x}\right] \tag{6.8}$$

This yields the overall RCS of the dipole array for a unit magnitude incident wave.

$$\sigma(\theta,\phi) = \lim_{R\to\infty} 4\pi R^2\left|\sum_{n=1}^{N}\left[\left\{\frac{j\eta_o}{4\lambda Z_a} h^2\left(\cos\theta\right)\vec{E}_n^r(\theta,\phi)\right\}\frac{e^{-j\vec{k}\vec{R}}}{R}\hat{x}\right]\right|^2$$

$$= 4\pi\left|\sum_{n=1}^{N}\left\{\frac{j\eta_o}{4\lambda Z_a} h^2\left(\cos\theta\right)\vec{E}_n^r(\theta,\phi)\right\}\right|^2 \tag{6.9}$$

The effective height of an *x*-polarised antenna element is given by (Jenn 1995)

$$\vec{h} = h\hat{x} = \left(\frac{1}{I(0)}\int_{\Delta l} I(l)\,dl\right)\hat{x} \tag{6.10}$$

where $I(0)$ is the current at the feed terminal of the dipole element and l is the dipole length. Assuming the current distribution on the single dipole antenna to be cosine distribution, i.e. $I(l) = I(0)\cos(kl)$, the expression for the effective height becomes

$$\vec{h} = \left(\frac{1}{I(0)}\int_{\Delta l} I(0)\cos(kl)\,dl\right)\hat{x} = \left(\int_{\Delta l}\cos(kl)\,dl\right)\hat{x} \tag{6.11}$$

Substitution of (6.11) in (6.9) yields

$$\sigma(\theta,\phi)=4\pi\left|\sum_{n=1}^{N}\left\{\frac{j\eta_o}{4\lambda Z_a}\left(\int_{\Delta l}\cos(kl)\,dl\right)^2(\cos\theta)\vec{E}_n^r(\theta,\phi)\right\}\right|^2$$

$$=4\pi\left|\frac{j\eta_o}{4\lambda Z_a}\left(\int_{\Delta l}\cos(kl)\,dl\right)^2\cos\theta\sum_{n=1}^{N}\vec{E}_n^r(\theta,\phi)\right|^2$$

$$\sigma(\theta,\phi)=4\pi\left|F\sum_{n=1}^{N}\vec{E}_n^r(\theta,\phi)\right|^2 \tag{6.12}$$

where $$F=\frac{j\eta_o}{4\lambda Z_a}\left(\int_{\Delta l}\cos(kl)\,dl\right)^2\cos\theta \tag{6.13}$$

$\vec{E}_n^r(\theta,\phi)$ represents the total scattered field returning to the aperture after being reflected from the mismatches prevailing within the feed network. This factor can be computed by traversing the path of the signal as it travels through the different components of the feed network. Thus

$$\sigma(\theta,\phi)=4\pi\left\{\left|\sigma_r(\theta,\phi)\right|^2+\left|\sigma_p(\theta,\phi)\right|^2+\left|\sigma_c(\theta,\phi)\right|^2+\left|\sigma_s(\theta,\phi)\right|^2\right\} \tag{6.14}$$

The total RCS of phased array is the summation of individual contribution from radiating elements, phase-shifters and coupling port of the couplers.

6.3 Impedance at different levels of the feed network

The signal from the antenna aperture enters into the feed network, passing through its different levels before reaching the receive port. The feed network is a combination of various components, *viz.* radiators, phase-shifters, couplers and terminating loads. Each of these components offers finite impedance to the travelling signal (Figure 6.2). Thus, in the feed network, a significant impedance mismatch can occur at the junction. This results in the reflection of the incoming signal depending upon the value of reflection coefficient, which in turn affects the total RCS of array. The following sub-sections discuss the impedances at each level of the feed network.

6.3.1 Impedance at the terminals of the dipole antenna

A dipole antenna in general will have finite impedance at its feed terminals. The impedance depends on the location of the antenna, the angle of incident signal and the mutual coupling. In this paper, the effect of mutual coupling is analysed for three different configurations of dipole array, *viz.* side-by-side, collinear and parallel-in-echelon (Figure 6.3).

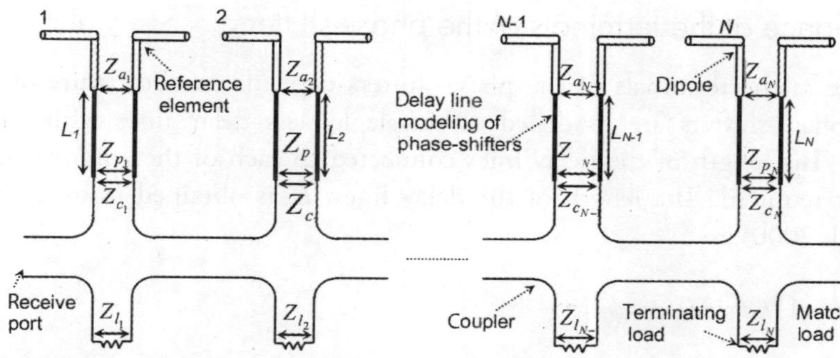

Figure 6.2 Impedances at different stages in a series-fed network

The antenna impedance is expressed as $Z_{a_n} = R_{a_n} + jX_{a_n}$. The corresponding impedance matrix (Gupta and Ksienski 1983) may be written as

$$z_{a_{x,y}} = \begin{pmatrix} z_{a_{1,1}} & z_{a_{1,2}} & \cdots & z_{a_{1,N}} \\ z_{a_{2,1}} & z_{a_{2,2}} & \cdots & z_{a_{2,N}} \\ \vdots & \vdots & \ddots & \vdots \\ z_{a_{N,1}} & z_{a_{N,2}} & \cdots & z_{a_{N,N}} \end{pmatrix} \tag{6.15}$$

where $z_{a_{i,j}}$ represents the self impedance of the ith element and $z_{a_{i,j}}$ $(i \neq j)$ is the mutual impedance between i^{th} and j^{th} element. The expressions for the self and mutual impedance between the dipoles for different configurations can be referred to in Appendix C. This yields the total antenna impedance at the nth element (Balanis 2005) as

$$Z_{a_n} = \sum_{y=1}^{N} z_{a_{x,y}} \frac{I_y}{I_x} \tag{6.16}$$

where I_n is the current at the feed terminals of nth antenna element.

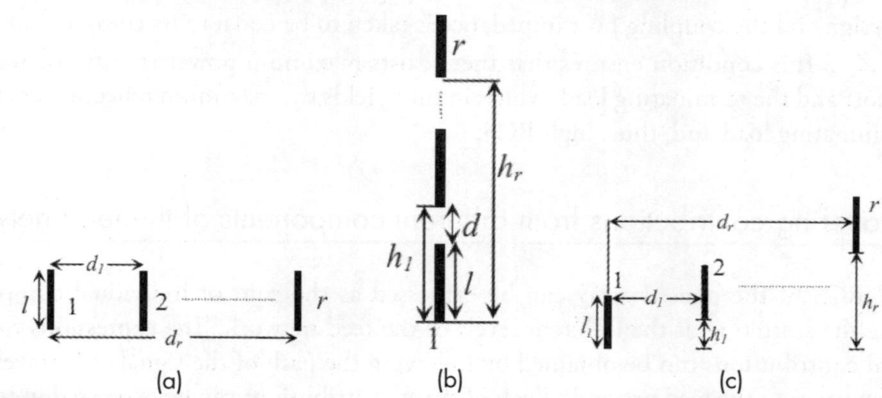

Figure 6.3 Schematic of dipole array (a) Side-by-side configuration, (b) Collinear configuration, (c) Parallel-in-echelon configuration

6.3.2 Impedance at the terminals of the phase-shifters

The impedance at the terminals of the phase-shifters depends on the nature of the phase-shifters. The phase-shifters are modelled as simple lossless delay lines with characteristic impedance Z_o. The length of the delay lines connected to each of the antennas depends on the phase-shift required. The length of the delay lines, L_n is obtained from the phase-shift (Batchelor et al. 2000) as

$$L_n = \frac{\lambda}{2\pi}\left[(n-1)kd\sin\theta_s\cos\phi_s\right] \tag{6.17}$$

These delay lines are connected to the antenna terminals at one end and to the coupling port at the other. Thus, the antenna impedance, i.e. the input impedance for the phase-shifter, is translated along its length (Liao 1990) as

$$Z_{p_n} = Z_o\left[\frac{Z_{a_n} + jZ_o\tan\left(\frac{2\pi}{\lambda}L_n\right)}{Z_o + jZ_{a_n}\tan\left(\frac{2\pi}{\lambda}L_n\right)}\right] \tag{6.18}$$

This yields the impedance Z_{p_n} at the other end of the phase-shifter (Figure 6.2).

6.3.3 Impedance at the coupler terminals

The impedances at different ports of the couplers depend on the type of couplers. The feed network is assumed to have lossless four-port couplers (Figure 6.4). Port 1 and Port 2 are the input and transmission ports, respectively, and form a part of the main feed line. Port 3 represents the coupling port, which connects to the antenna via phase-shifter. Port 4 represents the isolated port, which is terminated with a load. In general, the terminating load is chosen as per the design and the coupling port impedance is taken to be equal to its complex conjugate, i.e. $Z_{c_n} = Z_{l_n}^*$. This condition ensures that there exists maximum power transfer between the coupler port and the terminating load, which in turn yields the maximum reflection coefficient at the terminating load and, thus, high RCS.

6.4 Scattering contributions from different components of the feed network

The total RCS of the phased array can be expressed as the sum of individual components indicating the scattering at the different levels of the feed network. The expressions for these individual contributions can be obtained by following the path of the signal as it travels from antenna aperture to the feed network. Each of these contributions can be expressed in terms of reflection and transmission coefficients of feed network elements.

6.4.1 RCS component due to scattering from dipoles

The first source of scattering of the incident signal is the radiating element (e.g. a dipole). A well-designed antenna is expected to radiate/receive the entire energy incident on it. However, there exists a mismatch between the antenna impedance, Z_{a_n}, and the characteristic impedance of the delay line, Z_o, connected to it. This can be modelled as the antenna connected to a load representing the remaining part of the feed network (Figure 6.5). Further, the reflection of the incident signal can be expressed in terms of reflection coefficient ρ_{r_n} of the nth dipole. Mathematically

$$\rho_{r_n} = \left| \frac{Z_{a_n} - Z_o}{Z_{a_n} + Z_o} \right| \tag{6.19}$$

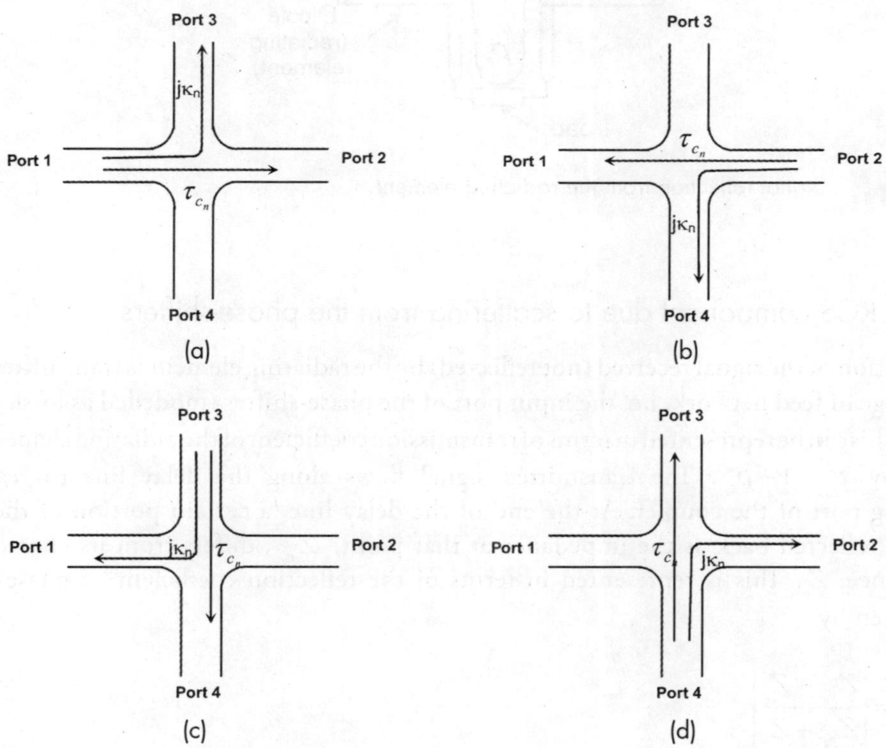

Figure 6.4 A four-port coupler with the transmission coefficient, τ_{c_n}, and the coupling, $j\kappa_n$, coefficient

This re-radiated portion of the signal from adjacent antenna elements exhibits a linear phase variation along the array. In terms of the reflection coefficient ρ_{r_n} and the phase of the nth element, the scattered electric field from n^{th} antenna element is given by

$$\vec{E}^r_{r_n}(\theta,\phi) = e^{j(n-1)\alpha} \rho_{r_n} e^{j(n-1)\alpha} = \rho_{r_n} e^{j2(n-1)\alpha} \tag{6.20}$$

The equation (6.20) represents the scattered field due to a single radiating element. Summation of (6.20) over all the elements of the array yields the total scattered field due to the dipole array. Thus, the corresponding RCS component is given by

$$\sigma_r(\theta,\phi)= F\sum_{n=1}^{N}\vec{E}^r_{r_n}(\theta,\phi) = F\sum_{n=1}^{N}\rho_{r_n}e^{j2(n-1)\alpha} \tag{6.21}$$

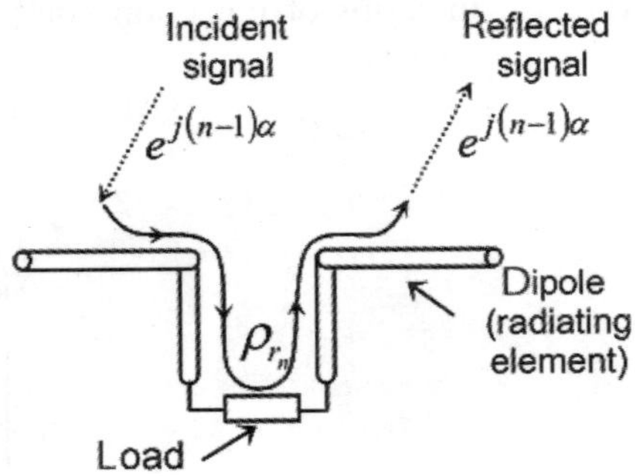

Figure 6.5 Signal reflection from the radiating element

6.4.2 RCS component due to scattering from the phase-shifters

The portion of the signal received (not reflected) by the radiating element is transmitted to the next stage of feed network, i.e. the input port of the phase-shifter (modelled as lossless delay lines). This can be represented in terms of transmission coefficient of the radiating elements, τ_{r_n}, given by $\tau^2_{r_n} = 1-\rho^2_{r_n}$. The transmitted signal flows along the delay line to reach the coupling port of the coupler. At the end of the delay line, a certain portion of the signal may be reflected back as the impedance at that point, Z_{p_n}, differs from its characteristic impedance, Z_o. This is represented in terms of the reflection coefficient of phase-shifter, ρ_{p_n} given by

$$\rho_{p_n} = \left|\frac{Z_{p_n} - Z_o}{Z_{p_n} + Z_o}\right| \tag{6.22}$$

This reflected signal propagates back towards the radiating element and suffers reflection and transmission for the second time. However, these higher order reflections can be neglected if their magnitude is negligible. Thus, following the path of the signal as shown in Figure 6.6, the scattered field due to n^{th} phase-shifter may be expressed as

$$\vec{E}^r_{p_n}(\theta,\phi) = \left(e^{j(n-1)\alpha}\tau_{r_n}\rho_{p_n}\tau_{r_n}e^{j(n-1)\alpha}\right) = \tau^2_{r_n}\rho_{p_n}e^{j2(n-1)\alpha} \tag{6.23}$$

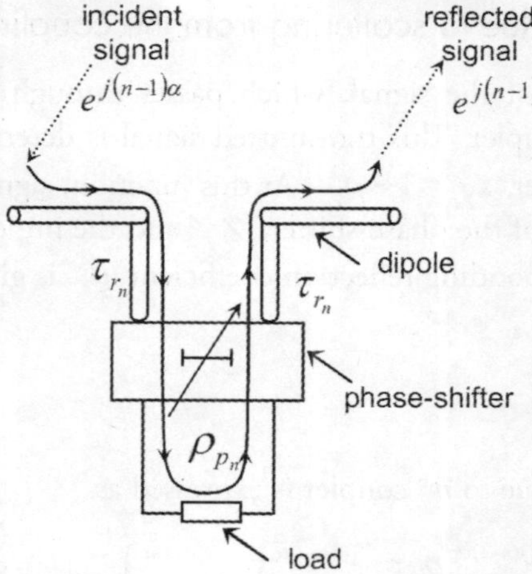

Figure 6.6 Signal reflection at the phase-shifter

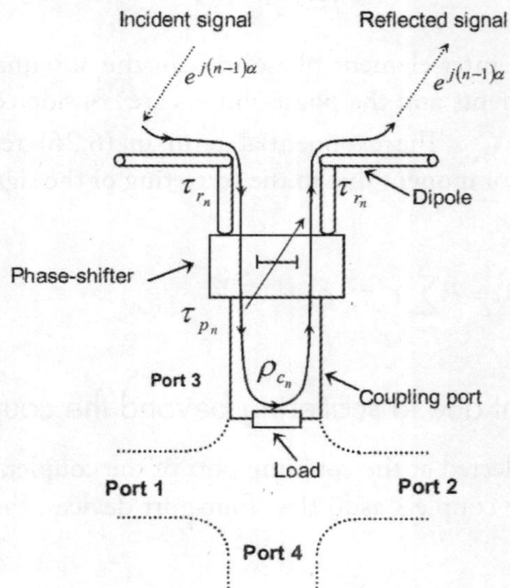

Figure 6.7 Signal reflection at the coupling port of the coupler

The antenna is assumed to be a reciprocal device, which results in the term $\tau_{r_n}^2$ in (6.23). Here, the load represents the levels of feed network beyond the phase-shifters. The exponential term in (6.23) signifies linear phase variation similar to that in the case of antenna elements. Likewise, the summation of (6.23) over all the elements in phased array yields the total scattered field due to phase-shifters.

The corresponding RCS component is given by

$$\sigma_p(\theta,\phi) = F\sum_{n=1}^{N} \vec{E}^r_{p_n}(\theta,\phi) = F\sum_{n=1}^{N} \tau_{r_n}^2 \rho_{p_n} e^{j2(n-1)\alpha} \tag{6.24}$$

6.4.3 RCS component due to scattering from the coupling port of the couplers

Figure 6.7 shows the path of the signal, which passes through the phase-shifters towards the coupling port of the coupler. This transmitted signal is determined by the transmission coefficient of the phase-shifter, $\tau_{p_n}^2 = 1 - \rho_{p_n}^2$. At this junction, significant reflection can occur if the impedance at the end of the phase-shifter, Z_{p_n} and the impedance at the coupler input port, Z_{c_n} differs. The corresponding reflection coefficient ρ_{c_n} is given by

$$\rho_{c_n} = \left| \frac{Z_{c_n} - Z_{p_n}}{Z_{c_n} + Z_{p_n}} \right| \tag{6.25}$$

Thus, the scattered field due to n^{th} coupler is expressed as

$$\vec{E}^r_{c_n}(\theta,\phi) = \left(e^{j(n-1)\alpha} \tau_{r_n} \tau_{p_n} e^{j(n-1)\alpha_s} \rho_{c_n} \tau_{p_n} e^{j(n-1)\alpha_s} \tau_{r_n} e^{j(n-1)\alpha} \right)$$

$$= \left(\tau_{r_n}^2 \tau_{p_n}^2 \rho_{c_n} e^{j2(n-1)\alpha_s} e^{j2(n-1)\alpha} \right) = \tau_{r_n}^2 \tau_{p_n}^2 \rho_{c_n} e^{j2(n-1)(\alpha+\alpha_s)} = \tau_{r_n}^2 \tau_{p_n}^2 \rho_{c_n} e^{j2(n-1)\zeta} \tag{6.26}$$

where α_s represents the inter-element phase to scan the antenna beam along *x*-direction. As both the radiating elements and the phase-shifters are considered to be reciprocal devices, (6.26) has terms $\tau_{r_n}^2$ and $\tau_{p_n}^2$. The exponential term in (6.26) represents the linear phase variation. Thus, the RCS component due to the scattering of the signal at the coupling port of the couplers is given by

$$\sigma_c(\theta,\phi) = F \sum_{n=1}^{N} \vec{E}^r_{c_n}(\theta,\phi) = F \sum_{n=1}^{N} \tau_{r_n}^2 \tau_{p_n}^2 \rho_{c_n} e^{j2(n-1)\zeta} \tag{6.27}$$

6.4.4 RCS component due to scattering beyond the coupling port of couplers

The path of signal, not reflected at the coupling port of the coupler, depends on the nature of the coupler. Assuming the couplers as lossless four-port devices, the coupling coefficient can be expressed as

$$\kappa_n = \frac{Z_{a_n} i_n^2}{\sum_{p=1}^{N} Z_{a_p} i_p^2 - \sum_{q=1}^{n-1} Z_{a_q} i_q^2} \tag{6.28}$$

The detailed derivation of the coupling coefficient of coupler is given in Appendix D. The transmission coefficient of the coupler will be $\tau_{c_n}^2 = 1 - \kappa_n^2$.

The signal incident at each of the antenna elements is expected to move towards the receiving port only. However, it has the ability to move towards (i) the next antenna element in the array, (ii) the previous antenna element in the array, (iii) the load terminating its own coupler and (iv) the port (backward direction) through which it was received. Each of these signals gives rise to scattering and, hence, contributes to the RCS significantly.

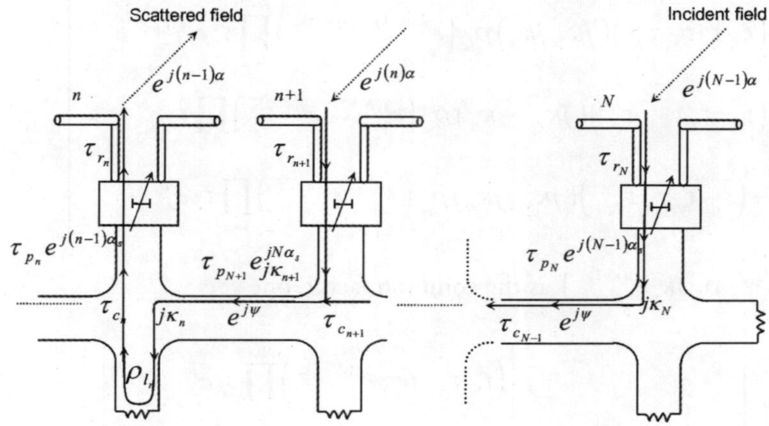

Figure 6.8 Path of the signal travelling towards the nth antenna element

The scattered field at the nth element of the array can be due to the signals incident on it from the next $(N\text{–}n)$ elements of the array. The path of such a signal is shown in Figure 6.8. Here, the reflection coefficient of the terminating load, ρ_{l_n}, can be expressed as

$$\rho_{l_n} = \left| \frac{Z_{l_n} - Z_{c_n}}{Z_{l_n} + Z_{c_n}} \right| \tag{6.29}$$

Following the path of the signals through each of the elements, the scattered electric field is given by

$$\vec{E}^{r}_{1_n}(\theta,\phi) = \begin{bmatrix} \begin{pmatrix} e^{j(N-1)\alpha}\tau_{r_N}\tau_{p_N} e^{j(N-1)\alpha_s} jK_N e^{j\psi}\tau_{c_{N-1}} e^{j\psi}\tau_{c_{N-2}} \ldots \\ e^{j\psi} jK_n \rho_{l_n}\tau_{c_n}\tau_{p_n} e^{j(n-1)\alpha_s}\tau_{r_n} e^{j(n-1)\alpha} \end{pmatrix} + \\ \begin{pmatrix} e^{j(N-2)\alpha}\tau_{r_{N-1}}\tau_{p_{N-1}} e^{j(N-2)\alpha_s} jK_{N-1} e^{j\psi}\tau_{c_{N-2}} \ldots \\ e^{j\psi} jK_n \rho_{l_n}\tau_{c_n}\tau_{p_n} e^{j(n-1)\alpha_s}\tau_{r_n} e^{j(n-1)\alpha} \end{pmatrix} + \ldots \\ + \begin{pmatrix} e^{j(n)\alpha}\tau_{r_{n+1}}\tau_{p_{n+1}} e^{j(n)\alpha_s} jK_{n+1} \\ e^{j\psi} jK_n \rho_{l_n}\tau_{c_n}\tau_{p_n} e^{j(n-1)\alpha_s}\tau_{r_n} e^{j(n-1)\alpha} \end{pmatrix} \end{bmatrix} \tag{6.30}$$

Regrouping of like factors yields

$$\vec{E}^{r}_{1_n}(\theta,\phi) = \begin{bmatrix} \left(\tau_{r_N}\tau_{r_n}\tau_{p_N}\tau_{p_n}\right)\left(jK_N jK_n\right)\rho_{l_n}\left(e^{j(N-1)\zeta} e^{j(n-1)\zeta}\right)\left(\tau_{c_{N-1}} e^{j\psi}\tau_{c_{N-2}} e^{j\psi}\ldots\tau_{c_n} e^{j\psi}\right) + \\ \left(\tau_{r_{N-1}}\tau_{r_n}\tau_{p_{N-1}}\tau_{p_n}\right)\left(jK_{N-1} jK_n\right)\rho_{l_n}\left(e^{j(N-2)\zeta} e^{j(n-1)\zeta}\right)\left(\tau_{c_{N-2}} e^{j\psi}\ldots\tau_{c_n} e^{j\psi}\right) + \ldots \\ + \left(\tau_{r_{n+1}}\tau_{r_n}\tau_{p_n}\tau_{p_{n+1}}\right)\left(jK_{n+1} jK_n\right)\rho_{l_n}\left(e^{j(n)\zeta} e^{j(n-1)\zeta}\right)\left(\tau_{c_n} e^{j\psi}\right) \end{bmatrix} \tag{6.31a}$$

$$
= \begin{bmatrix} \left(\tau_{r_N}\tau_{r_n}\tau_{p_N}\tau_{p_n}\right)\left(j\kappa_N j\kappa_n\right)\rho_{l_n}\left(e^{j(N-1)\zeta}e^{j(n-1)\zeta}\right)\prod_{i=n}^{N-1}\tau_{c_i}e^{j\psi} + \\ \left(\tau_{r_{N-1}}\tau_{r_n}\tau_{p_{N-1}}\tau_{p_n}\right)\left(j\kappa_{N-1} j\kappa_n\right)\rho_{l_n}\left(e^{j(N-2)\zeta}e^{j(n-1)\zeta}\right)\prod_{i=n}^{N-2}\tau_{c_i}e^{j\psi} + \ldots \\ +\left(\tau_{r_{n+1}}\tau_{r_n}\tau_{p_n}\tau_{p_{n+1}}\right)\left(j\kappa_{n+1} j\kappa_n\right)\rho_{l_n}\left(e^{j(n)\zeta}e^{j(n-1)\zeta}\right)\prod_{i=n}^{n}\tau_{c_i}e^{j\psi} \end{bmatrix} \tag{6.31b}
$$

Taking $\left(\tau_{r_n}\tau_{p_n}\rho_{l_n} j\kappa_n e^{j(n-1)\zeta}\right)$ as the common factor, one gets

$$
\vec{E}^{r}_{1_n}(\theta,\phi) = \left[\tau_{r_n}\tau_{p_n}\rho_{l_n} j\kappa_n e^{j(n-1)\zeta} \left\{ \begin{array}{l} \left(\tau_{r_N}\tau_{p_N} j\kappa_N e^{j(N-1)\zeta}\right)\prod_{i=n}^{N-1}\tau_{c_i}e^{j\psi} \\ +\left(\tau_{r_{N-1}}\tau_{p_{N-1}} j\kappa_{N-1} e^{j(N-2)\zeta}\right)\prod_{i=n}^{N-2}\tau_{c_i}e^{j\psi} \\ +\ldots+\left(\tau_{r_{n+1}}\tau_{p_{n+1}} j\kappa_{n+1} e^{j(n)\zeta}\right)\prod_{i=n}^{n}\tau_{c_i}e^{j\psi} \end{array} \right\} \right] \tag{6.32a}
$$

$$
\vec{E}^{r}_{1_n}(\theta,\phi) = \left[\tau_{r_n}\tau_{p_n}\rho_{l_n} j\kappa_n e^{j(n-1)\zeta} \sum_{m=n+1}^{N} \left(\tau_{r_m}\tau_{p_m} j\kappa_m e^{j(m-1)\zeta} \prod_{i=n}^{m-1}\tau_{c_i}e^{j\psi} \right) \right] \tag{6.32b}
$$

The scattered field at the same element n may also be affected by the signals, which travel from the previous n–1 elements of the array towards it. The path of such a signal is shown in Figure 6.9. Following the path of the signals through each of the antenna elements, one gets

$$
\vec{E}^{r}_{2_n}(\theta,\phi) = \begin{bmatrix} \begin{pmatrix} e^{j0\alpha}\tau_{r_1}\tau_{p_1}e^{j0\alpha'}\tau_{c_1}\rho_{l_1} j\kappa_1 e^{j\psi}\tau_{c_2}e^{j\psi}\ldots \\ \tau_{c_{n-1}}e^{j\psi} j\kappa_n\tau_{p_n}e^{j(n-1)\alpha'}\tau_{r_n}e^{j(n-1)\alpha} \end{pmatrix} + \\ \begin{pmatrix} e^{j1\alpha}\tau_{r_2}\tau_{p_2}e^{j1\alpha'}\tau_{c_2}\rho_{l_2} j\kappa_2 e^{j\psi}\tau_{c_3}e^{j\psi}\ldots \\ \tau_{c_{n-1}}e^{j\psi} j\kappa_n\tau_{p_n}e^{j(n-1)\alpha'}\tau_{r_n}e^{j(n-1)\alpha} \end{pmatrix} + \\ \ldots + \begin{pmatrix} e^{j(n-2)\alpha}\tau_{r_{n-1}}\tau_{p_{n-1}}e^{j(n-2)\alpha'}\tau_{c_{n-1}}\rho_{l_{n-1}} \\ j\kappa_{n-1}e^{j\psi} j\kappa_n\tau_{p_n}e^{j(n-1)\alpha'}\tau_{r_n}e^{j(n-1)\alpha} \end{pmatrix} \end{bmatrix} \tag{6.33}
$$

Regrouping of like factors yields

$$
\vec{E}^{r}_{2_n}(\theta,\phi) = \begin{bmatrix} \left(\tau_{r_n}\tau_{r_1}\tau_{p_n}\tau_{p_1}\right)\left(j\kappa_1 j\kappa_n\right)\rho_{l_1}\left(e^{j0\zeta}e^{j(n-1)\zeta}\right)\left(\tau_{c_1}e^{j\psi}\tau_{c_2}e^{j\psi}\ldots\tau_{c_{n-1}}e^{j\psi}\right)+ \\ \left(\tau_{r_n}\tau_{r_2}\tau_{p_n}\tau_{p_2}\right)\left(j\kappa_2 j\kappa_n\right)\rho_{l_2}\left(e^{j1\zeta}e^{j(n-1)\zeta}\right)\left(\tau_{c_2}e^{j\psi}\ldots\tau_{c_{n-1}}e^{j\psi}\right)+ \\ \ldots+\left(\tau_{r_n}\tau_{r_{n-1}}\tau_{p_n}\tau_{p_{n-1}}\right)\left(j\kappa_{n-1} j\kappa_n\right)\rho_{l_{n-1}}\left(e^{j(n-2)\zeta}e^{j(n-1)\zeta}\right)\left(\tau_{c_{n-1}}e^{j\psi}\right) \end{bmatrix} \tag{6.34a}
$$

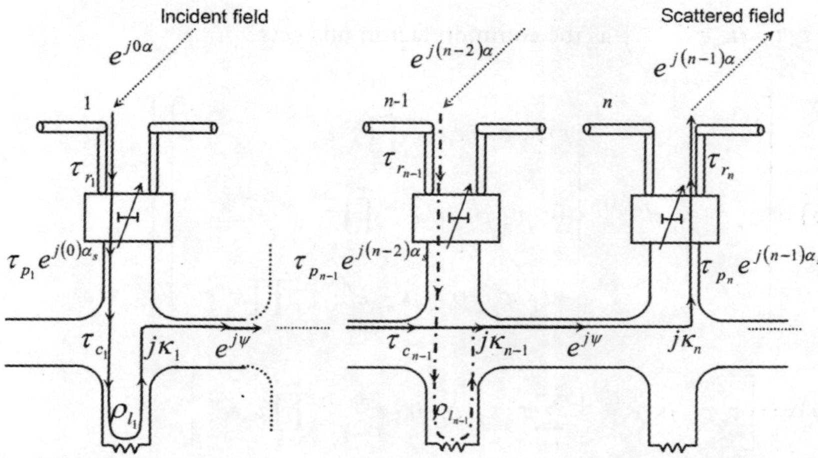

Figure 6.9 Path of the signal travelling towards the nth element from the previous $(n–1)$ elements

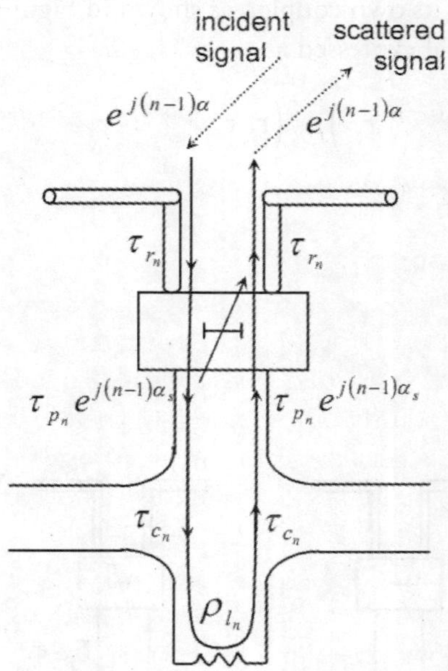

Figure 6.10 Path of the signal travelling towards the load terminating its own coupler

$$= \begin{bmatrix} \left(\tau_{r_n}\tau_{r_1}\tau_{p_n}\tau_{p_1}\right)\left(j\kappa_1 j\kappa_n\right)\rho_{l_1}\left(e^{j0\zeta}e^{j(n-1)\zeta}\right)\prod_{i=1}^{n-1}\tau_{c_i}e^{j\psi} + \\ \left(\tau_{r_n}\tau_{r_2}\tau_{p_n}\tau_{p_2}\right)\left(j\kappa_2 j\kappa_n\right)\rho_{l_2}\left(e^{j1\zeta}e^{j(n-1)\zeta}\right)\prod_{i=2}^{n-1}\tau_{c_i}e^{j\psi} + \\ \ldots + \left(\tau_{r_n}\tau_{r_{n-1}}\tau_{p_n}\tau_{p_{n-1}}\right)\left(j\kappa_{n-1} j\kappa_n\right)\rho_{l_{n-1}}\left(e^{j(n-2)\zeta}e^{j(n-1)\zeta}\right)\prod_{i=n-1}^{n-1}\tau_{c_i}e^{j\psi} \end{bmatrix}. \tag{6.34b}$$

Taking $\left(\tau_{r_n}\tau_{p_n} j\kappa_n e^{j(n-1)\zeta}\right)$ as the common factor, one gets

$$\vec{E}^r_{2_n}(\theta,\phi)=\left[\tau_{r_n}\tau_{p_n} jc_n e^{j(n-1)\zeta}\left\{\begin{array}{l}\tau_{r_1}\tau_{p_1}\rho_{l_1} j\kappa_1 e^{j0\zeta}\prod_{i=1}^{n-1}\tau_{c_i}e^{j\psi}\\ +\tau_{r_2}\tau_{p_2}\rho_{l_2} j\kappa_2 e^{j1\zeta}\prod_{i=2}^{n-1}\tau_{c_i}e^{j\psi}+\ldots\\ +\tau_{r_{n-1}}\tau_{p_{n-1}}\rho_{l_{n-1}} j\kappa_{n-1} e^{j(n-2)\zeta}\prod_{i=n-1}^{n-1}\tau_{c_i}e^{j\psi}\end{array}\right\}\right] \tag{6.35a}$$

$$\vec{E}^r_{2_n}(\theta,\phi)==\left[\tau_{r_n}\tau_{p_n} j\kappa_n e^{j(n-1)\zeta}\sum_{m=1}^{n-1}\tau_{r_m}\tau_{p_m}\rho_{l_m} j\kappa_m e^{j(m-1)\zeta}\prod_{i=m}^{n-1}\tau_{c_i}e^{j\psi}\right] \tag{6.35b}$$

The scattered field at the nth element can also be affected due to the signal, which moves towards the load, terminating its own coupler, as shown in Figure 6.10. This results in the self-scattering of the incident signal expressed as

$$\vec{E}^r_{3_n}(\theta,\phi)=\left(e^{j(n-1)\alpha}\tau_{r_n}\tau_{p_n}e^{j(n-1)\alpha_s}\tau_{c_n}\right)\rho_{l_n}\left(\tau_{c_n}\tau_{p_n}e^{j(n-1)\alpha_s}\tau_{r_n}e^{j(n-1)\alpha}\right) \tag{6.36a}$$

$$=\rho_{l_n}\tau^2_{r_n}\tau^2_{p_n}\tau^2_{c_n}e^{j2(n-1)\alpha}e^{j2(n-1)\alpha_s}$$

$$\vec{E}^r_{3_n}(\theta,\phi)=\rho_{l_n}\tau^2_{r_n}\tau^2_{p_n}\tau^2_{c_n}e^{j2(n-1)\zeta} \tag{6.36b}$$

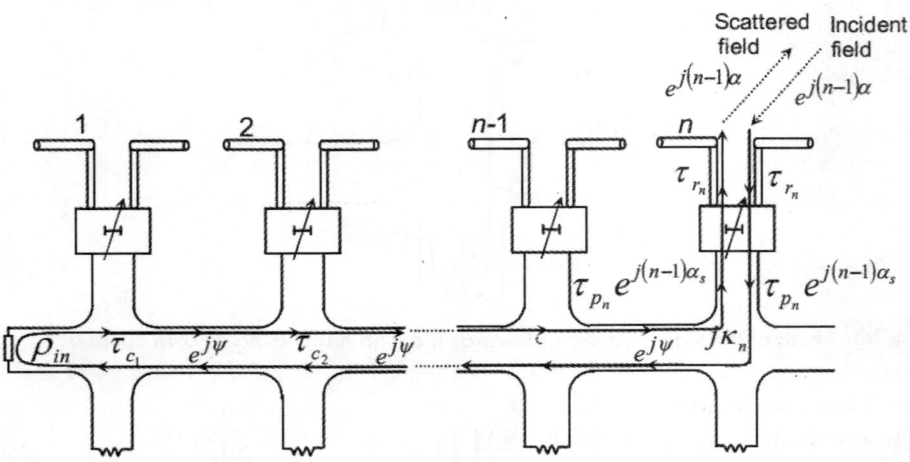

Figure 6.11 Path of the signal travelling towards the load terminating the input port

The signal, which succeeds in overcoming all the above-mentioned scattering sources, moves towards the receiving port (Figure 6.11). However, if there exists a mismatch between the receiving port and the input port of the coupler, significant scattering can occur even at this point. The reflection coefficient at this input port can be expressed as

$$\rho_{in} = \frac{Z_{12} - Z_o}{Z_{12} + Z_o} \tag{6.37}$$

where Z_{12} is the impedance at the input port of the first coupler in the array and Z_o is the characteristic impedance of the coaxial cable connected to the input port of the first coupler. The scattered field at the junction of receiving port and input port of coupler is expressed as

$$\vec{E}^r_{4_n}(\theta,\phi) = \begin{bmatrix} \left(e^{j(n-1)\alpha}\tau_{r_n}\tau_{p_n}e^{j(n-1)\alpha_s} j\kappa_n e^{j\psi}\tau_{c_{n-1}} \ldots e^{j\psi}\tau_{c_2}e^{j\psi}\tau_{c_1}\right)\rho_{in} \\ \left(\tau_{c_1}e^{j\psi}\tau_{c_2}e^{j\psi}\ldots\tau_{c_{n-1}}e^{j\psi} j\kappa_n\tau_{p_n}e^{j(n-1)\alpha_s}\tau_{r_n}e^{j(n-1)\alpha}\right) \end{bmatrix} \tag{6.38}$$

Rearrangement of like factors yields

$$\vec{E}^r_{4_n}(\theta,\phi) = \left[\tau^2_{r_n}\tau^2_{p_n}e^{j2(n-1)\alpha}e^{j2(n-1)\alpha_s}(j\kappa_n)^2\rho_{in}\left(\tau_{c_1}e^{j\psi}\tau_{c_2}e^{j\psi}\ldots\tau_{c_{n-1}}e^{j\psi}\right)^2\right] \tag{6.39a}$$

$$\vec{E}^r_{4_n}(\theta,\phi) = \left[\rho_{in}\tau^2_{r_n}\tau^2_{p_n}(j\kappa_n)^2 e^{j2(n-1)\zeta}\left(\prod_{i=1}^{n-1}\tau_{c_i}e^{j\psi}\right)^2\right] \tag{6.39b}$$

The total field due to the signal scattering beyond the coupling port of couplers is given by

$$\vec{E}^r_{s_n}(\theta,\phi) = \vec{E}^r_{1_n}(\theta,\phi) + \vec{E}^r_{2_n}(\theta,\phi) + \vec{E}^r_{3_n}(\theta,\phi) + \vec{E}^r_{4_n}(\theta,\phi) \tag{6.40}$$

The corresponding RCS component is expressed as

$$\sigma_s(\theta,\phi) = F\sum_{n=1}^{N}\vec{E}^r_{s_n}(\theta,\phi) \tag{6.41}$$

Substitution of (6.21), (6.24), (6.27) and (6.41) into (6.13) gives the total RCS for the phased array due to the mismatches of the feed network. Thus, the normalised RCS of phased array is expressed as

$$\sigma(\theta,\phi) = \frac{4\pi}{\lambda^2}\left\{\left|\sigma_r(\theta,\phi)\right|^2 + \left|\sigma_p(\theta,\phi)\right|^2 + \left|\sigma_c(\theta,\phi)\right|^2 + \left|\sigma_s(\theta,\phi)\right|^2\right\} \tag{6.42}$$

The RCS pattern of a linear series-fed phased array of infinitesimal dipoles ($l << \lambda$) has been analysed neglecting Mutual Coupling and edge effects. It must be borne in mind that although these assumptions simplify the RCS estimation of the array, they are unacceptable in a practical scenario. Figure 6.12 shows the RCS pattern of 50-element series-fed array with and without inclusion of impedance in the coupling coefficients formulation. The parameters are chosen as $\theta_s = 0°$, $\psi = \pi/2$, $d = 0.4\lambda$, $l = 0.5\lambda$ and $\rho_r = \rho_p = \rho_c = \rho_l = 0.2$ with uniform unit amplitude distribution. It is evident that both the RCS patterns are in excellent match. This derived formulation is generalised and accounts for both with and without mutual coupling case.

The effect of effective height of the dipole antenna element on the RCS of phased array is analysed next. Although an antenna is specified using its physical length (height), its effective aperture differs from the physical one. Thus, an accurate computation of the antenna RCS is expected to include the effective height rather than physical height of antenna element. The effective height of the antenna depends on the nature of the current distribution over its surface.

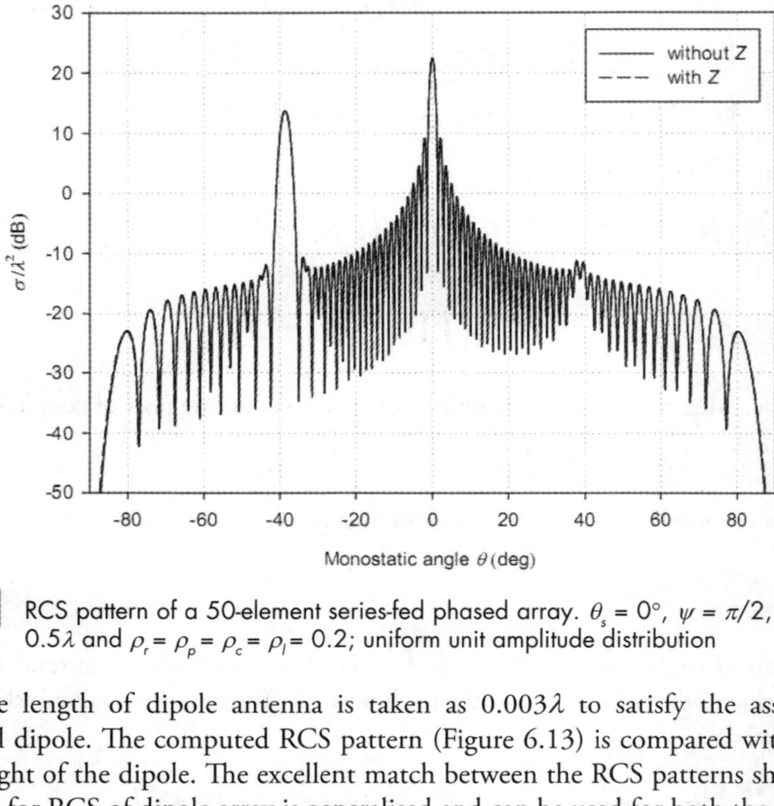

Figure 6.12 RCS pattern of a 50-element series-fed phased array. $\theta_s = 0°$, $\psi = \pi/2$, $d = 0.4\lambda$, $l = 0.5\lambda$ and $\rho_r = \rho_p = \rho_c = \rho_l = 0.2$; uniform unit amplitude distribution

Here, the length of dipole antenna is taken as 0.003λ to satisfy the assumption of infinitesimal dipole. The computed RCS pattern (Figure 6.13) is compared with that using physical height of the dipole. The excellent match between the RCS patterns shows that the formulation for RCS of dipole array is generalised and can be used for both the infinitesimal and finite dipoles.

The RCS of phased array is determined next using the antenna impedance instead of the antenna resistance. This is done to include the finite radius of the dipole antenna element in the expression of antenna impedance and, hence, the RCS of the phased array. It is necessary to choose the radius and length of the dipole element appropriately. Usually, the dipole is taken to be a thin wire. Here, the radius of the wire is fixed as $10^{-5}\lambda$. The antenna is assumed to be resonant and the antenna reactance is expected to vanish for a dipole of resonant length. Hence, the length of dipole is taken as 0.488λ.

This combination of length and radius of dipole antenna yields a radiation reactance of –7.566 × 10^{-5} Ω, which is almost equal to zero. The resultant RCS pattern is shown in Figure 6.14a.

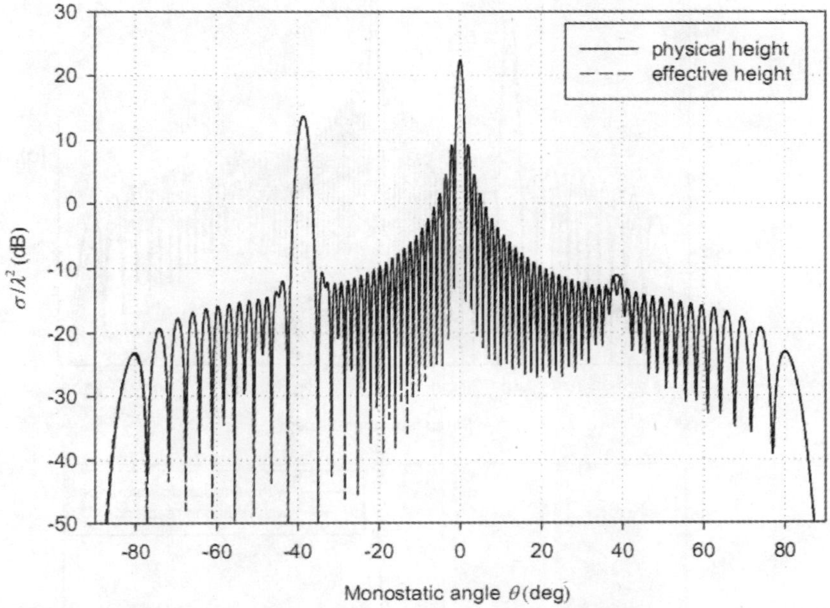

Figure 6.13 RCS of series-fed linear dipole array. $N = 50$, $\theta_s = 0°$, $\psi = \pi/2$, $d = 0.4\lambda$ and $l = 0.003\lambda$ and $\rho_r = \rho_p = \rho_c = \rho_l = 0.2$; uniform unit amplitude distribution

The RCS results are compared with and without using radiation reactance, keeping all other parameters same. It may be seen that the level of RCS pattern is significantly different for this choice of dipole length and inter-element spacing. The difference in the RCS patterns can be reduced for smaller inter-element spacing, keeping dipole length constant (Figure 6.14b). This observation leads to the conclusion that RCS of phased array can be controlled by multi-variant optimisation (Zhang et al. 2010, 2011). The parameters that can be optimised are the inter-element spacing, dipole length, or geometry.

In Figures 6.12 through 6.14, the electric field scattered due to the termination of the receive port of the array is taken as (Jenn and Lee 1995)

$$\vec{E}^r_{4_n}(\theta,\phi) = \rho_l\ \tau_r^2\tau_p^2 \left\{ \sum_{n=1}^{N} j\kappa_n e^{j(n-1)\zeta} \prod_{i=1}^{n-1} \tau_{c_i} e^{j\psi} \right\}^2 \tag{6.43}$$

However, if the scattered electric field is derived by tracing the signal path, one gets

$$\vec{E}^r_{4_n}(\theta,\phi) = \sum_{n=1}^{N} \left[\rho_{in}\ \tau_{r_n}^2 \tau_{p_n}^2 \left(j\kappa_n\right)^2 e^{j2(n-1)\zeta} \left(\prod_{i=1}^{n-1} \tau_{c_i} e^{j\psi} \right)^2 \right] \tag{6.44}$$

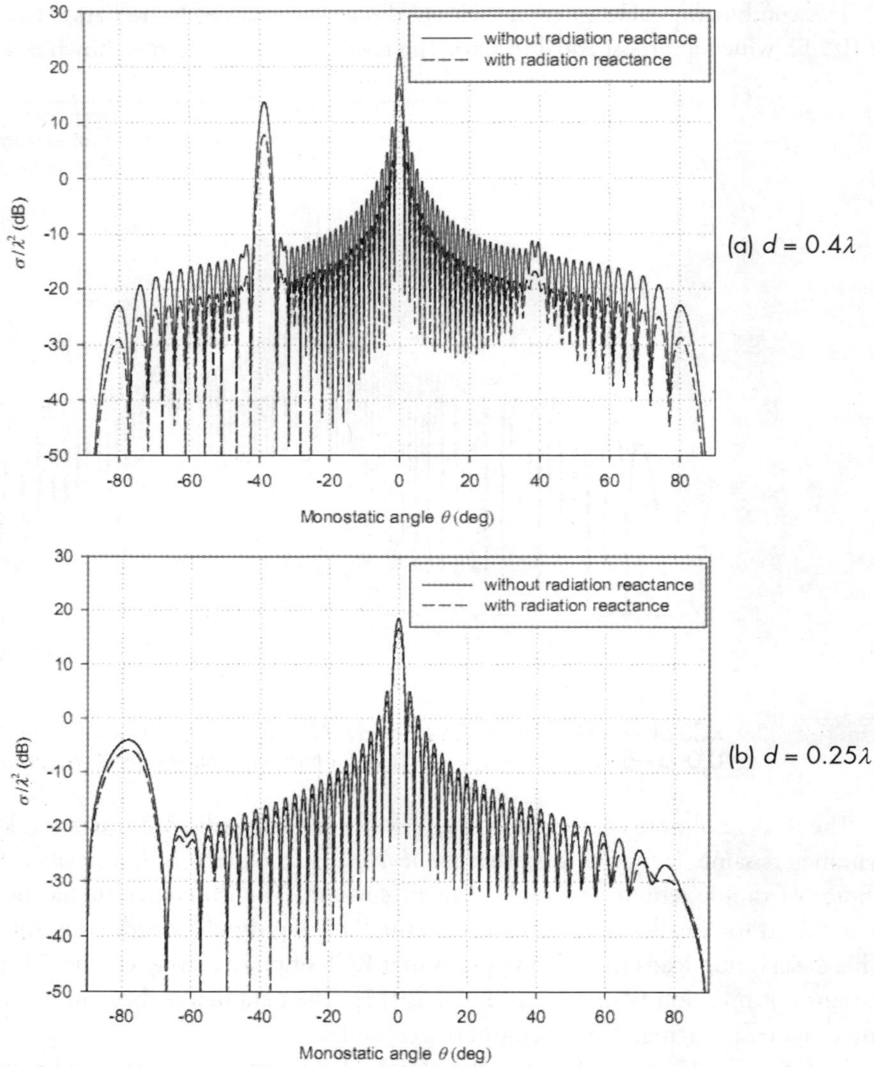

Figure 6.14 Effect of spacing between antenna elements on RCS of 50-element linear dipole array with series-fed network. $\theta_s = 0°$, $\psi = \pi/2$, $l = 0.488\lambda$ and $\rho_r = \rho_p = \rho_c = \rho_l = 0.2$; uniform unit amplitude distribution

The RCS pattern obtained using (6.44), i.e. through tracing the signal-path is shown in Figure 6.15. It may be observed that the load reflection lobe is absent in the RCS pattern as compared to the RCS pattern obtained using (6.43). Lu et al. (2009) reported the RCS pattern for a series-fed 60 × 60 planar dipole array by considering the effect of feed network. Here, the radiation resistance of dipole antenna is taken as $R_a = 24.7(kl)^{2.4}$. On the basis of Lu et al. formulation (2009), the RCS pattern for a 30-element series-fed linear dipole array is shown in Figure 6.16. It may be observed that the RCS patterns in Figures 6.15 and 6.16 are similar.

Next, Mutual Coupling effect is included in the estimation of RCS of dipole array in side-by-side configuration. Figure 6.17 presents the RCS pattern of 20-element dipole array with and without mutual coupling. Other parameters considered are $\theta_s = 0°$, $\psi = \pi/2$, $d = 0.1\lambda$, $l = 0.5\lambda$, $a = 10^{-5}\lambda$, $Z_o = 150\ \Omega$ and $Z_l = 225\ \Omega$. The amplitude distribution is taken as uniform distribution with unit amplitude. It is readily apparent that the mutual coupling changes the RCS of the dipole array significantly.

Figure 6.18 presents the variation in RCS pattern of dipole array (in collinear configuration) due to beam scanning. The beam scanning includes 0°, 45° and 85°. The other parameters considered are $N = 30$, $Z_0 = 75\ \Omega$ and $Z_l = 150\ \Omega$. It can be seen that the RCS pattern of collinear dipole array with and without mutual coupling are almost identical. However, the difference between the RCS patterns is noticeable with the increase in scan angle θ_s from 0° to 45° or 85°. This is in accordance with the reported inference (Lee and Chu 1988) that the mutual coupling affects both the array and RCS pattern significantly for large scan angles. The variation of the RCS pattern for different scan angles (0°, 45° and 85°) in case of a parallel-in-echelon and a side-by-side dipole array is shown in Figures 6.19 and 6.20, respectively. The parameters are taken to be same as in Figure 6.18 except for Z_o and Z_l. The values of Z_o and Z_l are chosen to be 125 Ω and 235 Ω in case of parallel-in-echelon and 150 Ω and 280 Ω for side-by-side configuration, respectively. It may be observed that the trend observed due to the the variation of scan angle is same for all three configurations. However, the difference in the RCS pattern with and without mutual coupling is least in the case of collinear configuration of dipole array.

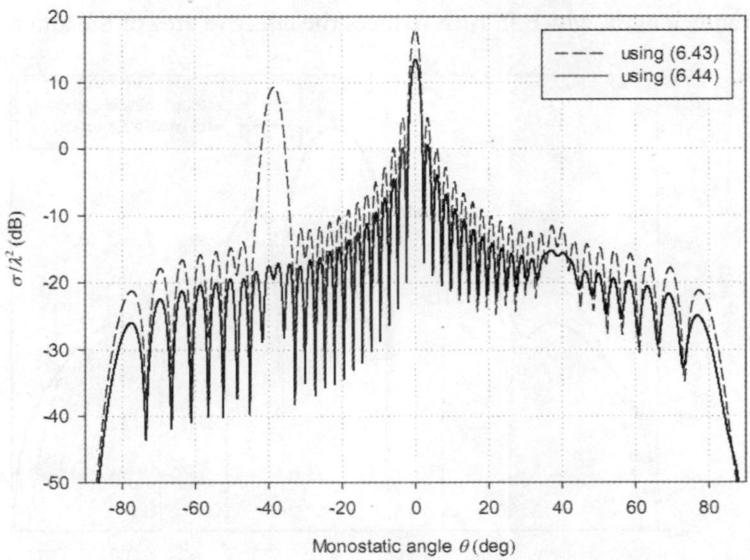

Figure 6.15 RCS of series-fed linear dipole array. $N = 30$, $\theta_s = 0°$, $d = 0.4\lambda$, $l = 0.5\lambda$ and $\rho = 0.2$

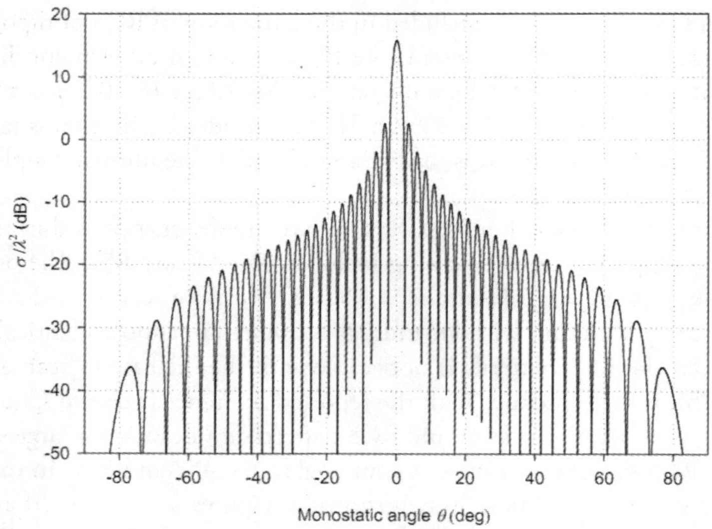

Figure 6.16 RCS of series-fed 30-element linear dipole array. $\theta_s = 0°$, $d = 0.4\lambda$, $l = 0.5\lambda$ and $\rho = 0.2$

The estimation of RCS can provide a useful insight in reducing the scattered field. Variation of certain design parameters of the feed network can reduce the RCS considerably. One such parameter is the number of antenna elements in the phased array. However, if the number of elements is less, the directivity of the array gets affected adversely. The other feasibility is to reduce the antenna length, which in turn reduces the effective area of the antenna array.

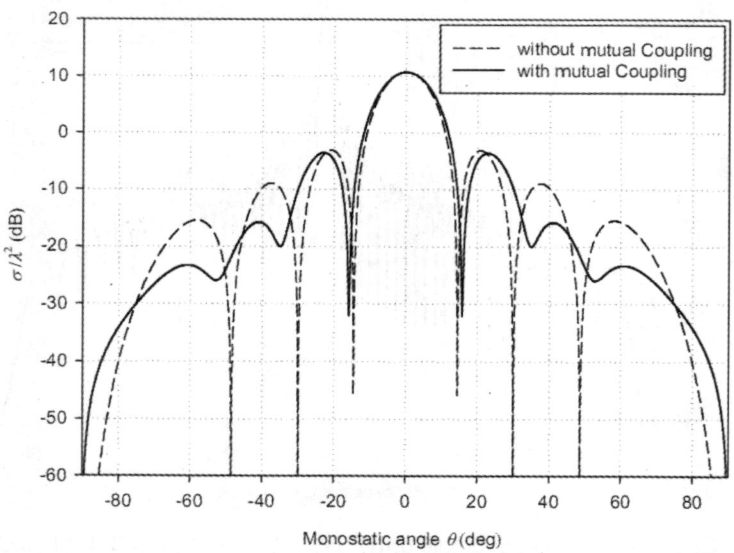

Figure 6.17 RCS of 20-element series-fed linear dipole array in side-by-side configuration

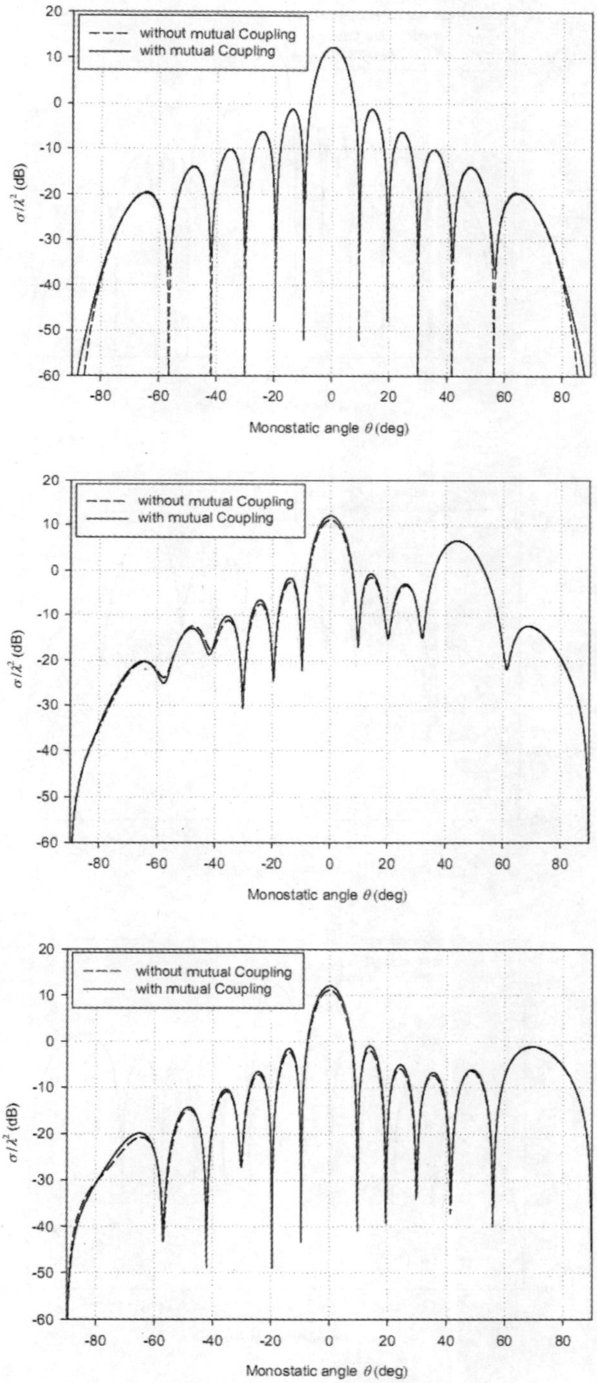

(a) $\theta_s = 0°$

(b) $\theta_s = 45°$

(c) $\theta_s = 85°$

Figure 6.18 RCS of series-fed linear collinear dipole array of $N = 30$, $\psi = \pi/2$, $d = 0.1\lambda$, $l = 0.5\lambda$, $a = 10^{-5}\lambda$, Z_0=75 Ω and Z_l=150 Ω; unit amplitude uniform distribution

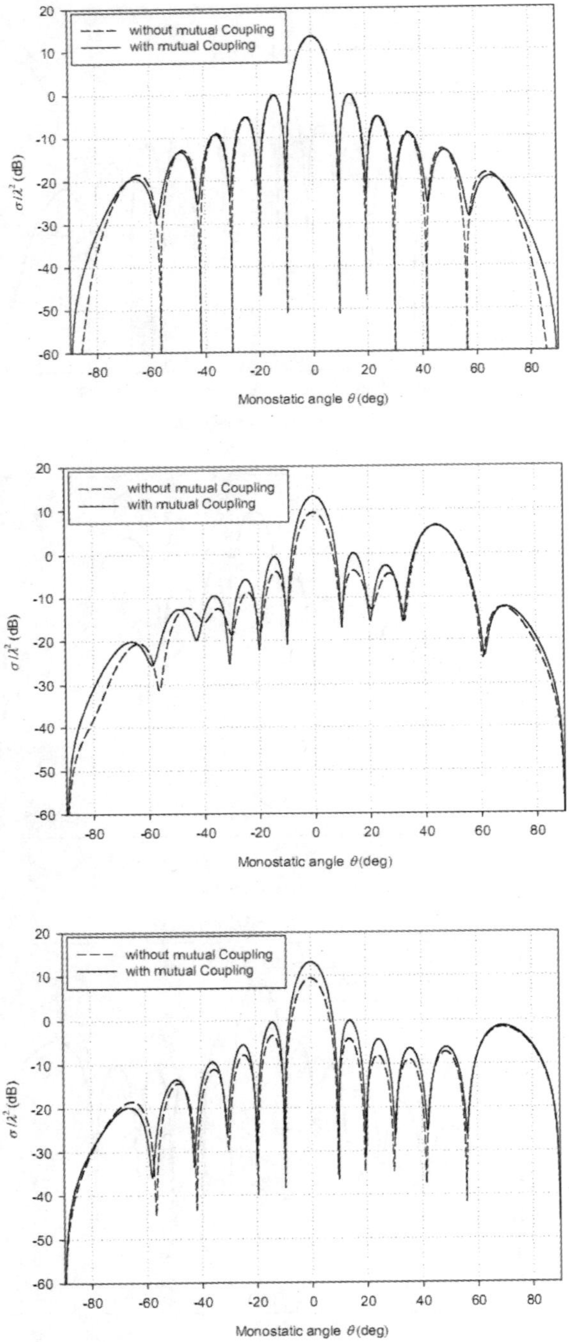

(a) $\theta_s = 0°$

(b) $\theta_s = 45°$

(c) $\theta_s = 85°$

Figure 6.19 RCS of series-fed linear parallel-in-echelon dipole array of $N = 30$, $\psi = \pi/2$, $d = 0.1\lambda$, $l = 0.5\lambda$, $a = 10^{-5}\lambda$, $Z_0 = 125\ \Omega$ and $Z_l = 235\ \Omega$; unit amplitude uniform distribution

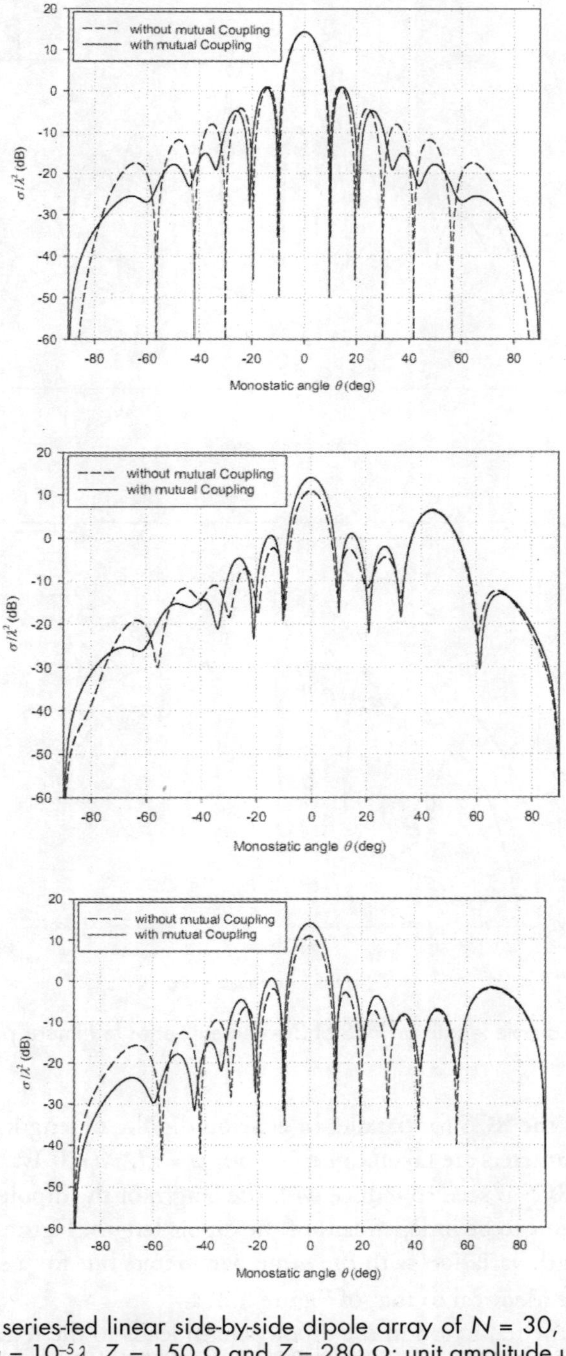

(a) $\theta_s = 0°$

(b) $\theta_s = 45°$

(c) $\theta_s = 85°$

Figure 6.20 RCS of series-fed linear side-by-side dipole array of $N = 30$, $\psi = \pi/2$, $d = 0.1\lambda$, $l = 0.5\lambda$, $a = 10^{-5}\lambda$, $Z_o = 150\ \Omega$ and $Z_l = 280\ \Omega$; unit amplitude uniform distribution

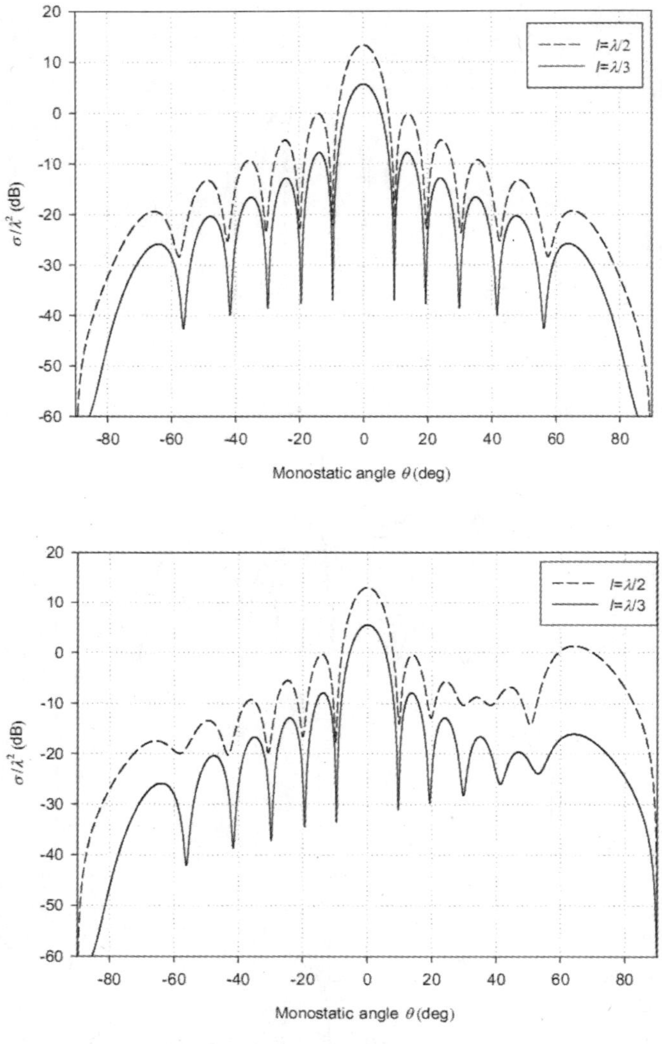

(a) $\theta_s = 0°$

(b) $\theta_s = 70°$

Figure 6.21 Effect of dipole length on RCS of 30-element series-fed linear parallel-in-echelon dipole array

Figure 6.21a shows the RCS for parallel-in-echelon dipoles of length $\lambda/2$ and $\lambda/3$ at a scan angle of 0°. Other parameters are taken to be $N = 30$, $\psi = \pi/2$, $d = 0.1\lambda$, $a = 10^{-5}\lambda$, $Z_o = 125\ \Omega$ and $Z_l = 235\ \Omega$. The RCS is seen to reduce with the length of the dipoles. However, the trend of the RCS pattern seems to be independent of the dipole length. Figure 6.21b shows the RCS pattern for dipole length variation with the same parameters but for a scan angle of 70°. The observations found are identical to that of Figure 6.21a.

This indicates that the reduction in the dipole length leads to the reduction in RCS for any scan angle. However, the variation in dipole length will affect the resonant frequency of the array. Therefore, reducing the RCS of a phased array just by reducing the dipole length is not recommended. Other factors such as the load termination or the antenna aperture distribution should also be chosen suitably.

The dependence of the array RCS on the amplitude distribution of the current feed exciting the couplers is studied next. The variations in the current at the antenna terminals affect the power distribution and, hence, vary the coupling coefficients of the couplers. Figure 6.22 shows the RCS pattern of dipole array (side-by-side configuration) for various amplitude distributions, *viz.* uniform, cosine squared on a pedestal, Dolph–Chebyshev and Taylor. The parameters chosen are of $N = 30$, $\theta_s = 0°$, $\psi = \pi/2$, $d = 0.1\lambda$, $l = 0.5\lambda$, $a = 10^{-5}\lambda$ for $Z_0 = 150\ \Omega$ and $Z_l = 280\ \Omega$. The mutual coupling factor is considered.

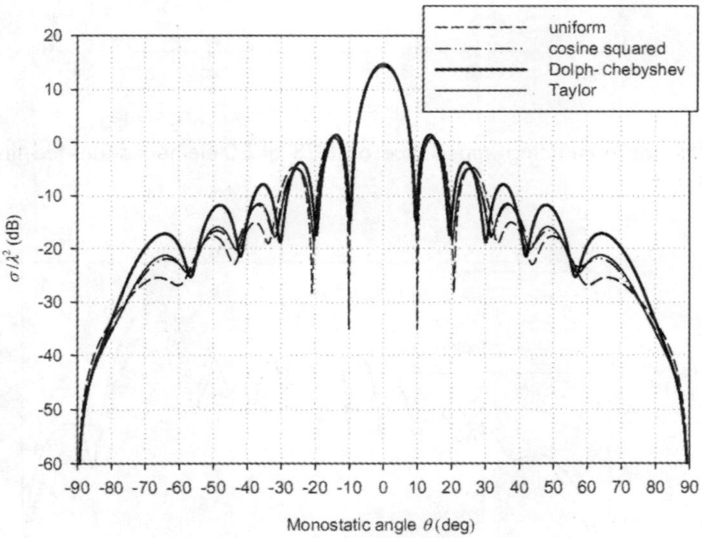

Figure 6.22 Effect of amplitude distribution on RCS of 30-element series-fed dipole array

It may be seen that the specular lobe has the same level irrespective of the amplitude distribution. However, the levels of the sidelobes vary considerably and are found to be maximum for Dolph–Chebyshev distribution. The termination of the isolated port of couplers varies the reflection coefficient at their respective terminal. This varies the amount of scattered field and, hence, the RCS of phased array. Figure 6.23 shows the RCS pattern for terminating impedances of $Z_l = 0\ \Omega$, $50\ \Omega$ and $280\ \Omega$. Other parameters are kept to be the same as in the previous case (Figure 6.22). It can be observed that the RCS pattern for the termination of 280 Ω is lower than that for the short circuit or for a termination of 50 Ω. However, the trend of the RCS pattern remains almost independent of the terminating load value. Figure 6.24 shows a similar comparison of the short and impedance terminated feed network for a scan angle of 50°. A similar trend is observed. This indicates that the termination of the coupler port with a suitable load will reduce the RCS irrespective of the scan angle.

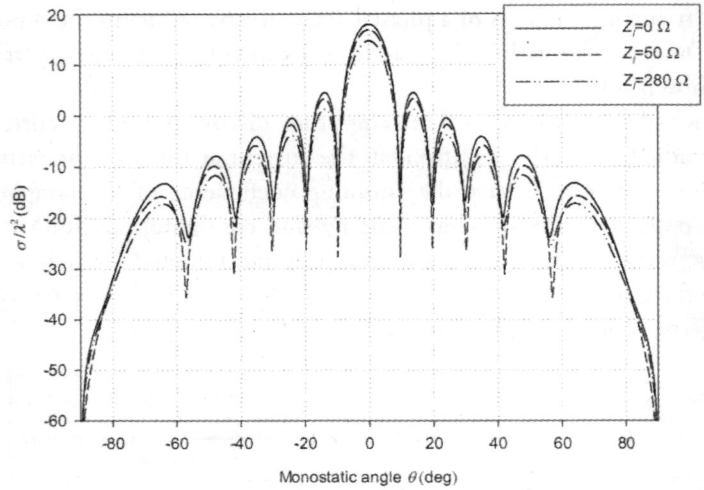

Figure 6.23 Effect of terminating impedance on RCS of 30-element series-fed linear dipole array at $\theta_s = 0°$

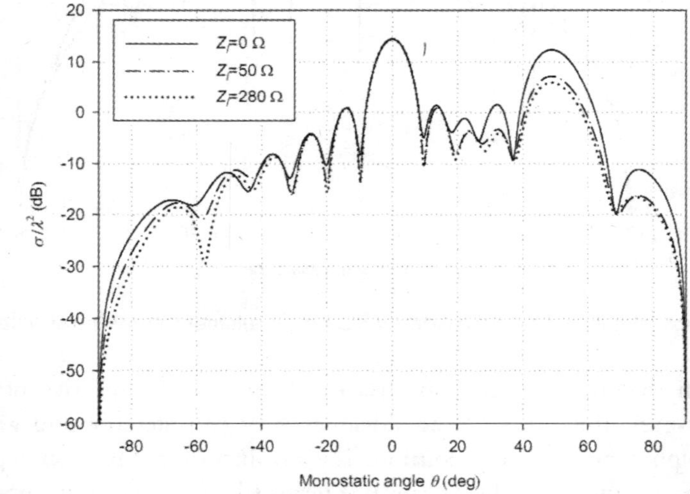

Figure 6.24 Effect of terminating impedance on RCS of 30-element series-fed linear array at $\theta_s = 50°$

6.5 Conclusion

The estimation of RCS of series-fed linear dipole array in the presence of mutual coupling is discussed. A step-by-step approach is followed to arrive at the equation of RCS for a phased array with the inclusion of Mutual Coupling factor. The formulation of the coupling coefficients is first derived in terms of impedance matrix, followed by the inclusion of (i) length of the dipole,

(ii) effective length of the dipole, and (iii) antenna impedance/radius of the dipole wire into the RCS equation. The summation term of the scattered field due to the terminating load is refined. Further, a detailed formulation is presented for the scattered field at each stage of the feed network. The simulation results are presented for the side-by-side, collinear and parallel-in-echelon configurations of linear dipole array. The RCS pattern obtained in the presence of mutual coupling is compared with the case where there is no mutual coupling.

It is shown that the mutual coupling significantly alters the impedance at the terminals of the radiating elements. The observed changes depend upon the configuration of dipole array considered, *viz.* side-by-side, collinear or parallel-in-echelon configuration. The results indicate that the mutual coupling affects the RCS pattern significantly for large scan angles. It is inferred that the RCS can be reduced by varying certain feed network parameters. The RCS of the dipole array can be reduced by (i) reducing the length of the dipole, (ii) terminating the isolated port of the coupler with a suitable load or (iii) choosing a suitable amplitude distribution.

References

Aumann, H. M., A. J. Fenn, and F. G. Willwerth. 1989. 'Phased array antenna calibration and pattern prediction using mutual coupling measurements.' *IEEE Transactions on Antennas and Propagation* 37: 844–50.

Batchelor, J. C., L. Economou, and R. J. Langley. 2000. 'Scanned microstrip arrays using simple integrated ferrite phase shifters.' *IEE Proceedings of Microwaves, Antennas and Propagation* 147(3): 237–41.

Balanis, C. A. 2005. *Antenna Theory, Analysis and Design.* New Jersey: John Wiley & Sons, 1117.

Chu, R. S. 1991. 'Analysis of an infinite phased array of dipole elements with RAM coating on ground plane and covered with layered radome.' *IEEE Transactions on Antennas and Propagation* 39: 164–76.

Dikmen, F., A. A. Ergin, B. Terzi, and L. Sevgili. 2010. 'Implementation of an efficient shooting and bouncing rays scheme.' *Microwave and Optical Technology Letters* 52: 2409–13.

Fan, G. -X. and J. -M. Jin. 1997. 'Scattering from a cylindrically conformal slotted waveguide array antenna.' *IEEE Transactions on Antennas and Propagation* 45: 1150–59.

Gupta, I. J. and A. A. Ksienski. 1983. 'Effect of mutual coupling on the performance of adaptive arrays.' *IEEE Transactions on Antennas and Propagation* 31: 785–91.

Hui, H. T. 2007. 'Decoupling methods for the mutual coupling effect in antenna arrays: A review.' *Recent Patents Engineering* 1(2): 187–93.

Hui, H. T. 2004. 'A practical approach to compensate for the mutual coupling effect of an adaptive dipole array.' *IEEE Transactions on Antennas and Propagation* 52: 1262–69.

Jenn, D. C. and S. Lee 1995. 'In-band scattering from arrays with series feed networks.' *IEEE Transactions on Antennas and Propagation* 43: 867–73.

Jenn, D. C. 1995. *Radar and Laser Cross Section Engineering.* Washington, DC: AIAA Education Series, 476.

Knott, E. F., J. F. Shaeffer, and M. T. Tuley. 1985. *Radar Cross Section.* Dedham, MA: Artech House Inc, 462.

Lee, K. C. and T. H. Chu. 2005. 'Mutual coupling mechanisms within arrays of nonlinear antennas.' *IEEE Transactions on Electromagnetic Compatibility* 47: 963–70.

Lee, K. M. and R. S. Chu. 1988. 'Analysis of mutual coupling between a finite phased array of dipoles and its feed network.' *IEEE Transactions on Antennas and Propagation* 36: 1681–99.

Li, W. -R., C. -Y., Chu, and S. -F. Chang. 2004. 'Switched-beam antenna based on modified Butler matrix with low sidelobe level.' *Electronics Letters* 40(5): 290–92.

Liao, S. Y. 1990. *Microwave Devices and Circuits.* Englewood Cliffs, NJ: Prentice Hall, 542.

Lu, B., S. X. Gong, S. Zhang, and J. Ling. 2009. 'A new method for determining the scattering of linear polarised element arrays.' *Progress in Electromagnetics Research M* 7: 87–96.

Ma, W., M. R. Rayner, and C. G. Parini. 2005. 'Discrete Green's function formulation of the FDTD method and its application in antenna modeling.' *IEEE Transactions on Antennas and Propagation* 53: 339–46.

Pozar, D. M. 1992. 'RCS of microstrip antenna on normally biased ferrite substrate.' *IEEE Microwave and Guided Wave Letters* 2(5): 1079–80.

Tao, Y., H. Lin, and H. Bao. 2010. 'GPU-based shooting and bouncing ray method for fast RCS prediction.' *IEEE Transactions on Antennas and Propagation* 58: 494–502.

Volakis, J. L. 2007. *Antenna Engineering Handbook.* Fourth edition. New York, USA: McGraw Hill, 1773.

Wang, W. T., S. X. Gong, Y. J. Zhang, F. T. Zha, and J. Ling. 2009. 'Low RCS dipole array synthesis based on MoM-PSO hybrid algorithm.' *Progress in Electromagnetics Research* 94: 119–32.

Wang, W. T., Y. Liu, S. X. Gong, Y. J. Zhang, and X. Wang. 2010. 'Calculation of antenna mode scattering based on method of moments.' *PIERS Letters* 15: 117–26.

Yu, Y. and H. T. Hui. 2011. 'Design of a mutual coupling compensation network for a small receiving monopole array.' *IEEE Transactions on Microwave Theory and Techniques* 59: 2241–45.

Zdunek, A. and W. Rachowicz. 2008. 'Cavity radar cross section prediction.' *IEEE Transactions on Antennas and Propagation* 56: 1752–62.

Zhang, S., S. X. Gong, Y. Guan, and B. Lu. 2011. 'Optimised element positions for prescribed radar cross section pattern of linear dipole arrays.' *International Journal of RF and Microwave Computer-Aided Engineering* 21(6): 622–28.

Zhang, S., S. X. Gong, Y. Guan, J. Ling, and B. Lu. 2010. 'A new approach for synthesising both the radiation and scattering patterns of linear dipole antenna array.' *Journal of Electromagnetic Waves and Application* 24(7): 861–70.

7

Performance of Sidelobe Cancellers in Active RCSR

7.1 Introduction

Optimum array processors find applications in diverse areas as versatile sensors. They are used in radar and communication systems that are subject to various types of interference and jamming. The sidelobe cancellers mainly perform the function of separating an unwanted signal from the desired signal to extract the relevant information from the signal while simultaneously reducing the noise power level. Various schemes of cancellers being proposed and implemented have better interference suppression capabilities and faster convergence (Li and Stoica 2005). This interference suppression capability of sidelobe cancellers can be used towards active radar cross section reduction (RCSR). Moreover, the application of an efficient adaptive algorithm like least mean square (LMS) in sidelobe cancellers can reduce the implementation complexity. In other words, sidelobe cancellers implementing an appropriate algorithm for weight optimisation can effectively suppress correlated and uncorrelated interfering signals. These interfering signals can be narrowband sources with single frequency, i.e. zero bandwidth or broadband sources distributed over a spectral band with some finite bandwidth.

Generalised sidelobe canceller (GSC) is especially useful when the information regarding the direction of arrival (DOA) of the interfering signal is not available *a priori*. The GSC is quite effective in reducing the computational burden when used with proper weight adaptation algorithms. The GSC employs linearly constrained minimum variance (LCMV) criterion to suppress the hostile sources incident from different directions but at the same time retains high output signal-to-interference noise ratio (SINR) towards the sources looking from desired directions. However, the performance of the GSC degrades considerably in case of discrepancy between the calculated and actual DOA (Figure 7.1). Moreover, the input signal of GSC is included in the stochastic gradient. Thus, the gradient becomes big and requires small step-size for weight adaptation for achieving converged solution.

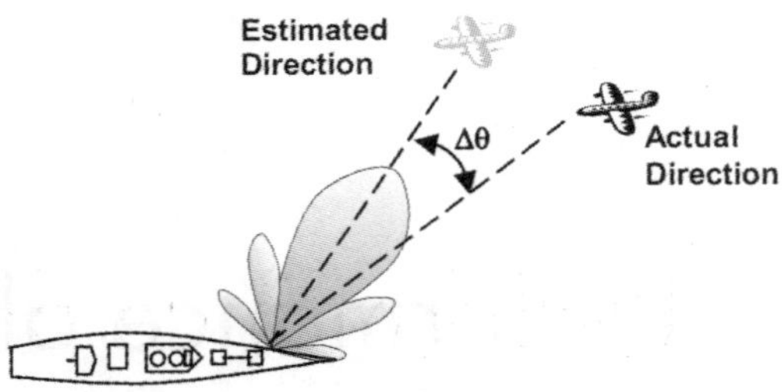

Figure 7.1 Direction of arrival (DOA) mismatch in adaptive arrays

This reduces the speed of convergence and, hence, the response of array towards the signal environment. To provide sufficient robustness to the sidelobe cancellers, an additional block consisting of filter and blind equaliser is included, which is called the decision feedback-generalised sidelobe canceller, i.e. DF–GSC. The blind equaliser of DF–GSC equalises the channel and, hence, the mismatch in DOA. The feedback filter helps in the cancellation of desired signal component, if any, in the error signal, thereby avoiding the self-cancellation problem. The inclusion of blind equaliser and decision feedback filter in the conventional GSC design facilitates the enhancement in robustness towards DOA mismatch errors, faster rate of convergence and higher output SINR.

There are different categories of constraints that may be incorporated in sidelobe cancellers for the enhancement of robustness. Point constraints are the minimum constraint, which upon addition sets the constraint matrix to a scalar. Directional constraints (Steele 1983) are very effective in suppressing interferences, whereas derivative constraints (Buckley and Griffiths 1986) provide the highest SINR.

In this chapter, the weight adaptation in GSC, DF–GSC, DF–GSC with blind equalisation is discussed. The performance of these sidelobe cancellers along with different LMS algorithms is analysed in terms of convergence rate and accurate placement of deep nulls towards the impinging angles of probing sources in the adapted pattern. Various signal scenarios comprising multiple desired and probing sources are also considered. The objective of analysis is to exploit the efficient interference suppression towards active RCSR.

7.2 Generalised sidelobe canceller (GSC)

The conventional schematic of GSC comprises an unconstrained weight w_{uc} and a signal cancelling branch that consists of the blocking matrix B and a constrained weight w_c (Figure 7.2). An important key element in the GSC model is the blocking matrix B. The transpose of this matrix is required to have N–1 linearly independent rows each of which sums up to zero. Numerous matrices can be generated that possess this property.

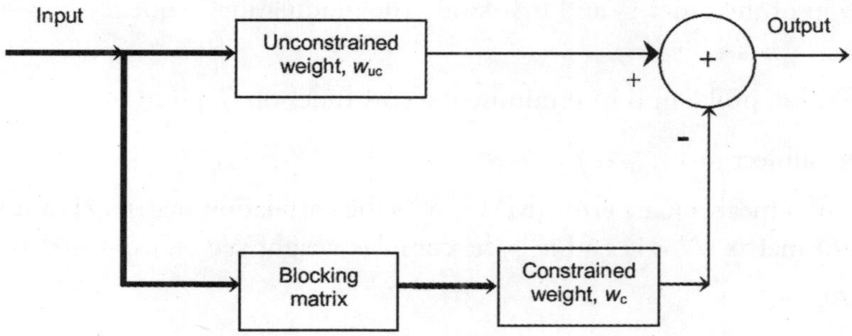

Figure 7.2 Generalised sidelobe canceller (GSC)

Here, the B matrix is generated by making use of Hadamard-ordered Walsh function (Harmuth 1969) given by

$$Had(i,j) = (-1)^{\sum_{n=0}^{3} b_n(i) b_n(j)} \tag{7.1}$$

where $b_o(n)$ represents the transmitted symbol.

The *Hadamard* matrix has Hermitian structure. Each column here is linearly independent. If a point constraint is included, one column gets deleted from the matrix, generating a blocking matrix of 16 × 15 dimension. The sum of each column elements of this matrix is zero. The Hadamard matrix may be calculated recursively as

$$Had_{n+1} = \begin{bmatrix} Had_n & Had_n \\ Had_n & -Had_n \end{bmatrix} \tag{7.2}$$

When the input signal travels through the blocking matrix, the desired signal component gets completely nullified. The constraint used is that the desired signal should have distortionless response. This is a zeroth-order point constraint and is the only constraint used in this chapter for the sake of convenience. The constrained weight is made to converge to a steady-state value using LMS adaptive algorithm (Godara 2004). Here, the linearly constrained minimum power (LCMP) norm is followed (Widrow and Stearns 1985). For sake of convenience, a quadrature phase-shift keying (QPSK)-based modulation is taken into consideration for flat-fading channel environments. A channel with flat fading characteristics is assumed, i.e. $s_o(n) = h(n).b_o(n)$, where $h(n)$ is the channel coefficient. Noise is modelled as white Gaussian noise. For simulations, random integers in the range [0,5] are generated with equal probability. These integers are the representative of the unmodulated discrete symbols. The modulated sequence $b_o(n)$ is obtained using $b_o(n) = \sqrt{2}\, e^{j(\frac{\pi}{4}+\frac{\pi}{2}m)}$, where m is the random integer generated at the nth snapshot. A signal transmission channel is subject to a lot of attenuation that can be modelled using fading coefficients (Anderson and Svensson 2003). Rayleigh flat fading channel coefficients are given by the formula $h(n) = J_{1/4}(2\pi fmn) / \sqrt[4]{n} \quad (n \neq 0)$. $J_{1/4}(x)$ is a Bessel function of the order ¼ and first kind. The modulating frequency f_m is subject to the constraint $-4 \le f_m^* n \le 4$.

The optimisation problem is to minimise the cost function, J, given by

$$J = w^H R_x w \text{ subject to } C^H w = f \tag{7.3}$$

where J denotes mean square error (MSE), R_x is the correlation matrix, f is a scalar quantity, C is a constraint matrix ($N \times 1$) and w is the complex weight vector, expressed as

$$w = w_{uc} - Bw_c \tag{7.4}$$

The steering vector for any source, $a(\theta_m)$ is given by

$$a(\theta_m) = \left[1, e^{i\tau_{\theta_m}}, e^{i2\tau_{\theta_m}}, \ldots, e^{i(N-1)\tau_{\theta_m}}\right]^T \quad (m = 0, 1, 2, 3) \tag{7.5}$$

where $\tau_{\theta_m} = (2\pi d / L_\lambda)\sin\theta_m$, where d is the element spacing, L_λ is the wavelength of incoming signal and θ_m is the DOA of the mth signal. The nth snapshot of the ($N \times 1$) received signal vector can be written as

$$\begin{aligned} x(n) &= a(\theta_o)s_o(n) + \sum_{m=1}^{3} a(\theta_m)s_m(n) + N(n) \\ &= s(n) + i(n) + N(n) \end{aligned} \tag{7.6}$$

Here, $s(n)$ is the received signal that impinges the array from the desired direction, $s_o(n)$ is the signal transmitted by the desired source, $i(n)$ is the received signal from the hostile probing sources and $N(n)$ is the random noise.

The output of the GSC block (Widrow and Stearns 1985) in Figure 7.2 is given by

$$y(n) = (w_{uc} - Bw_c)^H x(n) \tag{7.7}$$

Since C and f are known quantities, the unconstrained weight (w_{uc}) can be initially determined using

$$w_{uc} = C(C^H C)^{-1} f \tag{7.8}$$

The constrained weight w_c is calculated either iteratively using the adaptive algorithm or analytically.

The iterative approach involves the weight adaptation, e.g. LMS algorithm

$$w_c(n+1) = w_c(n) + \mu_c v(n) e^*(n) \tag{7.9}$$

where μ_c is the step-size that controls the rate of convergence and the output of blocking matrix block (Lee and Wu 2005) given by

$$v(n) = B^H x(n) \tag{7.10}$$

Since the GSC reuses the array output as the error signal $e(n)$, one has

$$e(n) = y(n) \tag{7.11}$$

Alternately, the constrained weight can be determined analytically. The optimum weight is expressed as

$$w_{opt} = w_{uc} - Bw_{c,opt} \tag{7.12}$$

with $w_{c,opt} = (B^H R_x B)^{-1} B^H R_x w_{uc}$ (7.12a)

The minimum mean squared error (MMSE), J_{min}, is given by

$$J_{\min} = w_{uc}^H R_x w_{opt} \tag{7.13}$$

where R_x is the array correlation matrix, expressed as

$$R_x = \sigma_{s_o}^2 a(\theta_o) a^H(\theta_o) + \sum_i \sigma_{s_i}^2 a(\theta_i) a^H(\theta_i) + \sigma_n^2 I \tag{7.14}$$

where σ_{sj}^2 represents the variance of $s_j(n)$.

Considering only the interference and noise, the correlation matrix R_{i+N} is given by

$$R_{i+N} = \sum_i \sigma_{s_i}^2 a(\theta_i) a^H(\theta_i) + \sigma_n^2 I \tag{7.15}$$

The minimum output power of GSC, i.e. $P_{o,min}$ is equal to minimum cost function,

$$P_{o,\min} = J_{\min} \tag{7.16}$$

The output power associated with impinging desired signal, P_s is given by

$$P_s = \sigma_{s_o}^2 \left| w_{uc}^H a(\theta_o) \right|^2 \tag{7.17}$$

where $\sigma_{s_o}^2$ is the input power of desired signal.

The optimum SINR of GSC is expressed as (Godara 2004)

$$\text{SINR}_{opt} = \frac{P_s}{P_{o,\min} - P_s} \tag{7.18}$$

The array gain corresponding to the adapted pattern is given by

$$\text{Gain}(\theta) = w_{opt}^H a(\theta) \tag{7.19}$$

This is analytical estimation of SINR and, hence, the adapted pattern.

The conventional approach of GSC implementation is susceptible to the mismatch in DOA estimation, which is a common observation in practical cases. The stochastic gradient in (7.9), i.e. $v(n)y^*(n)$, is never zero even when optimal weights are obtained. Thus, iterative calculations give rise to MSE. To reduce MSE, a smaller step-size is preferred, which in turn slows down the rate of convergence.

In case of DOA mismatch, the array tends to confuse in between the desired and probing signals, which may result in the suppression of desired signal. This situation is identified as self-cancellation. To avoid such a situation, one of the potential solutions is to include notch filter in the conventional design of GSC. Another way is to enforce constraints such as derivative constraints, linear constraints and directional constraints on the weight adaptation. However, enforcing additional constraints have a tendency to degrade the interference suppression in an array, owing to the consumption of its degrees of freedom (DOF) (Lee et al. 2005). It is known that lesser the number of DOF available, worse will be the degree of suppression in adaptive arrays.

Another approach towards the enhancement in robustness against various mismatch errors is to include blind equaliser in the processing unit of GSC. Such inclusion in the GSC structure is called decision feedback-generalised sidelobe canceller (DF–GSC), (Perreau et al. 2000). Blind equalisers are used especially for digital signals so as to circumvent the inter-symbol interference (ISI).

7.3 Decision feedback-generalised sidelobe canceller (DF–GSC)

The DF–GSC consists of an additional block with blind equaliser and decision feedback filter so as to alter the conventional GSC output. The modified GSC output is then used as the error signal (Figure 7.3).

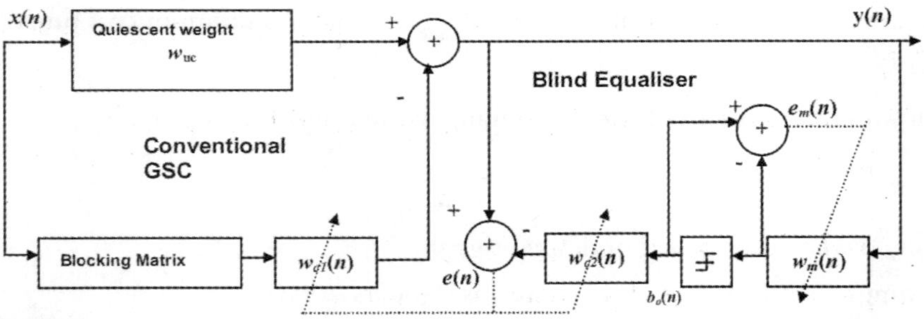

Figure 7.3 Schematic of decision feedback-generalised sidelobe canceller (DF–GSC)

The blind equaliser and the feedback filter operate with different error signals and, thus, have distinct weights, w_m and w_{c2}. The blind equaliser can recuperate the transmitted bits, $b_o(n)$ within acceptable phase uncertainty. The blind equaliser is trained by the error signal $e_m(n) = \hat{b}_o(n) - w_m^* y(n)$, where $\hat{b}_o(n)$ is the received bits and w_m is the tap weight of the equaliser.

Lee and Wu (2005) have suggested the weight updation for w_m as

$$w_m(n+1) = w_m(n) + \mu_m y(n) e_m^*(n) \tag{7.20}$$

where μ_m is the step-size governing the rate of convergence and * represents the complex conjugate of the vector. It is presumed that the blind equaliser detects the symbol accurately (Lee et al. 2005) and, hence, $s_o(n) = \hat{s}_o(n) = b_o(n) = \hat{b}_o(n)$.

The DF–GSC extracts the desired signal from the error signal, given by

$$e(n) = (w_{uc} - Bw_{c1})^H x(n) - w_{c2}^* b_o(n) \tag{7.21}$$

This error signal is employed to update the weights w_{c1} and w_{c2},

$$\begin{aligned} w_{c1}(n+1) &= w_{c1}(n) + \mu_{c1} v(n) e^*(n) \\ w_{c2}(n+1) &= w_{c2}(n) + \mu_{c2} b_o(n) e^*(n) \end{aligned} \tag{7.22}$$

The estimation of SINR remains the same as that in GSC except the computation of w_{opt}. The above approach is an iterative approach for the estimation of SINR. The alternative approach of analytical estimation involves the decomposition of w_{opt} into two weights, $w_{cl,opt}$ and $w_{c2,opt}$ for the estimation of SINR.

$$w_{c1,opt} = \frac{B^H R_x w_{uc}}{B^H R_x B} \tag{7.23}$$

and $w_{c2,opt} = S^H(\theta_o) w_{uc}$ (7.24)

Once optimum weights $w_{opt} = \begin{bmatrix} w_{c1} \\ w_{c2} \end{bmatrix}$ are obtained, adapted pattern can be generated.

7.4 Performance analysis

The performance analysis of GSC and DF–GSC is carried out considering a 16-element uniform linear array with λ/2 spacing. The signal scenario comprises one desired and three probing sources, incident at 0°, 30°, 60°, and –25°, respectively. The signal-to-noise ratio (SNR) for desired source is taken as 0 dB while interference-to-noise ratio (INR) as 20 dB per probing source.

The signals are assumed to be random QPSK signal having flat fading channel characteristics, such that $s_o(n) = h(n) \times b_o(n)$ and h is the coefficient of channel. The noise is AWGN with 0.5 variance. The DOAs of the impinging signals are assumed to be known *a priori*. Figure 7.4 presents the output SINR of GSC and DF–GSC versus snapshots. The weight adaptation is done using LMS algorithm.

It can be inferred from Figure 7.4 that the DF–GSC performs better in terms of SINR as well as convergence rate. The corresponding adapted pattern (Figure 7.5) shows that DF–GSC places deeper nulls towards the impinging angles of probing sources and maintains main lobe towards the desired direction. In other words, there is no signal transmitted towards any of the probing sources, thereby contributing towards low observability.

The capability of sidelobe cancellers to suppress the probing sources has potential applications in low-observable techniques. When such sidelobe cancellers are mounted in appropriate configurations over the platforms such as aircraft and missiles, they permit only the desired signal reception, suppressing each of hostile sources attempting to probe the array. In other words, these sidelobe cancellers along with an efficient weight adaptation algorithm have the potential of making them 'invisible' to the radars both on the hostile territory and aircraft.

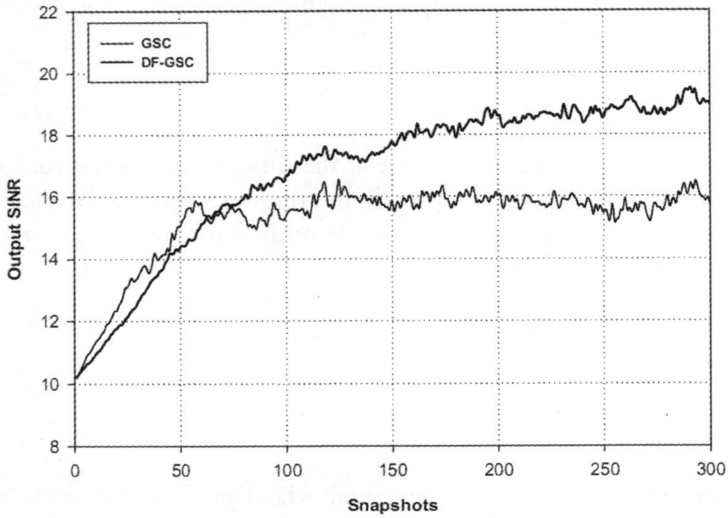

Figure 7.4 Output SINR of GSC and DF–GSC in signal environment consisting of three probing sources (30°, 60°, –25°; 20 dB each)

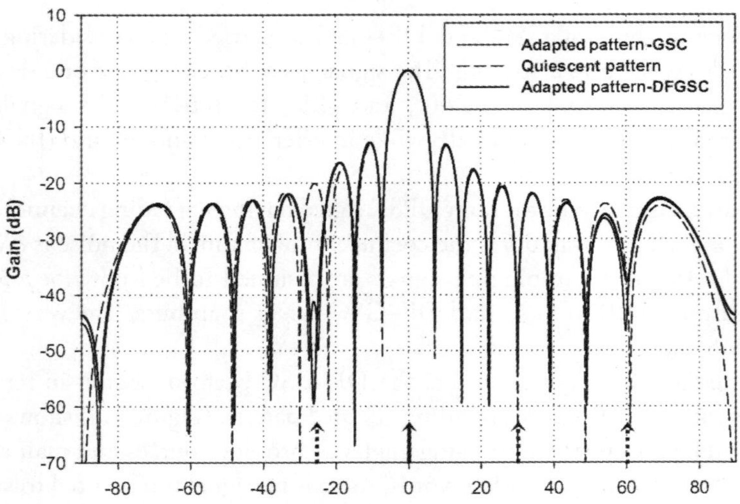

Figure 7.5 Adapted pattern for GSC and DF–GSC with one desired source (0°, 0 dB) and three probing sources (30°, 60°, –25°; 20 dB each). The desired source is marked as a solid arrow, while probing sources are shown as dashed arrows along the *x*-axis

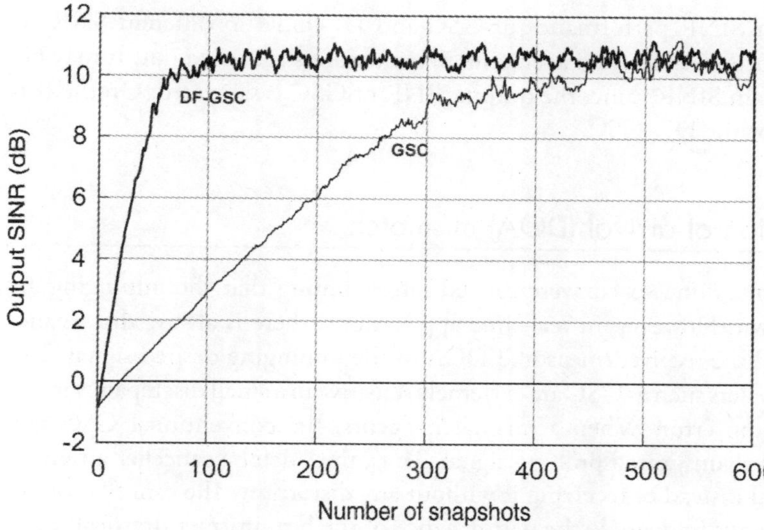

Figure 7.6 Learning curves for GSC and DF–GSC for the same SINR

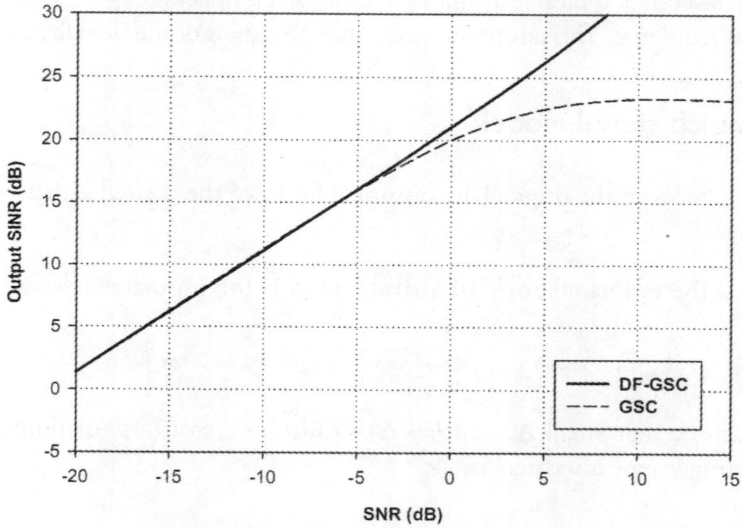

Figure 7.7 Steady-state SINR performance in different SNR environments

For the sake of comparing the rate of convergence of both GSC and DF–GSC schemes, the learning curves of the schemes for same SINR are presented in Figure 7.6. The signal environment consists of one desired source (0°) and two monochromatic radar sources probing at 45° and 20°. The power level of the sources is different (10 and 100). It is apparent that the rate of convergence of DF–GSC is higher than that of GSC to attain the target of optimal output SINR.

The output SINR performance of GSC and DF–GSC in different SNR environments is shown in Figure 7.7. Here, no averaging in simulations is carried out. It may be observed that beyond a certain SINR value, the output SINR of GSC is saturated. On the contrary, no such trend exists for the DF–GSC.

7.5 Direction of arrival (DOA) mismatch

The simulations done so far were carried out assuming that the impinging angle of desired signal is known. However, in real-time applications, there is always discrepancy between the estimated and the received/measured DOA of the impinging desired signal. The conventional sidelobe cancellers such as GSC are extremely sensitive to a small discrepancy in DOA estimation due to pointing error. When a mismatch occurs, the conventional GSC misconstrues the desired signal as unwanted probing signal. Thus, the sidelobe canceller attempts to nullify the probing signal instead of receiving it without any distortion. The cancellation of desired signal component tends to degrade the performance of the beamformer drastically.

The conventional approach of the beamformer assumes the absence of desired signal for the training period. This makes it sufficiently robust against any kind of DOA mismatches (Li and Stoica 2005). However, in typical real-time applications, it is not easy to get the training snapshots without the desired signal. This adversely affects the robustness of the sidelobe cancellers.

7.5.1 Mismatch signal model

For a mismatch between the original and assumed DOA of the desired signal,

$$\theta_o = \hat{\theta}_o + \Delta \tag{7.25}$$

where, $\hat{\theta}_o$ is the estimated angle of arrival and Δ is the mismatch, the steering vector is given by

$$S(\theta_o) = S(\hat{\theta}_o + \Delta) = \left[1,\ e^{i\alpha\sin(\hat{\theta}_o+\Delta)},\ e^{i2\alpha\sin(\hat{\theta}_o+\Delta)},\dots,\ e^{i(N-1)\alpha\sin(\hat{\theta}_o+\Delta)}\right]^T \tag{7.26}$$

with $\alpha = 2\pi d/\lambda$. For small Δ, $\sin(\hat{\theta}_o+\Delta) \approx \sin(\hat{\theta}_o) + \Delta\cos(\hat{\theta}_o)$. Assuming $\hat{\theta}_o = 0°$, the mismatch steering vector is written as

$$\begin{aligned} S(\theta_o) &= S(\Delta) \\ &= \left[1, e^{i\alpha\Delta}, e^{i2\alpha\Delta}, \dots\dots\dots, e^{i(N-1)\alpha\Delta}\right]^T \end{aligned} \tag{7.27}$$

Thus $s(n)$ can be re-written as $s(n) = S(\Delta)s_o(n)$.

7.5.2 DOA mismatch with GSC

If there is error in the estimation of the DOA of desired signal, the unconstrained weight vector w_{uc} does not match with the array response towards the desired signal. This mismatch affects the blocking matrix response. The matrix will not be able to block the desired signal in passing through the interference-cancelling filter. Thus, the filter would cancel the desired signal provided enough DOF are available. This signal cancellation hinders the sidelobe canceller to maximise SINR even with optimal weights. The output power associated with the desired signal in the conventional GSC (Lee and Wu 2005) becomes

$$P_s = \sigma_{s_o}^2 \left|(w_{uc} - Bw_{c1.opt})^H S(\Delta)\right|^2 \tag{7.28}$$

The extent of signal cancellation is governed by the signal power and the magnitude of the mismatch.

7.5.3 DOA mismatch with DF–GSC

In DF–GSC, the signal cancellation problem does not exist. Due to correlation between two signal paths in GSC system, the constrained weights $w_{c1,opt}$ and $w_{c2,opt}$ get coupled in the presence of DOA mismatch (Lee and Wu 2005). The coupled weights may be expressed as

$$w_{c1,opt} = \frac{B^H R_{i+N} w_{uc}}{B^H R_{i+N} B} \tag{7.29}$$

$$w_{c2,opt} = S^H(\Delta)(w_{uc} - Bw_{c1,opt}) \tag{7.30}$$

In the presence of DOA mismatch, the MMSE of DF–GSC is given by

$$J_{\min} = w_{uc}^H R_{i+N} w_{opt} \tag{7.31}$$

The output power is given by

$$P_{o,\min} = w_{opt}^H R_x w_{opt} \tag{7.32}$$

7.6 Constraints in adaptive array processing

In conventional design of sidelobe cancellers, such as GSC, a notch filter is added into the processor for separating the desired signal and the probing signals. The main objective is to enhance the robustness of sidelobe cancellers. Moreover, this addition reduces the divergence of adaptive weights from the optimal weights and, hence, enhances the rate of convergence (Lee et al. 2005). Alternative approach is to impose the constraints in array processing. An optimum adaptive beamforming with multiple constraints has become a well-known method in adaptive antenna arrays. There are several methods for imposing constraints to main lobe array response, constraining the processor of adaptive array to respond to the main beam signals, gain maintenance towards desired look direction and achieve a desired quiescent pattern.

7.6.1 Point constraints

This is the most common constraint that is imposed to the weight adaptation to generate the desired adapted pattern (Buckley and Griffiths 1986). If $C = D$ and $F = N$; C being the constraint vector, D is $N \times 1$ vector of 1s and N is the count of antenna elements, then the constraint equation can be written as

$$W^* D = N \tag{7.33}$$

A constraint is imposed on the weight vector to respond towards the main lobe direction. In other words, the weight vector tends to reduce the output power at all angles except towards the main lobe. As the probing directions (targets) and the look directions usually do not coincide, it is preferred that the multiple constraints are imposed to retain the angular extent of the main lobe and efficient nulling towards the probing sources.

7.6.2 Derivative constraints

It is known that for a desired signal with high power ratio, the signal suppression is less. Moreover, it affects the main lobe width. This can be minimised if the derivative constraints are imposed on the main lobe direction. Furthermore, by having control over the first few derivatives of the main lobe, adaptive arrays may circumvent null within the angular extent of the main lobe (Buckley and Griffiths 1986). These derivative constraints control the beam pattern sensitivity towards the discrepancy in between the beam steering angle and the angle at which the desired source impinges. On setting these derivatives to zero (Huarng and Yeh 1992), the constraint equation will be

$$\tilde{V} = \begin{bmatrix} v(\theta) & \dot{v}(\theta) & \ddot{v}(\theta) \ldots \end{bmatrix} \tag{7.34}$$

where V is a matrix constraint vector as columns, $v(\theta)$ is the column of point constraint, $\dot{v}(\theta)$ and $\ddot{v}(\theta)$ are the columns of the first and second derivative constraints, respectively.

The performance of adaptive array with derivative constraint does not depend on the phase origin and, thus, achieves high SINR provided the steering is error free. The main beam of the pattern will be broader for higher order derivatives. Such broader beam is helpful especially when the presumed and actual directions of impinging signal do not coincide precisely.

7.6.3 Directional constraints

If constraint of look direction is imposed in weight adaptation, it becomes easy to achieve probe suppression and high output SINR. However, if there is a mismatch in the presumed and actual look angle, the array may cancel the impinging desired signal considering it to be an interfering signal. The robustness of beamformers against beam steering angle errors can be improved by using either multiple directional or derivative constraints in the weight adaptation and, hence, the adapted pattern. In practical situations, the directional constraints are prone towards ill-conditioning, which is not true in case of derivative constraints (Takao et al. 1976). Furthermore, the derivative constraints are stated to be well-conditioned. It all depends on the choice of appropriate derivative constraint for achieving a better-conditioned array processing system.

7.6.4 Simulation results

If one assumes the mismatch of 2°, the SINR output of GSC and DF–GSC would depend upon the constraints considered.

Point constraint in GSC: If only point constraint is considered in GSC, the output SINR variation with snapshots (Figure 7.8) shows that the output SINR degrades drastically and converges around –8 dB.

Point constraint and first derivative constraint in GSC: On including derivative constraint and point constraint in weight adaptation, the performance of GSC improves further (Figure 7.8). The additional constraint in sidelobe canceller results in enhanced robustness. The output SINR is increased considerably. The plot stabilises much faster towards the value of 2.5 dB.

Point constraint in DF–GSC: Considering point constraints in DF–GSC, the output SINR of DF–GSC in the presence of DOA mismatch is shown in Figure 7.9a. The training of first 50 snapshots is used. The SINR converges beyond 2,500 snapshots and stabilises to 9 dB approximately.

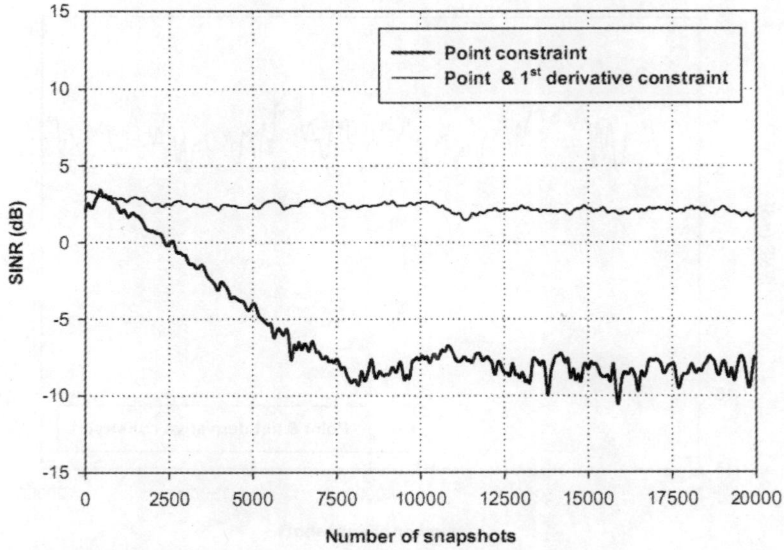

Figure 7.8 Learning curves for GSC in the presence of DOA mismatch utilising point and first-order derivative constraint

First derivative and point constraint in DF–GSC: The learning curve for DF–GSC in the presence of DOA mismatch is shown in Figure 7.9b. The weight adaptation is done imposing both point and first derivative constraints. Moreover, no training snapshots are required. The inclusion of derivative constraint in array processing reduces the SINR achieved; this may be due to lesser number of available DOF.

These simulations prove that DF–GSC scheme provides better robustness against mismatch errors as compared to the conventional sidelobe canceller such as GSC.

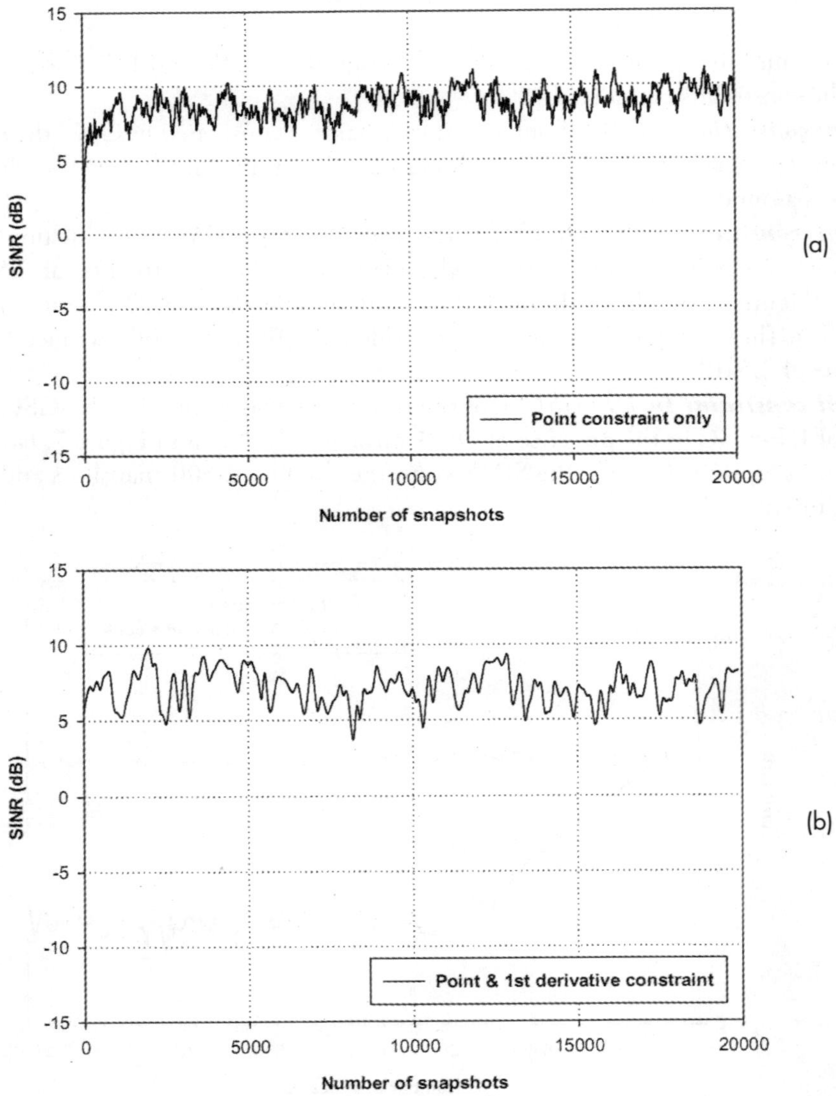

Figure 7.9 Learning curves for DF–GSC with DOA mismatch. (a) Point constraint only (b) First-order derivative constraint

7.7 Blind equalisation in sidelobe cancellers

Blind channel identification and equalisation have been widely studied from various perspectives to combat multipath effects in the absence of training samples. Blind methods are those methods in which no initial training is required. The performance of the adaptive array is enhanced while simultaneously maintaining the robustness against any mismatch errors. A major advantage of this approach is that it is unaffected due to error data being incorrectly received (Li and Liu 1998). This enhances the robustness of the sidelobe cancellers.

The blind adaptation techniques are associated with the poor rate of convergence as compared to the classical methods based on training bits. It is still a huge task to formulate a robust, computationally effectual subroutine for blind equalisation.

7.7.1 Theoretical background

A digital system that transmits data over a linearly distorting channel usually contains an equaliser to compensate for the channel distortion. The quantitative characteristics of the channel response are often not known *a priori*. There are three basic approaches for the equalisation of the unknown channel: (i) transmit and analyse a known training sequence, (ii) try to detect the reliable data and adjust equaliser accordingly and (iii) use statistical property of the transmitted signal for adjusting the equaliser (Leou et al. 2014).

Since the initial training snapshots are difficult to acquire, it becomes necessary to find a way, which can result in sufficient robustness without the use of initial training snapshots. Figure 7.10 shows the scheme for blind adaptation that utilises the data from the equalising block. The performance of blind adaptation is not affected by incorrectly received data.

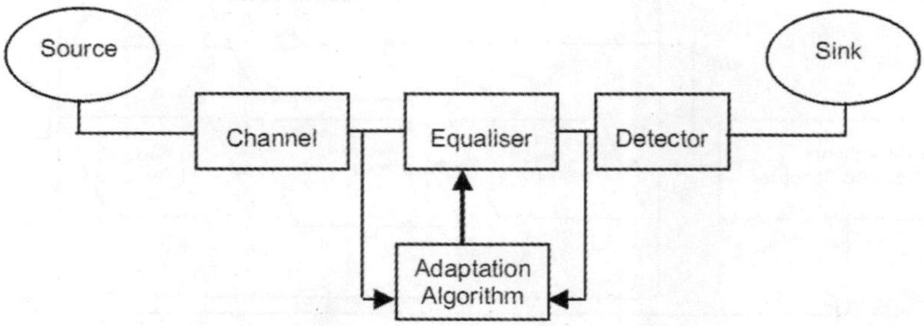

Figure 7.10 Blind adaptation scheme

7.7.2 Steps of algorithm

The principle involves initialisation of approximate derivative constraints in DF–GSC. These are released progressively. To begin with, the main lobe is broadened due to mismatch in DOA. However, the algorithm ensures that sufficient strength of desired signal exists in the output of quiescent filter. The equaliser will then proceed to convergence even in the absence of training bits. Once the convergence is achieved, the adaptive array dispenses the constraints sequentially. This in turn results in increase in the available number of DOF and, hence, accurate and deep null placement. Figure 7.11 describes the computation methodology of the blind approach.

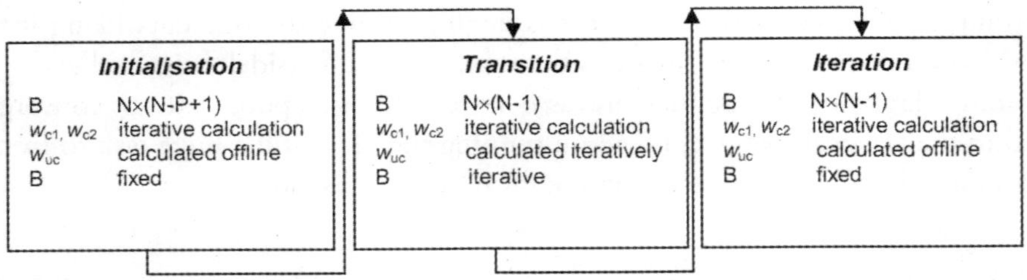

Figure 7.11 Steps in blind adaptation in sidelobe canceller

Blind approach: In blind adaptation methods, tracking of fast channel variations and re-capturing the functional conditions are based on information data. However, in case any specific information related to transmitted data is known *a priori*, it should be utilised. In many scenarios, the desired signal may be too frail for the initialisation of blind equaliser. This makes it difficult to extract accurate decision in initial stages in the presence of DOA mismatch (Xie et al. 2012). When blind adaptation is included in DF–GSC, the usage of training bits in the initial stage may be skipped and the required output SINR can be obtained. Lee and Wu (2005) proposed the inclusion of a blind equaliser to DF–GSC, shown in Figure 7.12. In this approach, three stages are included for the weight estimation process.

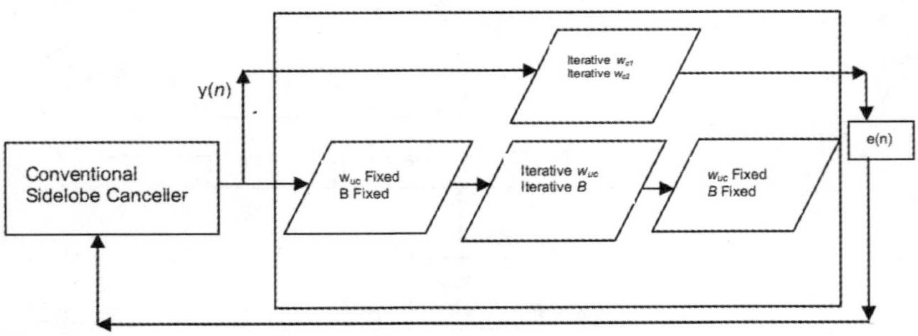

Figure 7.12 Algorithm for Blind DF–GSC

The first stage of initialisation involves the calculation of projection matrix. The vector dimension of weight coefficients is increased. The second stage is transition. Here, blocking matrix and quiescent weights are iteratively calculated. The number of point constraints along with derivative constraints is varied from p to $(p - 1)$. This increases the number of DOF from $N - p$ to $N - (p - 1)$. The third stage involves weight adaptation.
These stages are outlined as:

A. *Initialisation*: The projection matrix P of the array is calculated from the blocking matrix, B.

$$P = BB^H \tag{7.35}$$

The corresponding projection vector, p' is calculated including the derivative constraint as

$$p' = P.c_P \tag{7.36}$$

where c_p is the constraint vector.

A difference vector, d is given by (Lee and Wu 2005) as

$$d_{N-(P-1)} = w_{uc,N-(P-1)} - w_{uc,N-p} \tag{7.37}$$

An element of zero value is added into adaptive weight w_{c1}, for circumventing the mismatch between the new blocking matrix and the weight vector, i.e.

$$w_{c1,N-(P-1)} = [w_{c1,N-P} \quad 0] \tag{7.38}$$

B. *Transition*: New weight vector $w_{uc,N-(p-1)}$ and blocking matrix are calculated using

$$\begin{aligned} B_{N-(p-1)} &= [B_{N-p} \quad \gamma_n p'_{N-(p-1)}] \\ w_{uc,N-(p-1)} &= w_{uc,N-p} + \gamma_n d_{N-(p-1)} \end{aligned} \tag{7.39}$$

where $\gamma_n = 1 - \{(T-1) - n\}/(T-1), \quad 0 \le n \le (T-1)$

Steps are repeated till $\gamma_n = 1$. In this approach, DF–GSC skips the usage of training snapshots without any performance degradation.

For the performance analysis of DF–GSC with blind adaptation, a 16-element array with $\lambda/2$ uniform spacing is considered. The power level of desired signal is 0.2. The three probing sources have the power ratio of 100 each. The DOA of three probing signals is 20°, 50° and –35°, respectively. The variation of output noise power of the blind DF–GSC in the presence of 2° direction-of-arrival mismatch is presented in Figure 7.13. The weight adaptation involves three stages as discussed above. In the initialisation stage, the point and first derivative constraints are imposed on iterative weight estimation. After 110 snapshots, the second stage of weight estimation is carried out till 190th snapshot, beyond which the last stage begins keeping the blocking matrix dimension fixed. The length of transition period is taken as 100. Figure 7.14 shows the corresponding adapted pattern of Blind DF–GSC. The DF–GSC pattern is also included for the sake of comparison. It may be observed that the suppression capability of Blind DF–GSC is superior to that of DF–GSC and suppresses all the three interfering sources efficiently. Moreover, the DF–GSC along with blind adaptation shows better robustness than normal DF–GSC towards DOA mismatch.

Next, the suppression capability of Blind DF–GSC is demonstrated for three closely spaced interfering signals at 35°, 48° and 65° with 20-dB power level each. Figure 7.15 presents the adapted and quiescent pattern of Blind DF–GSC. It may be seen that all the three interfering signals are suppressed effectively without disturbing the main lobe in the direction of desired source.

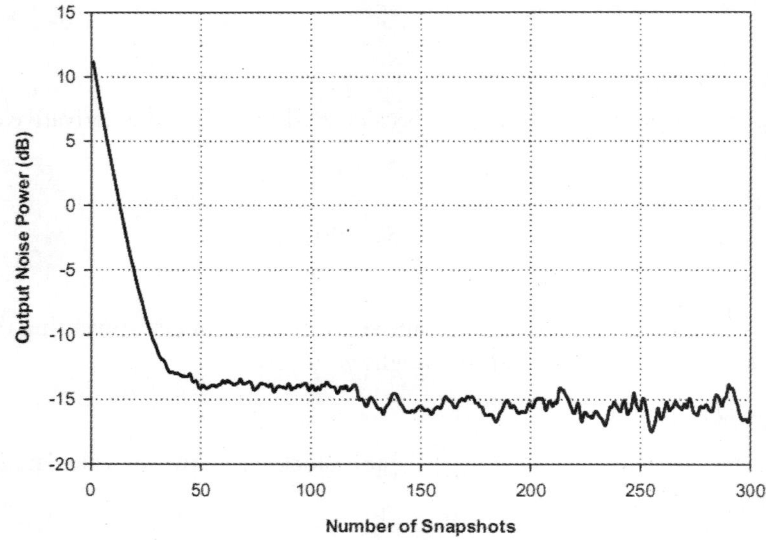

Figure 7.13 Output noise power of Blind DF–GSC

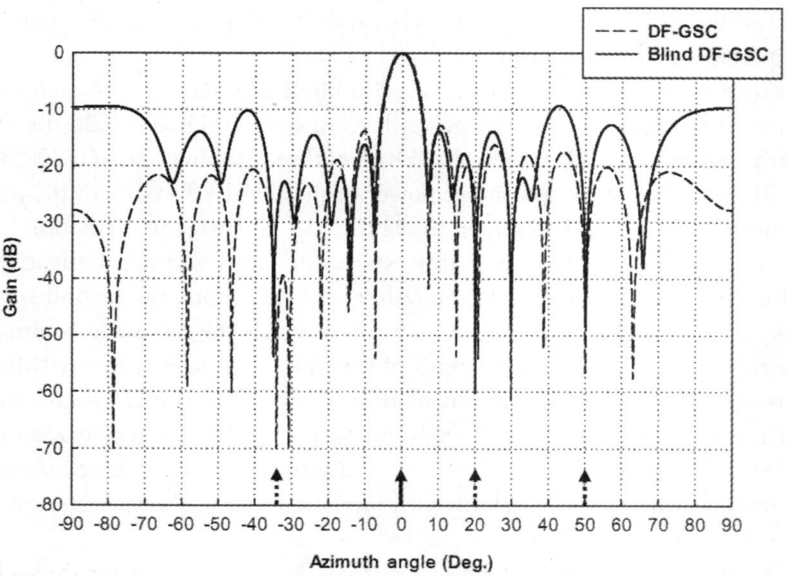

Figure 7.14 Adapted pattern of Blind DF–GSC for three probing signals (50°, 20°, –35°). The desired source is shown as a solid arrow, while probing sources are marked as dashed arrows along the *x*-axis

Semi-blind approach: The DF–GSC with blind adaptation is although superior to DF–GSC, but has high-computational cost. Thus, it is simplified to a two-staged Semi-blind approach in which only two stages, namely, initialisation and transition, are used for weight adaptation. In Figure 7.16, the output noise power of Semi-blind DF–GSC is shown. The signal scenario considered consists of single desired source (0°; 0 dB) and three hostile sources (50°, 20° and –35°; 20 dB). Since random signal generation is involved, the curve is plotted after averaging the results. The corresponding adapted pattern of semi-blind DF–GSC is presented in Figure 7.17. Both the quiescent and the adapted patterns of DF–GSC are included. It may be observed that suppression capabilities of semi-blind DF–GSC are superior to DF–GSC.

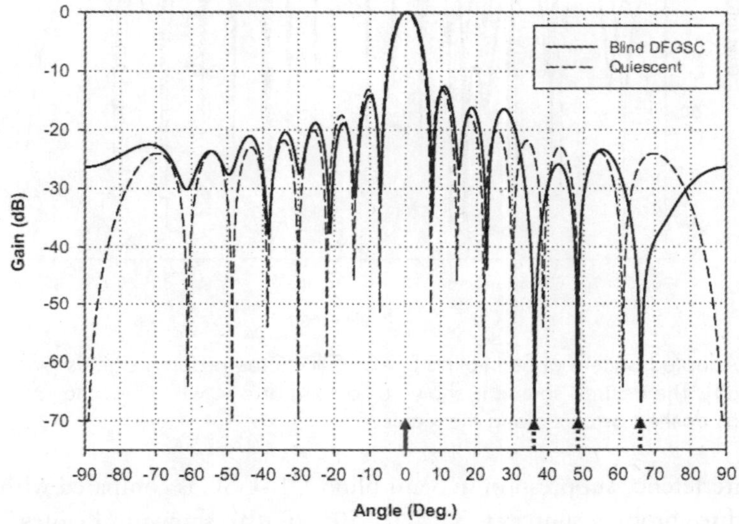

Figure 7.15 Adapted pattern of Blind DF–GSC for three closely spaced probing signals (35°, 48°, 65°). The desired source is shown as a solid arrow, while probing sources are marked as dashed arrows along the *x*-axis

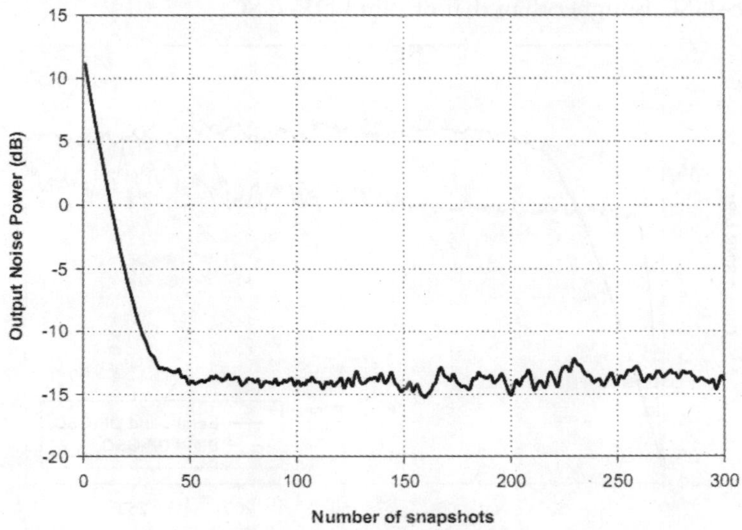

Figure 7.16 Output noise power of Semi-blind DF–GSC

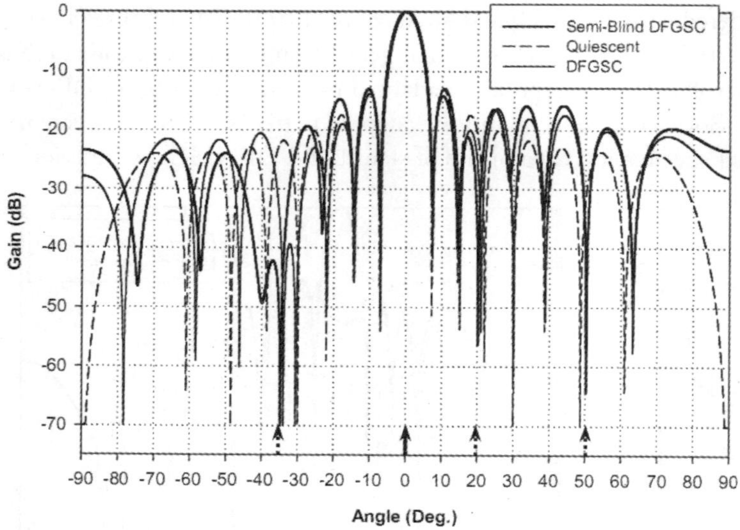

Figure 7.17 Adapted pattern of Semi-blind DF–GSC for three probing signals (–35°, 20°, 50°; 20 dB). The desired source is shown as a solid arrow, while probing sources are marked as dashed arrows along the *x*-axis

Next, the interference suppression in Semi-blind DF–GSC is compared with that of Blind DF–GSC for three probing sources (–35°, 20°, 50°; 20 dB), shown in Figures 7.18 and 7.19. As discussed in the previous sub-section, the computational burden of Semi-blind DF–GSC is less due to the two stages of computation of weight adaptation as compared to the three stages of computation in Blind DF–GSC. This is apparent from Figure 7.18. The output SINR of Semi-blind DF–GSC is higher than that of Blind DF–GSC.

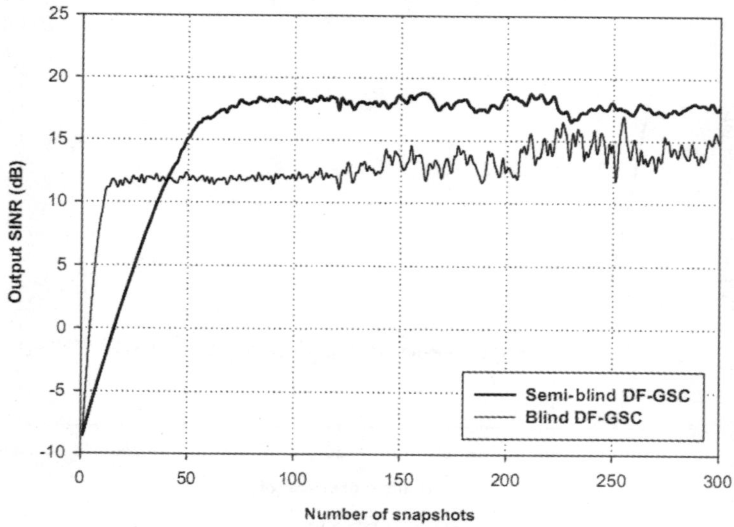

Figure 7.18 Comparison of performance of Semi-blind DF–GSC and Blind DF–GSC for single desired signal (0°) and three probing sources(–35°, 20°, 50°; 20 dB)

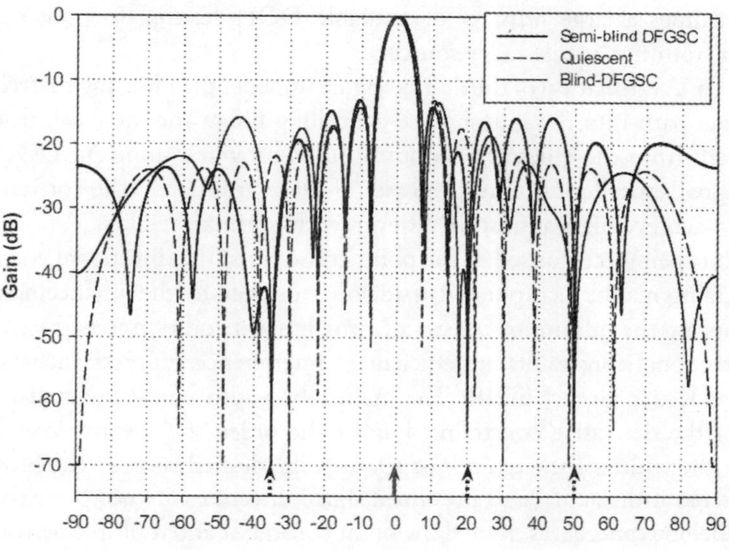

Figure 7.19 Comparison of suppression capabilities of Semi-blind DF–GSC and Blind DF–GSC. The desired source is shown as a solid arrow, while probing sources are marked as dashed arrows along the *x*-axis

The weight adaptation is done using the improved LMS algorithm. Moreover, the performance of Semi-blind DF–GSC is better than Blind DF–GSC in view of accurate placement of deep nulls in the probing directions. This underscores the capability of Semi-blind DF–GSC in transmitting no energy towards the directions of hostile sources and simultaneously maintaining the communication with the desired sources.

7.8 Conclusion

The generalised sidelobe canceller (GSC) schemes for adaptive arrays employ linearly constrained minimum variance (LCMV) criterion to suppress the hostile sources incident from different directions, while simultaneously maintaining high output SINR towards the desired ones. The GSC is especially useful when the direction of arrival (DOA) of the impinging signal is not known *a priori*. It is quite effective in reducing computational cost when used with proper adaptive algorithms. However, the performance of the GSC degrades considerably in case of a discrepancy between the estimated signal direction and the actual one.

To provide sufficient robustness, an additional block consisting of filter and blind equaliser is included, which is called DF–GFC. The DF–GSC scheme also avoids the signal cancellation in case of DOA mismatch. Another way for enhancing the robustness against errors is to impose additional constraints in the processing. However, this deteriorates the capability of the array to suppress the probing signals and thermal noise. This is because imposing multiple constraints consumes a large number of available DOFs leading to poorer suppression of signals incident from the hostile radar source.

The error in DOA estimation of the desired signal impedes the blocking matrix in obstructing the desired signal path into the interference-cancelling filter. The filter will delete the desired signal component from the quiescent output of filter, provided sufficient DOFs are available. This leads to signal cancellation. Thus, in cases of DOA mismatch, the conventional sidelobe cancellers fail to achieve high output SINR even with optimal weights.

The constraints being categorised as the point constraints, the directional constraints and the derivative constraints may be incorporated in sidelobe cancellers for the enhancement of robustness. Point constraints are the minimum constraint, which upon addition, sets the constraint matrix to a scalar. Directional constraints are efficient in interference suppression, whereas derivative constraints provide the highest SINR. The main lobe width of the adapted pattern depends on the order of the derivative constraint. Higher the order of derivative constraint imposed, broader will be the width of the main lobe towards the desired source. The broader main lobe of array is preferred if the actual and presumed signal directions are not precisely matched. The robustness of sidelobe canceller is more if the point constraint and multiple derivative constraints are imposed in weight adaptation. The convergence rate of output SINR is also improved. It is observed that although the addition of derivative constraints in the array processing enhances robustness, the output SINR deteriorates. Moreover, these imposed additional constraints need training in initial steps, which is a challenging task in real scenario.

To achieve higher output SINR while maintaining the increased robustness, a new approach known as blind approach for DF–GSC is used. In this technique, all the simulations are divided into three parts. Instead of using either of the constraints separately, both the constraints are employed simultaneously. In the first part, weights are determined iteratively and after that a new period called the transition period starts where the columns of blocking matrix are reduced gradually. In this stage, both the blocking matrix and weights are calculated recursively. This provides additional robustness to the DF–GSC. Thus, DF–GSC can converge faster without any use of initial training snapshots. A novel technique of blind adaptation that has only two stages called the Semi-blind DF–GSC is also discussed. The performance of sidelobe cancellers along with blind adaptation is analysed in terms of output SINR and accurate placement of deep nulls in probing directions.

The study carried out is intended to corroborate the potential of sidelobe cancellers for active RCSR applications.

References

Anderson, J. B. and A. Svensson. 2003. *Coded Modulation Systems*. New York: Kluwer Academic/Plenum Publishers, 485.

Buckley, K. M. and L. J. Griffiths. 1986. 'Adaptive generalised sidelobe canceller with derivative constraints.' *IEEE Transactions on Antennas and Propagation* 34: 311–319.

Godara, L. C. 2004. *Smart Antennas.* Washington DC: CRC Press, 448.

Harmuth, H. F. 1970. *Transmission of Information by Orthogonal Functions.* New York: Springer-Verlag, 325.

Huarng, K. -C. and C. -C. Yeh. 1992. 'Performance analysis of derivative constraint adaptive arrays with pointing errors.' *IEEE Transactions on Antennas and Propagation* 40: 975–81.

Lee, J. -H., K. -P. Cheng, and C. -C. Wang. 2005. 'Robust adaptive array beamforming under steering angle mismatch.' *Signal Processing* 86: 296–309.

Lee, Y. and W. -R. Wu. 2005. 'A robust adaptive generalized sidelobe canceller with decision feedback.' *IEEE Transactions on Antennas and Propagation* 53: 3822–32.

Leou, M. -L., C. -M. Wu, Y. -C. Liaw, and H. -K. Su. 2014. 'An orthogonalized blind algorithm for hybrid of adaptive array and equalizer.' *International Journal of Communication Systems* 27: 201–15.

Li, J. and P. Stoica. 2005. *Robust Adaptive Beamforming.* Hoboken, NJ: John Wiley & Sons, 422.

Li, Y. and K. J. R. Liu. 1998. 'Adaptive blind source separation and equalisation for multiple-input/multiple-output systems.' *IEEE Transactions on Information Theory* 44: 2864–76.

Perreau, S., L. B. White, and P. Duhamel. 2000. 'A blind decision feedback equalizer incorporating fixed lag smoothing.' *IEEE Transactions on Signal Processing* 48: 1315–28.

Steele, A. K. 1983. 'Comparison of directional constraints for beamformers subject to multiple linear constraints.' *IEE Proceedings, Pt. F and H* 130(1): 41–45.

Takao, K., M. Fujita, and T. Nishi. 1976. 'An adaptive antenna array under directional constraint.' *IEEE Transactions on Antennas and Propagation* 24: 662–69.

Widrow, B. and S. D. Stearns. 1985. *Adaptive Signal Processing.* Englewood Cliffs, NJ: Prentice-Hall, 474.

Xie, N., H. Hu, and H. Wang. 2012. 'A new hybrid blind equalization algorithm with steady-state performance analysis.' *Digital Signal Processing* 22(2): 233–37.

8

Emerging RCSR Techniques

8.1 Introduction

The active radar cross section reduction (RCSR) and the algorithms discussed in this book relate to the real-time RCSR and control of aerospace structures. The antenna array mounted over a platform actively adapts the pattern in coherence with the overall structural radar cross section (RCS) of the platform. The pattern adaptation depends on parameters such as the number of radar sources, their frequencies, the angle of arrival, the power levels, and the bandwidth. The RCS of aerospace platform with phased array mounted over it is reduced towards each of the hostile probing sources while simultaneously maintaining communication links with friendly radars.

Apart from the concept of active RCSR, several other techniques and ideas have been proposed towards achieving low observability through passive means. These include the usage of engineered materials, i.e. frequency selective surfaces (FSS), artificial magnetic conductor (AMC), metamaterials, conformal load-bearing antennas, etc. The research is emphasised towards the miniaturisation, bandwidth enhancement and conformality of the structures. This is especially true for the antennas mounted over the surface of aerospace structures. Typically, RCS of a non-stealthy aircraft may vary from 10 dBsm to 30 dBsm, whereas missiles have relatively lower RCS (–20 dBsm to 10 dBsm) (Doan et al. 1995) (Paterson 1999). Moreover, various sensors and antennas are mounted over these platforms owing to applications such as surveillance, tracking and reconnaissance and, hence, contribute to the overall RCS of the platform. Extensive studies have, therefore, been done to reduce the antenna RCS (Volakis et al. 1992), (Zhang et al. 2012). The solutions proposed include the shaping of the surface, application of radar absorbing material (RAM) (Pozar 1992), (Wang et al. 2011), using tapered resistive sheets in antenna (Gustafsson 2006), and (Genovesi et al. 2012) in radome, passive RCSR (Knott et al. 2004) and active RCSR. Scattering at the edges can be suppressed effectively by using tapered loading. The average backscattering reduction with serrated loading exceeds 30 dB for the aspect angles between 30° and 85° (Chen et al. 2011). The RCSR is, however, associated with high cost, weight and system maintenance. Apart from the structural RCS of antenna or antenna array, it is the scattering within the feed network of antenna system that requires rigorous analysis. This makes the design of a low RCS antenna array, with a good radiation characteristic, a big challenge. When the terminating port of feed network is matched, the antenna scattering is called the structural mode of scattering. Otherwise, a portion of the received wave is reflected and, hence, reradiated. This is called antenna mode of scattering (Hu et al. 2011). The structural RCS of antenna can be reduced by 10–20 dB for out-of-band frequencies if the ground plane is replaced with hybrid structures like FSS or AMC. This concept had been applied to passive antennas like reflectarray (Misran et al. 2003). The concept cannot be extended directly for printed antenna like microstrip patch antenna. The FSS-based ground plane permits quasi-TEM mode of propagation. Nevertheless, this idea has been employed for a monopole array (Wang et al. 2009). The hybrid ground plane may contribute significantly in RCSR for the frequencies of FSS transmission band. However, the robustness and RCSR in all directions is uncertain. The in-band scattering can be controlled by the optimised design of the antenna systems including its feed network, which reduces the internal reflections of the incoming signal at the component level (Sneha et al. 2012).

This chapter discusses the advancements in design and development in a few frontier areas towards RCSR of aerospace vehicles. These include embedded antennas, conformal load-bearing antennas, FSS, metamaterials and plasma-based RCSR.

8.2 Embedded antennas

Antenna designs are considered to be effective when they realise the desired radiation pattern and polarisation diversity, i.e. the antenna should be capable of radiating linear, circular or elliptical polarisation. The antenna/array can be fabricated over the surface by flush mounting. In addition, the antenna element should be immune to failure, rugged and reliable. In real situations, none of the radiating elements can satisfy every requirement. However, a trade-off can be done for hardware realisation of array system. One should not forget that choice of radiating antenna element should be based on the radiation characteristics of a group of elements constituting array and not on that of the isolated element.

The geometry of the array has important significance in controlling the array performance owing to the coupling effect. Further, the mounting of the array on the platform affects the radiation characteristics drastically. The role of surface becomes dominant, which includes the curvature and the material. This makes the location of array a critical issue. The field of view, i.e. scan area of the array should be such that the antenna can perform the desired task. The other components of the airframe should not hinder the field of view of the antenna array. The array should also not face electromagnetic interference (EMI) from other active parts of the aircraft. Other factors for antenna location are the airframe, impact of ice, dust, water and temperature. Thus, one can infer that the performance of antenna system is affected due to airframe hindrance, non-optimal antenna siting or improper selection of antenna components. The sensor payload is known to pose difficulties in sustaining environmental loads.

In addition, the antenna and sensor design and its integration into the composite structure should be included right from the beginning of the aerospace vehicle's design and development. This includes the electromagnetic (EM) design, fabrication and installation of conformal load-bearing antennas on the platform, performance related to the sidelobe level, controlled beam steering, thermal balance due to circuitry of the feed network and the processing unit.

Planar printed antennas are ideal choice for use in aerial vehicles as they are easy to fabricate and bond onto the airframe, making them an integral part of the vehicle. The feed network of such an antenna array can be fabricated as a part of the antenna system owing to its planar structure, making it a compact and lightweight system. The most common location is the front or nose cone region of the aircraft. The nose cone has the greatest available frontal square footage in which antenna array can be placed along with dielectric covering. The hemispherical scan coverage can be easily achieved at this location. Another possible location is the leading edge of the wings, which provides the desired elevation scan coverage. In azimuth plane, however, there can be blockage by the fuselage if an array is mounted only on one wing. The UHF and X-band antenna array can be integrated into the structure for multiple frequency bands and full angular coverage. This eliminates the weight issues of conventional radars and radomes. High gain, electronically steerable UHF antennas mounted over the aircraft wings can locate and track both the stationary and moving targets. In general, the blade antenna employed in data-link applications is mounted over the fuselage. Since it protrudes out of the airframe, it is prone to the damage. Further, such protrusions contribute to EM scattering and, hence, the RCS. Such antennas can be replaced by flush-mounted slot antenna within circular composite radome. Similarly, dipole antenna can be embedded into the structure.

Recent advances in materials, electronics and novel designs proved that flexible, conducting and high-tensile threads can be weaved over the vehicle structure. Such textile-based antennas facilitate lowering the operational frequency (Bayram et al. 2010). Moreover, such antennas conform over the platform. These antennas are scalable and can be removed without any adverse effects. These antennas add to structural stiffness of the platform. These conformal, lightweight and load-bearing antennas have wide applications in the aerospace sector. However, embedding such antennas within a skin of the vehicle requires proper surface modelling. The antennas should be robust enough to withstand flexibility, and the algorithms implemented within the system must control the beam steering over a curved, moving surface.

In the current scenario, miniaturised conformal multifunctional antennas are being used on aerospace platforms. These antennas are developed using lightweight, high-contrast polymer-composite materials, which allow 3-D fabrication. This combination shows low electric loss tangent ($\tan \delta < 0.02$) for a wide range of frequency (GHz) and, hence, may be used as dielectric substrates for printed antennas. Carbon nanotube (CNT) is used for lowloss printing on the polymer substrate (Zhou et al. 2010), (Bayram et al. 2010). Such a CNT antenna exhibits light weight, flexibility, controlled permittivity and low electric loss tangent. These antennas can be a preferred choice for aerospace vehicles.

Further, degrees of conforming and structure compatibility are essential part of the antenna specifications. When the antenna conforms to the platform, it is either bent or stretched. In such situations, the antenna performance must be maintained besides withstanding mechanical forces like vibrations, pressure and temperature variations. These requirements are fulfilled by composite materials. Such materials may be processed at normal temperature, thereby making them favourable for onboard multilayered compact system packaging.

Another current trend in aerospace antennas is the use of ultra wideband (UWB) antennas, especially for wireless communication and defense applications. The wide frequency operation range makes the RCSR of UWB antenna a difficult task (Salman 2009). (Jiang et al. 2010) computed the RCS of UWB antenna in terms of the open-circuit and short-circuit RCS. Several circular and semicircular holes are made in the outer metallic covering of the antenna, thereby reducing its structural RCS by 3–5 dB, especially in the high-frequency range (6–20 GHz). However, these holes do not affect the radiation characteristics of the antenna.

Printed antennas are the preferred antenna for conformal arrays mounted on platforms such as aircraft and missiles. These antennas can be easily implemented over curved surfaces, reducing the overall scattering cross section of the vehicle. The RCS of such an antenna can be reduced further by using lossy substrate above the microstrip patches (Jackson 1990), by using ferrite material as substrate and cover layer (Yang et al. 1992), by lumped loading (Yang and Gong 2004), (Ma et al. 2004), or by distributed impedance loading (Volakis 1992). These methods, however, affect the radiation characteristics of antennas. Zhao et al. (2009) used reconfigurable Yagi-Uda substrate on the top of the patch to reduce RCS by 20–25 dB. Cutting slots on patch or ground plane and using FSS as ground can also contribute in RCSR up to 15 dB (Hong et al. 2010), (Ren et al. 2011).

The antenna radiation pattern can be dynamically controlled by using electronically controllable switching devices such as the PIN diodes (Jamlos et al. 2010), (Bai et al. 2011). The antenna mode can be switched between low RCS and radiation state, making both the radar cross section reduction and optimal radiation performance possible. The low RCS state over a large angular span is obtained by RF resistance characteristics in forward bias ON condition of the diodes. During forward bias, the field energy induced in the PIN diode dissipates, thereby reducing the RCS by 15–20 dB (Shang et al. 2012). When an antenna is required in optimal radiation state, the capacitance characteristic of PIN diode is used by applying DC reverse bias voltage.

A dual-polarised conformal array operating in X-band has been proposed for unmanned aerial vehicles (Ahn et al. 2011). The antenna array has sixteen 2 × 2 sub-arrays conformal to the cylindrical surface. The radii of curvature can be altered for fitting payload container in the vehicle. The feasibility of mounting a conformal active electronically scanned antenna (AESA) over the airborne vehicles has been demonstrated (Gerini and Zappelli 2005), (Knott et al. 2012).

8.3 Conformal load-bearing antenna

Conformal antenna structures integrated into aircraft and missiles have the potential to reduce weight, size and observability. In general, commercial and military aircraft consist of antennas that contribute towards aerodynamic drag and weight, since these antennas protrude out of the surface. Even conventional conformal antenna structures involve housing the antenna within a radome, which is either attached to the skin of the aircraft or replaces part of the outer skin. Such designs also add to overall weight and drag of the platform. As per recent trends, these antenna structures can be replaced with conformal antenna patterns deposited directly onto or embedded within the composite aerodynamic surfaces. Such antennas are referred to as conformal load-bearing antenna structure (CLAS).

The main objective of such antennas is to avoid mounting of conventional antennas that protrude out from the airframe. Moreover, they conform the outer mould line (OML) of aircraft structure, support structural loads, and perform the transmission and reception of the signals. The antennas embedded within the composite structure have improved RF performance, low weight, low cost and low drag, and they do not disturb the aerodynamic characteristics (Callus 2008).

Since CLAS supports the structural load, the panel skin from outside should be made of high-strength materials such as carbon fibre reinforced polymer (CFRP) or alloys. However, these highly stiff materials do not allow EM propagation through them. Thus, a rectangular-shaped recess is created in the inner portion of skin such that it remains continuous along the outer mould line, providing support to the structural loads and space for fixing of antenna components. The outer face of CLAS conforms to OML of the platform.

The material of the cover should be transparent to RF. Thus, the cover may be of glass fibre reinforced polymer (GFRP) or quartz-based fibre composites. The transmission losses are minimised by keeping the material thickness up to few millimetre and distance from the antenna of the order of half-wavelength. The low stiffness of GFRP should not lead to secondary bending and, hence, reduction of fatigue period. Moreover, the deformation effect should not degrade the antenna radiation and structural characteristics. The radiating elements can be of different shapes and dimension. The material of the elements can be a good conductor but less dense in view of durability when embedded in composite structures. Most CLAS radiators are oriented along OML to minimise antenna thickness. However, recent designs include radiators oriented in the through-thickness direction. All orientations that satisfy the volume/size constraints and produce the desired radiation pattern could be considered.

Next to the radiating elements, either honeycomb- or foam-based core is placed in between the load-bearing sheets. The honeycomb core rather than foams is used because of its lower density and electromagnetic losses. The CFRP sheet is bonded to the core so as to bear the load. The radiating element is placed in a bathtub-shaped recess. This conductive sheet serves as the ground plane. A lossy dielectric layer is put on the back face of CLAS to avoid back lobe radiations.

The CLASs are mainly in the form of printed antennas, i.e. microstrip patch antennas. The efficiency of microstrip antenna is known to depend on the transmission loss within the dielectric substrate and the energy coupled within the antenna. Finding appropriate material with the required permittivity and permeability is a challenge for antenna designers. The solution can be a metamaterial-based substrate. Another option may be to mix dielectric materials (GFRP or QFRP) with magnetic materials (ferromagnetic or paramagnetic strands). The mechanical strength of such antennas would satisfy the stiffness requirement of load-bearing structure and would be able to tolerate the strains within the structures. The problem is to integrate such materials within the conformal load-bearing structures. The manufacturing process will depend on the size and tolerance requirements. This issue is more critical for antenna design at the higher frequencies.

In a typical microstrip patch antenna, the dielectric substrate is thick (~10 mm for X-band). Since the composite aerospace vehicle skin thickness is of the same order, the skin at antenna location becomes doubly thick and increases further if supporting structures are added for stiffness. On the contrary, such type of thickness gets absorbed within the sandwich skin in CLAS due to the recessing of load-bearing skin. In general, the materials used for recess are resins or composite fibres. The stiffness of these materials is comparatively less than that of CFRP skin of the aerospace vehicle. Further, on applying load to such structures, even if along the skin, it creates secondary bending. Thus, the vehicle skin needs thickening for resisting any strain or bending, which on the contrary adds to the weight and complexity of the antenna system. However, an intelligent design might help solve this issue.

If the outer plies of CFRP skin are put off-axis with respect to the main load, the antenna would have least effect on the load-bearing capability of the platform skin (Callus 2008). The feed line inserted through hole should also not degrade the structural integrity. This is not an issue since the composite aircraft structure is often made with the strength of open or filled hole compression.

Another antenna that can be integrated on the airframe structure without additional reinforcement and acceptable performance degradation is the slot array. The composite skin itself can be used as a ground plane. To restore aerodynamic surfaces, the slots should be filled with materials like low modulus polymer resin or composites such as GFRP or QFRP. The windows with straight sides have low load-bearing capability and strain tolerance, which can be improved by scarfing the joints and putting tapered metallic bushes (Callus 2008).

Load-bearing aerospace structures are commonly made of hat- or blade-based stiffened skins. These stiffeners are similar to slot array in waveguides. However, the size and shape of such stiffeners should be properly designed so as to achieve structural support for the slot array over the desired frequency range. The bandwidth can be further increased by filling the slots with dielectric materials.

Instead of installing different antennas over the aircraft surface for operation at distinct frequencies, multiple waveguide stiffeners with varied dimensions can be designed, each tuned for a different frequency range. Another option is to have dispersed slot configuration in which the slots for higher frequency are designed in between the lower frequency slots. The critical issue in the slot design is the inter-element spacing. The frequency bands should be sufficiently separated so as to accommodate high-frequency slots in between the low-frequency slots. The radiation over the entire frequency range can be achieved by changing the shape and size of the slots, e.g. the spiral slot for bandwidth enhancement.

The composite UAV surface consisting of blades in internal design serves both as the stiffeners and the base for cutting slot arrays. This design reduces the skin thickness while maintaining sufficient structural strength. Materials like ferroelectrics are being actively used in the design and development of tunable load-bearing antennas. These materials are of interest because of the significant variation in their dielectric permittivity on the application of electric field. In particular, when used in load-bearing antenna subjected to mechanical strains, the permittivity of the material varies, which in turn affects its radiation characteristics.

8.4 FSS-based RCSR

The FSS consist of the periodic arrangement of metallic structures of arbitrary shape and size over dielectric substrate or their complimentary design with apertures on a metallic sheet. The most critical issue is the appropriate choice of geometry of the FSS element which could be rectangular/circular slots, annular slots, a single/multiloaded slot, loop, ring, etc. The performance of the FSS structure is governed by the type and geometry of the element, the constitutive parameters of dielectric substrate and the periodicity of elements. The performance indices of FSS structures include the frequency response, transfer function and its dependence on the angle of incidence and polarisation (Munk 2000). The choice of FSS element depends on the design specifications such as narrowband/multiband operation, occupancy and bandwidth requirements, etc. The FSS are essentially filters that transmit or reflect EM waves at the chosen frequency bands.

Metallic-patch-based FSS exhibit total reflection at the design frequency while aperture-based FSS allow total transmission at the resonant frequency. These FSS structures can be incorporated into radome or antenna systems without affecting the desired radiation characteristics of the antenna. Moreover, these structures facilitate the antenna RCSR for all out-of-band frequencies (Ren et al. 2011).

To reduce the antenna RCS, the recent trend is to incorporate FSS ground plane in the design of patch antennas. The main aim in such designs is to include a perfect electric conductor (PEC)-based ground plane valid for the antenna-operating frequency range and pass-band characteristics at other frequencies. Further, FSS prevent the reflection of energy back to the antenna aperture. This leads to significantly lower antenna RCS while simultaneously maintaining the radiation characteristics (Lu et al. 2009). An out-of-band RCSR of 20–30 dB can be achieved.

Loaded FSS structures such as single- and double-square loops loaded with chip can lead to wideband absorption owing to the ohmic losses in lumped resistors. Such structures have application as thin and low cost radar absorbers (Yang and Shen 2007). Further, the capabilities of the FSS can be extended by incorporating active devices within the FSS unit cells (Chang et al. 1996), (Khosravi and Abrishamian 2007). An active FSS can switch its frequency response from a rejecting structure to one which is virtually transparent but with a small insertion loss. This change in characteristics is electronically controlled and can be exploited for RCSR.

The FSS structure behind the low-profile printed antenna array allows the incident waves to travel through the antenna structure in the pass-band frequency range of FSS (Genovesi et al. 2012). If the reflecting structure exists behind such an antenna, the transmitted energy into the antenna structure through FSS might be reflected back, limiting the efficiency of FSS layer in the antenna system. The solution to this problem is obtained by placing the absorbing screen behind the antenna structure in the vicinity of the FSS layer. This helps in modifying the impedance experienced by the incident waves. This effect depends on the optimal separation between the antenna structure and the absorbing layer. A perfect impedance match is achieved when the reflection behaviour of the multilayered antenna system is close to that of free space.

The RCS of a low-profile microstrip patch array with FSS-based ground plane is reported to be 15–20 dB less within the transmission band of the FSS. The RCSR is achieved even for oblique angle of incidence for the same frequency band (Genovesi et al. 2012). The radiation characteristic of the array is maintained with significant reduction for the out-of-band radar signatures. The FSS-based design employed for RCSR must satisfy the conditions that (i) the total reflection frequency band should match with the antenna resonance frequency range and (ii) the resonant frequency of the antenna system should be stable within the range of incident angles of the incoming waves.

8.5 Metamaterial-based RCSR

Metamaterials owing to their peculiar propagation characteristics are used for RCSR (Xu et al. 2010), for antenna design with low sidelobe level and for avoiding EM interference. Multilayered structures composed of lossy dielectric materials and metamaterials have been used for the realisation of wideband RAM (Oraizi and Abdolali 2008). Oraizi et al. (2010) presented the applications of double zero metamaterial (DZR) coatings in dispersive and non-dispersive conditions. The reflected power from RAM coatings is optimised using the hybrid method (GA-CG) based on genetic algorithm and conjugate gradient technique.

The FSS can be used for the out-of-band RCSR of antennas. The challenge is to achieve in-band RCSR of antennas. The electromagnetic band-gap (EBG)-based absorbing material loaded with lumped components such as resistances (Li et al. 2008) or metamaterial-based absorbers (Yang et al. 2013) have been used for RCSR of slot array on a ridged waveguide.

Metamaterial-based absorbers are reported to show absorbance bandwidth above 90 per cent along with RCSR of 10 dB (Yang et al. 2013). The low RCS behaviour remains insensitive to the angle of incidence and both the parallel and perpendicular polarisations. If a metamaterial-based absorber instead of PEC is used as a ground plane of slot antenna on a waveguide, the in-band antenna RCS is reduced significantly (Liu et al. 2013). Overall RCSR reduction up to 8 dB can be achieved within the operational frequency band. Moreover, the frontal RCS can be reduced up to 15 dB, without affecting the radiation characteristics.

Recently, a dual-band metamaterial absorber has been proposed in microstrip antenna design (Zhang et al. 2014). This four-layered structure shows high absorption for wide incidence angles and is polarisation-insensitive. If such an absorber is loaded in a microstrip patch antenna, its in-band RCS reduces up to 9 dB for both the TE and TM polarisations without any adverse effect on the radiation pattern.

A structurally integrated metamaterial-based conformal antenna can provide the required communication capability without degradation in aerodynamic, structural and RF performance. The ground plane is made of metamaterial EBG material, reducing the antenna height. The EBG-based ground plane behaves as a perfect magnetic conductor (PMC), unlike conventional perfect electric conductor (PEC) ground. The EBG-based ground plane allows the antenna radiating element to be placed in close proximity to the ground, as compared to $\lambda/4$ spacing from a PEC ground, thus providing significant reduction in height. Moreover, the metamaterial EBG ground plane allows the antenna thickness to be approximately 10 per cent that of the conventional antenna design. Overall RCSR of up to 10–15 dB can be achieved without degrading the antenna performance. If magnetic materials such as polycrystalline ferrites are used in the antenna design, bandwidth of the antenna can be further improved.

A structurally integrated metamaterial-based conformal array being a low-profile array occupies less volume of the platform and, hence, provides extra space for other sub-systems and auxiliaries. There are minimal protrusions from the surface, thereby reducing aerodynamic drag and EM scattering from the antennas. Since the antenna is structurally integrated with the platform, there is no need for parasitic packaging. This facilitates weight reduction and enhanced flexibility. The system robustness will be more, with reduced maintenance and life cycle cost. Such structurally integrated antenna can be mounted over various platforms such as aircraft, missiles or unmanned vehicles. Further, the choice of materials ranges from metal, dielectrics to composites. This helps in preserving the radiation and structural performance of the antenna system.

Extensive research has been carried out for planar metamaterial-based absorbers towards RCSR and control. Recently, a conformal metamaterial absorber has been proposed for curved surfaces (Jang et al. 2013). It is known that the angle of incidence changes at different points on a curved surface. This requires a variational design of metamaterial unit cell at different angles. Based on three different metamaterial-based unit cells, a conformal absorber on a cylindrical surface has been proposed. An RCSR of 10–12 dB is achieved while maintaining high absorption characteristics for both the polarisations.

Apart from RCSR, metamaterials can also be exploited for antenna gain enhancement. (Pan et al. 2013) proposed a partially reflecting surface (PRS)-based microstrip patch antenna design. The patch antenna is embedded between the two parallel reflectors, one being the metallic ground plane and another being the PRS, consisting of metallic periodic structure placed at half-wavelength above a ground plane. This creates a Fabry–Perot (F–P) type cavity, thereby enhancing the antenna directivity. Due to different metallic patterns on both the sides of a dielectric slab, the upper and bottom surfaces exhibit different transmission and reflection behaviour. The upper surface shows low reflections, thereby reducing the out-of-band antenna RCS (10–25 dB). An in-band antenna RCSR of 3 dB is obtained. The proposed design shows a gain enhancement of 6 dB.

Apart from the antenna RCS, metamaterial-based coatings have also been used for the RCSR of structures (Bucinskas et al. 2010), Culhaoglu et al. (2013) demonstrated that an electrically thin metamaterial-based coating with low-reflection characteristics can achieve significant reduction of both monostatic and bistatic RCS of a metallic cube of electrically large dimensions, in Ka band. The proposed coating consists of a metallic strip array printed over a substrate. These strips are of sub-wavelength dimension with capacitive load behaviour. The reduction in scattering cross section of ≈10 dB is archived in specular as well as non-specular directions. This can be further explored for electrically large 3 D structures.

8.6 Plasma-based RCSR

Plasma stealth is yet another technique that has been tried for implementing low observability. The plasma, an ionised gas with different parameters (plasma density, temperature, collision frequency and applied magnetic field) and electron density profile (exponential, parabolic and Epstein) has different responses towards EM wave propagation. This peculiar feature of plasma has potential in stealth applications.

In plasma-based stealth, the electric field surrounding the vehicle is altered to reduce the RCS. If a plasma cloud could be created around the aircraft exterior, EM waves would be absorbed by the surface instead of reflecting back to the radar antenna. This plasma-based method represents one of the leading edge stealth technologies and is still being explored.

The motivation to use plasma stealth is similar to that of RAMs. When an EM wave propagates in weakly ionised plasma, it undergoes scattering and absorption. The absorption is mainly due to the transfer of energy from EM wave to charged particles of plasma and, then, to neutral atoms due to elastic/inelastic collisions (Yuan et al. 2011). Wave scattering takes place due to the variation of refractive index, when EM wave travels either from free space to plasma or propagates through a varying density within the plasma (Laroussi and Roth 1993).

The advantages of plasma absorber include being lightweight, having attenuation of EM wave over a wide frequency range and having provision of switching off the plasma generator when not required. However, the power supply required is bulky. The plasma absorber may be used to wrap the portions, which contribute significantly to the overall RCSR (20–30 dB) (Chaudhury and Chaturvedi 2005, 2009). Unmagnetised plasma can also be used as RAM over a range of frequencies (Stadler et al. 1992).

Sufficiently dense plasma exhibits conductor-like properties and, thus, can be used as transmitting and receiving antenna. The plasma behaves as a conductor when illuminated by an EM source. On the contrary, it acts as a non-conductive material once the source is switched off. This property can be exploited to have an RCS that virtually disappears for the probing radar. Further, a plasma antenna, compared to the conventional antenna, can serve as reconfigurable radiating element. Its radiation characteristics may be varied electrically instead of avoiding mechanical control.

Plasma-based RCSR has both advantages and disadvantages. The plasma cloud can replace RAM coatings, thereby reducing the weight issues and the cost of maintenance. Moreover, the plasma clouds provide heat shielding and improve the aerodynamic performance of the aircraft. However, there are a few challenges related to plasma stealth technology that need to be considered. The first issue is to generate the plasma cloud around the aircraft. The source requires high power and would increase the overall weight of the aircraft. The next problem is to avoid the blocking of pilot's radar by the plasma layer (Vidmar 1990). Finally, the plasma cloud generates a visible plasma trail which may defeat the very purpose of RCSR.

8.7 Conclusion

Demand for small-size and mitigation of RF congestion (communications, imaging and information gathering on small platforms) requires antennas to be reduced in size by a factor of 10 or more, while maintaining the desired gain and data rates. Moreover, the antennas/arrays are required to be multifunctional, covering wide frequency range (UWB). The entire aerospace vehicle surface might serve as an RF (smart) skin, implying structural integrity of the RF functional layers. The antennas mounted over the vehicle surface are subjected to deflections and aero-elastic deformations. This problem is reduced if the antenna array is embedded within the structure. This helps in weight and size control, apart from the RCSR.

The realisation of low RCS antenna can be done using tapered resistive sheets, plasma-based, FSS-based or metamaterial-based antenna design. The active RCSR is required to be integrated with the passive RCSR to get overall control over electromagnetic scattering. For this, conformal structurally integrated antennas would be preferred choice.

The physical size of antennas for low-frequency applications is an issue for aerospace vehicles. The textile thread-based load-bearing array can be attached to the structure. The power requirements of phased array vary according to the application and frequency range. The factors that affect antenna performance include airframe hindrance, non-optimal antenna siting or improper selection of antenna components. These conformal, lightweight and load-bearing antennas serve as communication, navigation and surveillance systems.

RCSR using FSS is achieved by virtue of its pass-band and stop-band frequency characteristics. Transparency can be achieved within the operating frequency range while reflecting the out-of-band signals in different directions. This method is referred to as counterfeit RCSR, effectual in monostatic radars. However, for bistatic radars, the incident signal reflected in other directions can also be detected.

The use of lossy paints over radomes could cause high-insertion loss. Metamaterial-based absorber can be one of the options to circumvent this problem. However, these absorbers also have limited efficiency owing to narrowband absorption characteristics. Another possibility is plasma stealth for aerospace vehicles. Although several technical issues remain to be sorted out, researchers expect that substantial RCSR is possible to achieve by covering the reflective portions of the structure by radar-absorbent plasma. These reflective portions include vertical stabilisers, turbojet engine fan blades and engine air intakes of the aircraft.

References

Ahn, C. -H., Y. -J. Ren, and K. Chang. 2011. 'A dual-polarised cylindrical conformal array antenna suitable for unmanned aerial vehicles.' *International Journal of RF and Microwave Computer-Aided Engineering* 21: 91–98.

Bai, Y. -Y., S. Q. Xiao, and M. -C. Tang. 2011. 'Wide-angle scanning phased array with pattern reconfigurable elements.' *IEEE Transactions on Antennas and Propagation* 59: 4071–76.

Bayram, Y., Z. Yijun, B. S. Shim, S. Xu, J. Zhu, N. A. Kotov, and J. L. Volakis. 2010. 'E-textile conductors and polymer composites for load bearing antennas.' *IEEE Transactions on Antennas and Propagation* 58: 2732–36.

Bucinskas, J., L. Nickelson, and V. Shugurovas. 2010. 'Microwave scattering and absorption by a multilayered lossy metamaterial-glass cylinder.' *Progress in Electromagnetics Research*. 105: 103–18.

Callus, P. J. 2008. 'Novel concepts for conformal load-bearing antenna structure.' DSTO-TR-2096, Technical Report, DSTO Platforms Sciences Laboratory, Australia, 94.

Chang, T. K., R. J. Langley, and E. A. Parker. 1996. 'Active frequency selective surfaces.' *IEE Proceedings on Microwave Antennas Propagation* 143(1): 62–66.

Chaudhury, B. and S. Chaturvedi. 2009. 'Study and optimisation of plasma-based radar cross section reduction using three-dimensional computations.' *IEEE Transactions on Plasma Science* 37: 2116–27.

Chaudhury, B. and S. Chaturvedi. 2005. 'Three-dimensional computation of reduction in radar cross section using plasma shielding.' *IEEE Transactions on Plasma Science* 33: 2027–34.

Chen, H. -Y., L. -J. Deng, P. -H. Zhou, and J. -L. Xie. 2011. 'Tapered impedance loading for suppression of edge scattering.' *IET Microwave, Antennas and Propagation* 5(14): 1744–49.

Culhaoglu, A. E., A. V. Osipov, and P. Russer. 2013. 'Mono- and bistatic scattering reduction by a metamaterial low reflection coating.' *IEEE Transactions on Antennas and Propagation* 61: 462–66.

Doan, L. R., P. A. Day, and O. Brovko. 1995. 'Large dynamic range radar cross section parallel tracking.' Los Angeles CA: RL-TR-95-212, Hughes Aircraft Company, 192.

Genovesi, S., F. Costa, and A. Monorchio. 2012. 'Low-profile array with reduced radar cross section by using hybrid frequency selective surfaces.' *IEEE Transactions on Antennas and Propagation*. 60: 2327–35.

Gerini, G. and L. Zappelli. 2005. 'Multilayer array antennas with integrated frequency selective surfaces conformal to a circular cylindrical surface.' *IEEE Transactions on Antennas and Propagation* 53: 2020–30.

Gustafsson, M. 2006. 'RCS reduction of integrated antenna arrays with resistive sheets.' *Journal of Electromagnetic Waves and Applications* 20: 27–40.

Hong, T., L. -T. Jiang, Y. -X. Xu, S. -X. Gong, and W. Jiang. 2010. 'Radiation and scattering analysis of a novel circularly polarised slot antenna.' *Journal of Electromagnetic Waves and Applications* 24: 1709–20.

Hu, S., H. Chen, C. Law, Z. Shen, L. Zhu, W. Zhang, and W. Dou. 2007. 'Backscattering cross section of ultra wideband antennas.' *IEEE Antennas and Wireless Propagation Letters* 6: 70–73.

Jackson, D. R. 1990. 'The RCS of a rectangular microstrip patch in a substrate-superstrate geometry.' *IEEE Transactions on Antennas and Propagation* 38: 2–8.

Jamlos, M. F., O. A. Aziz, T. B. A. Rahman, M. R. B. Kamarudin, P. Saad, M. T. Ali, and M. N. Md Tan. 2010. 'A reconfigurable radial line slot array (RLSA) antenna for beam shape and broadside application.' *Journal of Electromagnetic Waves and Applications* 24: 1171–82.

Jang, Y., M. Yoo, and S. Lim. 2013. 'Conformal metamaterial absorber for curved surface.' *Optics Express* 21(20): 24163–70.

Jiang, W., T. Hong, Y. Liu, S. -X. Gong, Y. Guan, and S. Cui. 2010. 'A novel technique for radar cross section reduction of printed antennas.' *Journal of Electromagnetic Waves and Applications* 24: 51–60.

Khosravi, M. and M. S. Abrishamian. 2007. 'Reduction of monostatic RCS by switchable FSS elements.' *PIERS Online* 3(6): 770–73.

Knott, E. F., J. F. Shaeffer, and M. T. Tuley. 2004. *Radar Cross Section*. Raleigh, NC: SciTech, 611.

Knott, P., T. Bertuch, H. Wilden, O. Peters, A. R. Brenner, and I. Walterscheid. 2012. 'SAR experiments using a conformal antenna array radar demonstrator.' *International Journal of Antennas and Propagation* 2012: 7.

Laroussi, M. and J. R. Roth. 1993. 'Numerical calculation of the reflection, absorption, and transmission of microwaves by a non-uniform plasma slab.' *IEEE Transactions on Plasma Science* 21: 366–72.

Li, Y., H. Zhang, Y. Fu, and N. Yuan. 2008. 'RCS reduction of ridged waveguide slot antenna array using EBG radar absorbing material.' *IEEE Antennas Wireless Propagation Letters* 7: 473–76.

Liu, T., X. Cao, J. Gao, Q. Zheng, W. Li, and H. Yang. 2013. 'RCS reduction of waveguide slot antenna with metamaterial absorber.' *IEEE Transactions on Antennas and Propagation* 61: 1479–84.

Lu, B., X. Gong, J. Ling, and H. W. Yuan. 2009. 'Radar cross-section reduction of antennas using a stop-band frequency selective surface.' *Microwave Journal* 52: 6.

Ma, H., C. Xu, and H. Zheng. 2004. 'Effect of impedance load on radiation and scattering of microstrip antenna.' *Modern Electronic Technology* 6: 6–7.

Misran, N., R. Cahill, and V. F. Fusco. 2003. 'RCS reduction technique for reflectarray antennas.' *Electronic Letters* 39(23): 1630–31.

Munk, B. A. 2000. *Frequency Selective Surfaces: Theory and Design*. New York: Wiley-Interscience, 440.

Oraizi, H. and A. Abdolali. 2008. 'Design and optimisation of planar multilayer antireflection metamaterial coatings at Ku band under circularly polarised oblique plane wave incidence.' *Progress in Electromagnetics Research C* 3: 1–18.

Oraizi, H., A. Abdolali, and N. Vaseghi. 2010. 'Application of double zero metamaterials as radar absorbing materials for the reduction of radar cross section.' *Progress in Electromagnetics Research* 101: 323–37.

Pan, W., C. Huang, P. Chen, X. Ma, C. Hu, and X. Luo. 2014. 'A low-RCS and high-gain partially reflecting surface antenna.' *IEEE Transactions on Antennas and Propagation* 62: 945–49.

Paterson, J. 1999. 'Overview of low observable technology and its effect on combat aircraft survivability.' *Journal of Aircraft* 36(2): 380–88.

Pozar, D. M. 1992. 'RCS reduction for a microstrip antenna using a normally biased ferrite substrate.' *IEEE Microwave Guided Wave Letters* 2: 196–98.

Ren, L. -S., Y. -C. Jiao, J. -J. Zhao, and F. Li. 2011. 'RCS reduction for a FSS-backed reflectarray using a ring element.' *Progress in Electromagnetics Research Letters* 26: 115–23.

Salman, N. 2009. 'Low profile and compact size coplanar UWB antenna working from 2.8 GHz to over 40 GHz.' *Microwave and Optical Technology Letters* 51: 408–11.

Shang, Y. P., S. Q. Xiao, J. L. Li, and B. -Z. Wang. 2012. 'An electronically controllable method for radar cross section reduction for a microstrip antenna.' *Progress in Electromagnetics Research* 127: 15–30.

Sneha, H. L., H. Singh, and R. M. Jha. 2012. 'Mutual coupling effects for radar cross section (RCS) of a series-fed dipole antenna array.' *International Journal of Antennas and Propagation* 2012: 20.

Stadler, K. R., R. J. Vidmar, and D. J. Eckstrom. 1992. 'Observations of strong microwave absorption in collisional plasmas with gradual density gradients.' *Journal of Applied Physics* 72: 5089–94.

Vidmar, R. J. 1990. 'On the use of atmospheric pressure plasmas as electromagnetic reflectors and absorbers.' *IEEE Transactions on Plasma Science* 18: 733–41.

Volakis, J. L., A. A. Lexanian, and J. M. Lin. 1992. 'Broadband RCS reduction of rectangular patch by using distributed loading.' *Electronic Letters* 28(25): 2322–23.

Wang, F. W., S. X. Gong, S. Zhang, X. Mu, and T. Hong. 2011. 'RCS reduction of array antennas with radar absorbing structures.' *Journal of Electromagnetic Waves and Applications* 25: 2487–96.

Wang, W. -T., S. -X. Gong, X. Wang, H. -W. Yuan, J. Ling, and T. -T. Wan. 2009. 'RCS reduction of array antenna by using bandstop FSS reflector.' *Journal of Electromagnetic Waves and Applications* 23: 1505–14.

Xu, H. -Y., H. Zhang, X. Yin, and K. Lu. 2010. 'Ultra-wideband Koch fractal antenna with low backscattering cross section.' *Journal of Electromagnetic Waves and Applications* 24: 2615–23.

Yang, H., X. Cao, J. Gao, W. Li, Z. Yuan, and K. Shang. 2013. 'Low RCS metamaterial absorber and extending bandwidth based on electromagnetic resonances.' *Progress in Electromagnetics Research* 33: 31–44.

Yang, H. and S. Gong. 2004. 'RCS reduction technique out of operation of microstrip antennas.' *Journal of Microwaves* 20: 35–39.

Yang, H. -Y., J. A. Castaneda, and N. G. Alexopoulos. 1992. 'Multifunctional and low RCS nonreciprocal microstrip antennas.' *Electromagnetics* 12(1): 17–31.

Yang, J. and Z. X. Shen. 2007. 'A thin and broadband absorber using double-square loops.' *IEEE Antennas and Wireless Propagation Letters* 6: 388–91.

Yuan, C. X., Z. X. Zhou, J. W. Zhang, X. L. Xiang, Y. Feng, and H. G. Sun. 2011. 'Properties of propagation of electromagnetic wave in a multilayer radar absorbing structure with plasma and radar absorbing material.' *IEEE Transactions on Plasma Science* 39: 1768–75.

Zhao, S. -C., B. -Z. Wang, and W. Shao. 2009. 'Reconfigurable Yagi-Uda substrate for RCS reduction of patch antenna.' *Progress in Electromagnetics Research B* 11: 173–87.

Zhou, Y., Y. Bayram, F. Du, L. Dai, and J. L. Volakis. 2010. 'Polymer-carbon nanotube sheets for conformal load bearing antennas.' *IEEE Transactions on Antennas and Propagation* 58: 2169–75.

Zhang, H., X. -Y. Cao, J. Gao, H. Yang, and Q. Yang. 2014. 'A novel dual-band metamaterial absorber and its application for microstrip antenna.' *Progress in Electromagnetics Research Letters* 44: 35–41.

Zhang, J., J. Wang, M. Chen, and Z. Zhang. 2012. 'RCS reduction of patch array antenna by electromagnetic band-gap structure.' *IEEE Antennas and Wireless Propagation Letters* 11: 1048–51.

Epilogue

The main objective of this book is to establish the feasibility of a novel concept of active radar cross section reduction (RCSR). In this context, the phased antenna array theory and algorithms are discussed.

A discerning reader may conclude that this book does not directly lead to the realisation of low-observable platforms, which is correct. The book in hand is not a handbook or a reference manual for achieving active RCSR over low-observable platforms.

Design and realisation of active RCS low-observable platforms through this route (i.e. the theory and algorithms of phased antenna arrays) would also involve on-board/airborne antenna analysis. As mentioned in the Preface of this book, the methods and techniques for on-board/airborne antenna analysis are now well-established across all the three (low, resonance and high) frequency regions. References are numerous and readers may have their own preferences.

Another important aspect is that of conformal arrays. Again, the modern conformal antenna theory now appears to have been systematised (Josefsson and Persson, 2006). Incidentally, the analytical core of the conformal array theory in this reference (Josefsson and Persson, 2006) relies extensively on the work of one of the authors (Jha et al. 1993), (Jha et al. 1994) or (Jha and Wiesbeck, 1995). Analysis of conformal phased arrays over the airborne platform would be required towards achieving the low-observable platform. The conformal antenna array RCS is summed up with the structural RCS of the platform to obtain a quiescent RCS pattern. Following the sensing of the direction of arrival (DOA) of the incoming radar wave, the overall structural plus antenna mode RCS pattern is adapted to obtain the active RCSR in the DOA, thereby leading to the active low-observable platform.

It may be noted that other techniques to realise low-observable platforms are also prevalent and have been discussed in this book. Broadbanding of RCSR is important and the basic theory of RAM design (Vinoy and Jha 1996) should not be lost sight of. Likewise, plasma stealth is emerging as yet another feasible option in the context of low-observable platforms.

Finally, it must be understood that the phased array may not be the only route to adapting the RCS pattern. Since active elements are involved, one is well-advised to expand the scope and look beyond to the FSS antennas and metamaterials as well.

References

Jha, R. M. and W. Wiesbeck. 1995. 'Geodesic constant method: A novel approach to analytical surface-ray tracing on convex conducting bodies.' *IEEE Antennas and Propagation Magazine* 37(2): 28–38.

Jha, R. M., P. R. Mahapatra, and W. Wiesbeck. 1994. 'Surface-ray tracing on hybrid surfaces of revolution for UTD mutual coupling analysis.' *IEEE Transactions on Antennas and Propagation* 42, 1167–75.

Jha, R. M., S. A. Bokhari, and W. Wiesbeck. 1993. 'A novel ray-tracing on general paraboloids of revolution for UTD applications.' *IEEE Transactions on Antennas and Propagation* 41, 934–39.

Josefsson L. and P. Persson. 2006. *Conformal Array Antenna: Theory and Design.* Hoboken, NJ: John Wiley & Sons, 472.

Vinoy, K. J. and R. M. Jha. 1996. *Radar Absorbing Materials: From Theory to Design and Characterisation.* Norwell, Boston: Kluwer Academic Publishers, 209.

Appendices

Appendix A

Calculation of self and mutual impedance between two antennas

When two antenna elements are close to each other, i.e. $d_{spacing} < \lambda/2$, the impedance of each antenna depends on the self impedance of that antenna and the mutual impedance between the two antennas. To illustrate, a two-element array can be represented by a general four terminal network (Figure A.1).

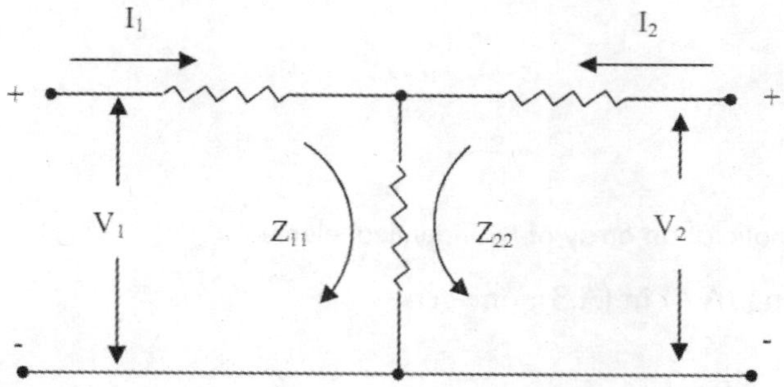

Figure A.1 Network equivalent of an array of two identical antenna elements

The voltages at input and output ports will be

$$V_1 = I_1 Z_{11} + I_2 Z_{12} \tag{A.1}$$

$$V_2 = I_1 Z_{21} + I_2 Z_{22} \tag{A.2}$$

where, Z_{11} and Z_{22} are self impedance of Antenna 1 and Antenna 2, respectively. Z_{12} and Z_{21} are mutual impedance between the antennas. The mutual impedance of Antenna 2 due to Antenna 1 is expressed as

$$Z_{21} = \frac{V_{21}}{I_1(0)} \tag{A.3}$$

where, V_{21} is the open-circuit voltage at the terminals of Antenna 2 produced by the base current $I_1(0)$ in Antenna 1. The open-circuit voltage may be expressed as

$$V_{21} = \frac{-1}{I_2(0)} \int_0^h E_{Z_{21}} I_2(z) dz \tag{A.4}$$

h is the height of Antenna 1, $E_{Z_{21}}$ is the parallel component of E-field along the axis of Antenna 2, $I_2(z)$ is the current along Antenna 2.

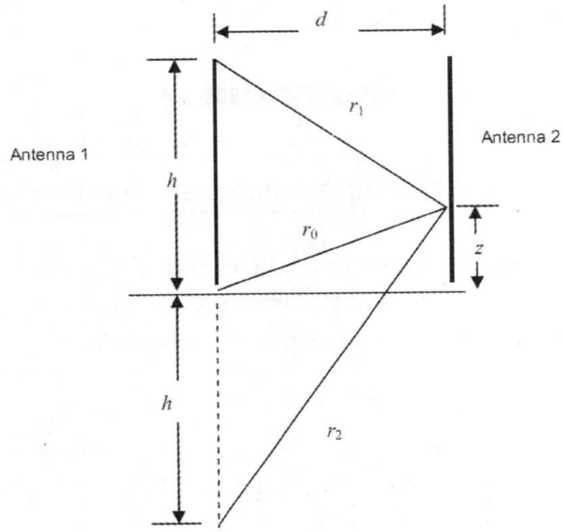

Figure A.2 Schematic of an array of two identical elements

Thus, substituting (A.4) in (A.3), one gets

$$Z_{21} = \frac{-1}{I_1(0) I_2(0)} \int_0^h E_{Z_{21}} I_2(z) dz \tag{A.5}$$

where, $I_2(z) = \dfrac{-I_2(0)}{\sin kh} \sin k(h-z) = I_{2m} \sin k(h-z)$

I_{2m} is the maximum current at Antenna 2.

For two antennas of height h mounted on the ground, as shown in Figure A.2

$$r_0 = \sqrt{d^2 + z^2} \quad r_1 = \sqrt{d^2 + (h-z)^2} \quad r_2 = \sqrt{d^2 + (h+z)^2}$$

where d is the spacing between antennas.

The parallel component of E-field along the axis of Antenna 2 is given by

$$E_{Z_{21}} = -j30 I_{m1} \left(\frac{e^{-jkr_1}}{r_1} + \frac{e^{-jkr_2}}{r_2} - 2\cos kh \frac{e^{-jkr_0}}{r_0} \right) \tag{A.6}$$

Thus, (A.5) becomes

$$Z_{21} = \frac{-30 I_{1m} I_{2m}}{I_1(0) I_2(0)} \int_0^H \begin{pmatrix} -\dfrac{je^{-jkr_1}}{r_1} - \dfrac{je^{-jkr_2}}{r_2} \\ +2j\cos kh \dfrac{e^{-jkr_0}}{r_0} \end{pmatrix} \sin k(h-z)dz \tag{A.7}$$

The mutual impedance of Antenna 2 with respect to the base of Antenna 1.

The mutual impedance with respect to loop current is obtained by multiplying the above equation by the term $\dfrac{I_1(0)I_2(0)}{I_{1m}I_{2m}}$.

Thus, the mutual impedance of Antenna 2 with respect to loop current is

$$Z_{21} = -30 \int_0^H \left(-\frac{je^{-jkr_1}}{r_1} - \frac{je^{-jkr_2}}{r_2} + 2j\cos kh \frac{e^{-jkr_0}}{r_0} \right) \sin k(h-z)d \tag{A.8}$$

$$z = R_{mutual} + jX_{mutual}$$

where R_{21} is the mutual resistance, X_{21} is the mutual inductance of Antenna 2, expressed as

$$R_{21} = 30 \int_0^H \sin k(h-z) \begin{pmatrix} \dfrac{\sin k\sqrt{d^2+(h-z)^2}}{\sqrt{d^2+(h-z)^2}} + \dfrac{\sin k\sqrt{d^2+(h+z)^2}}{\sqrt{d^2+(h+z)^2}} \\ -2\cos kh \dfrac{\sin k\sqrt{d^2+z^2}}{\sqrt{d^2+z^2}} \end{pmatrix} dz \tag{A.9}$$

$$X_{21} = 30 \int_0^H \sin k(h-z) \begin{pmatrix} \dfrac{\cos k\sqrt{d^2+(h-z)^2}}{\sqrt{d^2+(h-z)^2}} + \dfrac{\cos k\sqrt{d^2+(h+z)^2}}{\sqrt{d^2+(h+z)^2}} \\ -2\cos kh \dfrac{\cos k\sqrt{d^2+z^2}}{\sqrt{d^2+z^2}} \end{pmatrix} dz \tag{A.10}$$

For a pair of $\lambda/4$ monopoles spaced at distance d, the mutual resistance and mutual reactance are given by

$$R_{21} = 15[2C_i(kd) - C_i(\sqrt{(kd)^2+\pi^2} - \pi) - C_i(\sqrt{(kd)^2+\pi^2} + \pi)] \tag{A.11}$$

$$X_{21} = 15[S_i(\sqrt{(kd)^2+\pi^2} - \pi) + S_i(\sqrt{(kd)^2+\pi^2} + \pi) - 2S_i(kd)] \tag{A.12}$$

where, $C_i(kd)$ and $S_i(kd)$ are cosine and sine integrals, expressed as

$$S_i(x) = \sum_{k=0}^{\infty} \frac{(-1)^k x^{2k+1}}{(2k+1)(2k+1)!} .. \tag{A.13}$$

$$C_i(x) = C + \ln(x) + \sum_{k=1}^{\infty} (-1)^k \frac{x^{2k}}{2k(2k)!} \tag{A.14}$$

The general expression for the mutual impedance between antennas of equal height H and a distance d apart is given as

$$R_{21} = \begin{pmatrix} 30[\sin(kh)\cos(kh)(S_i(u_2) - S_i(v_2) - 2S_i(v_1) + 2S_i(u_1)) \\ -\dfrac{\cos(2kh)}{2}(2C_i(u_1) - 2C_i(u_0) + 2C_i(v_1) - C_i(u_2) - C_i(u_2)) \\ -(C_i(u_1) - 2C_i(u_0) + C_i(v_1))] \end{pmatrix} \tag{A.15}$$

$$X_{21} = \begin{pmatrix} 30[\sin(kh)\cos(kh)(2C_i(v_1) - 2C_i(u_1) + C_i(v_2) - C_i(u_2)) \\ -\dfrac{\cos(2kh)}{2}(2S_i(u_1) - 2S_i(u_0) + 2S_i(v_1) - S_i(u_2) - S_i(u_2)) \\ -(S_i(u_1) - 2S_i(u_0) + S_i(v_1))] \end{pmatrix} \tag{A.16}$$

where $u_0 = kd \quad u_1 = k(\sqrt{d^2 + h^2} - h) \quad v_1 = k(\sqrt{d^2 + h^2} + h)$

$u_2 = k(\sqrt{d^2 + (2h)^2} + 2h) \quad v_2 = k(\sqrt{d^2 + (2h)^2} - 2h)$

For a pair of $\lambda/2$ dipoles placed side-by-side (Figure A.3), spaced at distance d, the self and mutual impedances are given by

$$R_{self} = \frac{\eta}{2\pi}\left\{C + \ln(kl) - C_i(kl) + \frac{1}{2}\sin(kl)\left[S_i(2kl) - 2S_i(kl)\right]\right. \\ \left. + \frac{1}{2}\cos(kl)\left[C + \ln(kl/2) + C_i(2kl) - 2C_i(kl)\right]\right\} \tag{A.17}$$

$$X_{self} = \frac{\eta}{4\pi}\left\{2S_i(kl) + \cos(kl)\left[2S_i(kl) - S_i(2kl)\right]\right. \\ \left. - \sin(kl)\left[2C_i(kl) - C_i(2kl) - C_i\left(\frac{2ka^2}{l}\right)\right]\right\} \tag{A.18}$$

$$R_{mutual} = \frac{\eta}{4\pi}\left[2C_i(kd) - C_i\left(k\left(\sqrt{d^2 + l^2} + l\right)\right) - C_i\left(k\left(\sqrt{d^2 + l^2} - l\right)\right)\right] \tag{A.19}$$

$$X_{mutual} = -\frac{\eta}{4\pi}\left[2S_i(kd) - S_i\left(k\left(\sqrt{d^2 + l^2} + l\right)\right) - S_i\left(k\left(\sqrt{d^2 + l^2} - l\right)\right)\right] \tag{A.20}$$

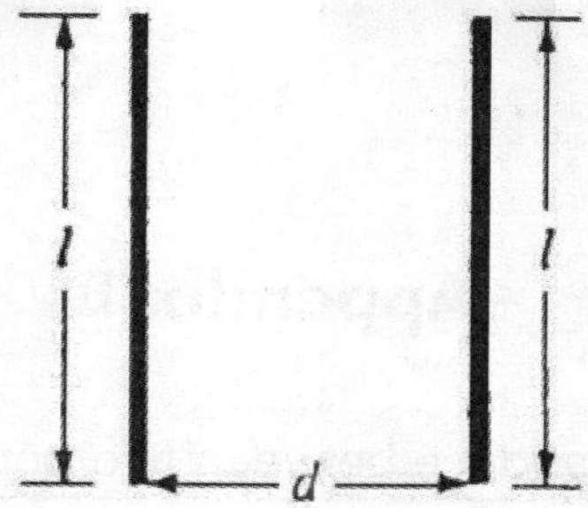

Figure A.3 Side-by-side configuration of two identical dipole antennas of length l

Appendix B

Calculation of mutual impedance between two antennas of unequal lengths

Let l_1 and l_2 be the half-lengths of two centre-fed dipole antennas separated by distance d and staggered by height h. Consider r_0, r_1 and r_2 be the distance of differential element of the second element from the centre, top and base of first dipole element, respectively. The mutual impedance between two antennas is given by

$$Z_{21} = -\frac{V_{21}}{I_{1b}} \tag{B.1}$$

where V_{21} is the open-circuit voltage at Antenna 2 terminal due to base current I_{1b} in Antenna 1. This open-circuit result from the voltages induced in all the elemental lengths of antenna may be found by reciprocity theorem and is given by

$$V_{21} = \frac{1}{I_{2b}}\left(\int_{h}^{l_2+h} E_{z1} I_2(z)dz + \int_{l_2+h}^{2l_2+h} E_{z1} I_2(z)dz \right) \tag{B.2}$$

where I_{2b} is the base current of Antenna 2 and E_{z1} is the electric intensity component parallel to the antenna axis at point z along Antenna 2 due to current in Antenna 1.

The antenna current distribution is given by

$$I_2(z) = I_{2m} \sin\beta(z-h) \qquad h < z < l_2 + h \tag{B.3}$$

$$I_2(z) = I_{2m} \sin\beta(2l_2 + h - z) \qquad l_2 + h < z < 2l_2 + h \tag{B.4}$$

where I_{2m} is the maximum current.

The expression for parallel component of electric field is given by

$$E_{z1} = 30 I_{1m}\left[\frac{-je^{-j\beta r1}}{r_1} + \frac{-je^{-j\beta r2}}{r_2} + \frac{2j\cos\beta l_1 e^{-j\beta r0}}{r_o} \right] \tag{B.5}$$

The relation between mutual impedance and loop currents is given by

$$Z_{12\,loop} = \frac{I_{1b} I_{2b}}{I_{1m} I_{2m}} Z_{12\,base} \tag{B.6}$$

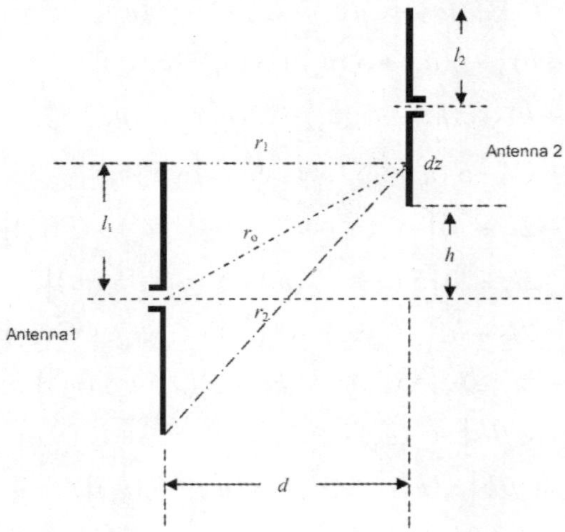

Figure B.1 Schematic of echelon dipole array

$$Z_{12} = -30\frac{1}{I_{2b}}\left\{\begin{array}{l}\left[\int_{h}^{l_2+h}\sin\beta(z-h)+\int_{l_2+h}^{2l_2+h}\sin\beta(2l_2+h-z)\right]\\ \left[\frac{-je^{-j\beta r1}}{r_1}+\frac{-je^{-j\beta r2}}{r_2}+\frac{2j\cos\beta l_1 e^{-j\beta r0}}{r_o}\right]dz\end{array}\right\} \tag{B.7}$$

where

$$\begin{aligned} r_0 &= \sqrt{d^2+z^2}\\ r_1 &= \sqrt{d^2+(l_1-z)^2}\\ r_2 &= \sqrt{d^2+(l_1+z)^2}\end{aligned} \tag{B.8}$$

(B.8) can be reduced to

$$Z_{12} = -30\frac{1}{I_{2b}}\left\{\begin{array}{l}\left[\int_{h}^{l_2+h}\sin\beta(z-h)+\int_{l_2+h}^{2l_2+h}\sin\beta(2l_2+h-z)\right]\\ \left[\frac{\sin\beta r_1}{r_1}+\frac{\sin\beta r_2}{r_2}-\frac{2\cos\beta l_1\sin\beta r_o}{r_o}\right]dz\end{array}\right\} \tag{B.9}$$

On solving the above integration, the real part gives the mutual resistance R_{12} and the imaginary part gives the mutual reactance X_{12}. Hence, the mutual resistance and mutual reactance is given by

$$R_{12} = 15 \left\{ \begin{array}{l} [\cos\beta(l_1 - h)\{C_i(u_0) + C_i(v_0) - C_i(u_1) - C_i(v_1)\} \\ + \sin\beta(l_1 - h)\{-S_i(u_0) + S_i(v_0) + S_i(u_1) - S_i(v_1)\} \\ + \cos\beta(l_1 + h)\{C_i(u_0') + C_i(v_0') - C_i(u_2) - C_i(v_2)\} \\ + \sin\beta(l_1 + h)[-S_i(u_0') + S_i(v_0') + S_i(u_2) - S_i(v_2)] \\ + \cos\beta(l_1 - 2l_2 - h)[-C_i(u_1) - C_i(v_1) + C_i(u_3) + C_i(v_3)] \\ + \sin\beta(l_1 - 2l_2 - h)[S_i(u_1) - S_i(v_1) - S_i(u_3) + S_i(v_3)] \\ + \cos\beta(l_1 + 2l_2 + h)[-C_i(u_2) - C_i(v_2) + C_i(u_4) + C_i(v_4)] \\ + \sin\beta(l_1 + 2l_2 + h)[S_i(u_2) - S_i(v_2) - S_i(u_4) + S_i(v_4)] \\ +2\cos\beta l_1 \cos\beta h[-C_i(w_1) - C_i(y_1) + C_i(w_2) + C_i(y_2)] \\ +2\sin\beta l_1 \sin\beta h[S_i(u_1) - S_i(v_1) - S_i(u_1) + S_i(v_3)] \\ +2\cos\beta l_1 \cos\beta(2l_2 + h)[C_i(w_2) + C_i(y_2) - C_i(w_3) - C_i(y_3)] \\ +2\cos\beta l_1 \sin\beta(2l_2 + h)[-S_i(w_2) + S_i(y_2) + S_i(w_3) - S_i(y_3)] \end{array} \right\} \tag{B.10}$$

where

$$S_i(u_0) = \int_0^{u_0} \frac{\sin u}{u} du \quad C_i(u_0) = \int_0^{u_0} \frac{\cos u}{u} du$$

$$u_0 = \beta(\sqrt{d^2 + (h - l_1)^2} + (h - l_1)); v_0 = \beta(\sqrt{d^2 + (h - l_1)^2} - (h - l_1));$$

$$u_0' = \beta(\sqrt{d^2 + (h + l_1)^2} - (h + l_1)); v_0' = \beta(\sqrt{d^2 + (h + l_1)^2} + (h + l_1));$$

$$u_1 = \beta(\sqrt{d^2 + (h - l_1 + l_2)^2} + (h - l_1 + l_2)); v_1 = \beta(\sqrt{d^2 + (h - l_1 + l_2)^2} - (h - l_1 + l_2));$$

$$u_2 = \beta(\sqrt{d^2 + (h + l_1 + l_2)^2} - (h + l_1 + l_2));$$

$$v_2 = \beta(\sqrt{d^2 + (h + l_1 + l_2)^2} + (h + l_1 + l_2));$$

$$u_3 = \beta(\sqrt{d^2 + (h - l_1 + 2l_2)^2} + (h - l_1 + 2l_2));$$

$$v_3 = \beta(\sqrt{d^2 + (h - l_1 + 2l_2)^2} - (h - l_1 + 2l_2));$$

$$u_4 = \beta(\sqrt{d^2 + (h + l_1 + 2l_2)^2} - (h + l_1 + 2l_2));$$

$$v_4 = \beta(\sqrt{d^2 + (h + l_1 + 2l_2)^2} + (h + l_1 + 2l_2));$$

$$w_1 = \beta(\sqrt{d^2 + h^2} - h); y_1 = \beta(\sqrt{d^2 + h^2} + h);$$

$$w_2 = \beta(\sqrt{d^2 + (h + l_2)^2} - (h + l_2)); y_2 = \beta(\sqrt{d^2 + (h + l_2)^2} + (h + l_2));$$

$$w_3 = \beta(\sqrt{d^2 + (h + 2l_2)^2} - (h + 2l_2)); y_3 = \beta(\sqrt{d^2 + (h + l_2)^2} + (h + l_2));$$

Likewise

$$X_{12} = 15 \left\{ \begin{aligned} &\cos\beta(l_1 - h)\left[-S_i(u_0) - S_i(v_0) + S_i(u_1) + S_i(v_1)\right] \\ &+\sin\beta(l_1 - h)\left[-C_i(u_0) + C_i(v_0) + C_i(u_1) - C_i(v_1)\right] \\ &+\cos\beta(l_1 + h)\left[-S_i(u_0') - S_i(v_0') + S_i(u_2) + S_i(v_2)\right] \\ &+\sin\beta(l_1 + h)\left[-C_i(u_0') + C_i(v_0') + C_i(u_2) - C_i(v_2)\right] \\ &+\cos\beta(l_1 - 2l_2 - h)\left[S_i(u_1) + S_i(v_1) - S_i(u_3) - S_i(v_3)\right] \\ &+\sin\beta(l_1 - 2l_2 - h)\left[C_i(u_1) - C_i(v_1) - C_i(u_3) + C_i(v_3)\right] \\ &+\cos\beta(l_1 + 2l_2 + h)\left[S_i(u_2) + S_i(v_2) - S_i(u_4) - S_i(v_4)\right] \\ &+\sin\beta(l_1 + 2l_2 + h)\left[C_i(u_2) - C_i(v_2) - C_i(u_4) + C_i(v_4)\right] \\ &+2\cos\beta l_1 \cos\beta h\left[S_i(w_1) + S_i(y_1) - S_i(w_2) - S_i(y_2)\right] \\ &+2\sin\beta l_1 \sin\beta h\left[C_i(u_1) - C_i(v_1) - C_i(u_1) + C_i(v_3)\right] \\ &+2\cos\beta l_1 \cos\beta(2l_2 + h)\left[-S_i(w_2) - S_i(y_2) + S_i(w_3) + S_i(y_3)\right] \\ &+2\cos\beta l_1 \sin\beta(2l_2 + h)\left[-C_i(w_2) + C_i(y_2) + C_i(w_3) - C_i(y_3)\right] \end{aligned} \right\} \qquad \text{(B.11)}$$

If both the antenna elements are equal in length, i.e. $l_1 = l_2 = l = \lambda/2$ with some height H, then the expressions for the mutual resistance R_{12} and mutual reactance X_{12} for two element parallel-in-echelon dipole array are given by

$$R_{12} = -\frac{\eta}{8\pi}\cos(kh)\begin{bmatrix} -2C_i\left(k\left(\sqrt{d^2+h^2}+h\right)\right) - 2C_i\left(k\left(\sqrt{d^2+h^2}-h\right)\right) + \\ C_i\left(k\left[\sqrt{d^2+(h-l)^2}+(h-l)\right]\right) + \\ C_i\left(k\left[\sqrt{d^2+(h-l)^2}-(h-l)\right]\right) + \\ C_i\left(k\left[\sqrt{d^2+(h+l)^2}+(h+l)\right]\right) + \\ C_i\left(k\left[\sqrt{d^2+(h+l)^2}-(h+l)\right]\right) \end{bmatrix}$$

$$+\frac{\eta}{8\pi}\sin(kh)\begin{bmatrix} -2S_i\left(k\left(\sqrt{d^2+h^2}+h\right)\right) - 2S_i\left(k\left(\sqrt{d^2+h^2}-h\right)\right) + \\ S_i\left(k\left[\sqrt{d^2+(h-l)^2}+(h-l)\right]\right) + S_i\left(k\left[\sqrt{d^2+(h-l)^2}-(h-l)\right]\right) + \\ S_i\left(k\left[\sqrt{d^2+(h+l)^2}+(h+l)\right]\right) + S_i\left(k\left[\sqrt{d^2+(h+l)^2}-(h+l)\right]\right) \end{bmatrix} \qquad \text{(B.12)}$$

$$X_{12} = -\frac{\eta}{8\pi}\cos(kh)\begin{bmatrix} -2S_i\left(k\left(\sqrt{d^2+h^2}+h\right)\right)+2S_i\left(k\left(\sqrt{d^2+h^2}-h\right)\right)- \\ S_i\left(k\left[\sqrt{d^2+(h-l)^2}+(h-l)\right]\right)- \\ S_i\left(k\left[\sqrt{d^2+(h-l)^2}-(h-l)\right]\right)- \\ S_i\left(k\left[\sqrt{d^2+(h+l)^2}+(h+l)\right]\right)- \\ S_i\left(k\left[\sqrt{d^2+(h+l)^2}-(h+l)\right]\right) \end{bmatrix}$$

$$+\frac{\eta}{8\pi}\sin(kh)\begin{bmatrix} -2C_i\left(k\left(\sqrt{d^2+h^2}+h\right)\right)-2C_i\left(k\left(\sqrt{d^2+h^2}-h\right)\right)- \\ C_i\left(k\left[\sqrt{d^2+(h-l)^2}+(h-l)\right]\right)+ \\ C_i\left(k\left[\sqrt{d^2+(h-l)^2}-(h-l)\right]\right)- \\ C_i\left(k\left[\sqrt{d^2+(h+l)^2}+(h+l)\right]\right)+ \\ C_i\left(k\left[\sqrt{d^2+(h+l)^2}-(h+l)\right]\right) \end{bmatrix} \tag{B.13}$$

where $C_i(x)$ and $S_i(x)$ are cosine and sine integrals.

Appendix C

Self and mutual impedance of dipole array

The self impedance of a dipole is given by

$$R_{self_n} = \frac{\eta}{2\pi}\left[\begin{array}{l} C + \ln(kl_n) - C_i(kl_n) + \frac{1}{2}\sin(kl_n)\{S_i(2kl_n) - 2S_i(kl_n)\} \\ + \frac{1}{2}\cos(kl_n)\left\{C + \ln\left(\frac{kl_n}{2}\right) + C_i(2kl_n) - 2C_i(kl_n)\right\} \end{array}\right] \tag{C.1}$$

$$X_{self_n} = \frac{\eta}{4\pi}\left[\begin{array}{l} 2S_i(kl_n) + \cos(kl_n)\{2S_i(kl_n) - S_i(2kl_n)\} \\ -\sin(kl_n)\left\{2C_i(kl_n) - C_i(2kl_n) - C_i\left(\frac{2ka_n^2}{l_n}\right)\right\} \end{array}\right] \tag{C.2}$$

where $C_i(kl_n)$ and $S_i(kl_n)$ are cosine and sine integrals expressed as

$$S_i(x) = \sum_{k=0}^{\infty}\frac{(-1)^k x^{2k+1}}{(2k+1)(2k+1)!} \qquad C_i(x) = C + 1n(x) + \sum_{k=1}^{\infty}(-1)^k \frac{x^{2k}}{2k(2k)!}$$

The expressions of the mutual impedances for different configurations are

Side-by-side configuration

$$R_{s_r} = \frac{\eta}{4\pi}\left[2C_i(u_{0_r}) - C_i(u_{1_r}) - C_i(u_{2_r})\right] \tag{C.3}$$

$$X_{s_r} = -\frac{\eta}{4\pi}\left[2S_i(u_{0_r}) - S_i(u_{1_r}) - S_i(u_{2_r})\right] \tag{C.4}$$

where $u_{0_r} = kd_r; \quad u_{1_r} = k\left(\sqrt{d_r^2 + l_n^2} + l_n\right); \quad u_{2_r} = k\left(\sqrt{d_r^2 + l_n^2} - l_n\right)$

Collinear configuration

$$R_{c_r} = -\frac{\eta}{8\pi}\cos\left(v_{0_r}\right)\left[-2C_i\left(2v_{0_r}\right)+C_i\left(v_{2_r}\right)+C_i\left(v_{1_r}\right)-ln\left(v_{3_r}\right)\right] + \frac{\eta}{8\pi}\sin\left(v_{0_r}\right)\left[2S_i\left(2v_{0_r}\right)-S_i\left(v_{2_r}\right)-S_i\left(v_{1_r}\right)\right] \tag{C.5}$$

$$X_{c_r} = -\frac{\eta}{8\pi}\cos\left(v_{0_r}\right)\left[2S_i\left(2v_{0_r}\right)-S_i\left(v_{2_r}\right)-S_i\left(v_{1_r}\right)\right] + \frac{\eta}{8\pi}\sin\left(v_{0_r}\right)\left[2C_i\left(2v_{0_r}\right)-C_i\left(v_{2_r}\right)-C_i\left(v_{1_r}\right)-ln\left(v_{3_r}\right)\right] \tag{C.6}$$

where $v_{0_r} = k\left(d+l_n\right)$; $v_{1_r} = 2k\left(d+2l_n\right)$; $v_{2_r} = 2kd$; $v_{3_r} = \left(d^2+2dl_n\right)/\left(d+l_n\right)^2$

Parallel-in-echelon configuration

$$R_{p_r} = -\frac{\eta}{8\pi}\cos\left(w_{0_r}\right)\left[-2C_i\left(w_{1_r}\right)-2C_i\left(w'_{1_r}\right)+C_i\left(w_{2_r}\right)+C_i\left(w'_{2_r}\right)+C_i\left(w_{3_r}\right)+C_i\left(w'_{3_r}\right)\right] + \frac{\eta}{8\pi}\sin\left(w_{0_r}\right)\left[2S_i\left(w_{1_r}\right)-2S_i\left(w'_{1_r}\right)-S_i\left(w_{2_r}\right)+S_i\left(w'_{2_r}\right)-S_i\left(w_{3_r}\right)+S_i\left(w'_{3_r}\right)\right] \tag{C.7}$$

$$X_{p_r} = -\frac{\eta}{8\pi}\cos\left(w_{0_r}\right)\left[2S_i\left(w_{1_r}\right)+2S_i\left(w'_{1_r}\right)-S_i\left(w_{2_r}\right)-S_i\left(w'_{2_r}\right)-S_i\left(w_{3_r}\right)-S_i\left(w'_{3_r}\right)\right] + \frac{\eta}{8\pi}\sin\left(w_{0_r}\right)\left[2C_i\left(w_{1_r}\right)-2C_i\left(w'_{1_r}\right)-C_i\left(w_{2_r}\right)+C_i\left(w'_{2_r}\right)-C_i\left(w_{3_r}\right)+C_i\left(w'_{3_r}\right)\right] \tag{C.8}$$

where $w_{0_r} = kh_r$; $w_{1_r} = k\left(\sqrt{d^2+h_r^2}+h_r\right)$; $w'_{1_r} = k\left(\sqrt{d^2+h_r^2}-h_r\right)$

$$w_{2_r} = k\left(\sqrt{d^2+\left(h_r-l_n\right)^2}+\left(h_r-l_n\right)\right);\quad w'_{2_r} = k\left(\sqrt{d^2+\left(h_r-l_n\right)^2}-\left(h_r-l_n\right)\right)$$

$$w_{3_r} = k\left(\sqrt{d^2+\left(h_r+l_n\right)^2}+\left(h_r+l_n\right)\right);\quad w'_{3_r} = k\left(\sqrt{d^2+\left(h_r+l_n\right)^2}-\left(h_r+l_n\right)\right)$$

(C.1) through (C.8) are valid for the dipoles of equal length l with identical wire radius a. Thus, $l_n = l$ and $a_n = a$ for all n elements of the array. However, d_r represents the relative distance between the pair of elements considered; h_r represents the offset distance between the elements under consideration and $r = (x,y)$ indicates the coordinate pair representation of these elements.

Appendix D

Coupling and transmission coefficients: Formulation

The coupling coefficient for a coupler is the ratio of amount of power coupled into the coupling port to the amount of power at its input or receiving port. Thus,

$$\text{Coupling coefficient} = \frac{\text{Power at the coupling port}}{\text{Power at its receiving port}}$$

For each coupler in the network, the power at its coupling port will be proportional to the product of current and the impedance at its terminals. However, the power at the receiving port of the nth coupler will be the total received power minus the power received due to the elements 1 to $n - 1$ of the array.

The expression for the coupling coefficient of first element is given by

$$\kappa_1 = \frac{P_1}{P_{rx}} \tag{D.1}$$

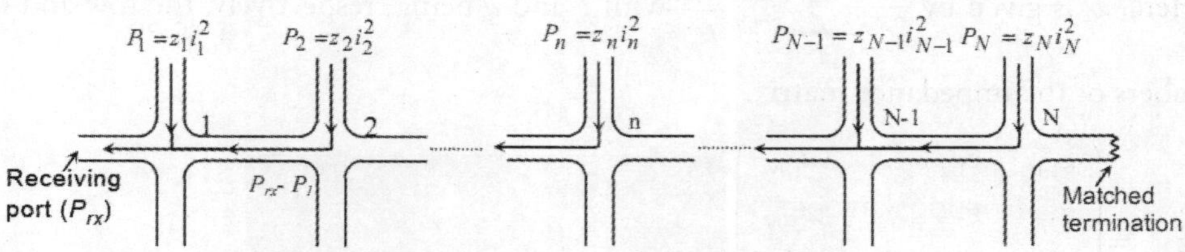

Figure D.1 Coupling and transmission coefficients of the couplers

Similarly, for other elements

$$\kappa_2 = \frac{P_2}{P_{rx} - P_1} = \frac{P_2}{P_{rx} - \sum_{q=1}^{1} P_q} \quad \text{(coupling coefficient of the second element)} \tag{D.2}$$

$$\kappa_3 = \frac{P_3}{P_{rx} - (P_1 + P_2)} = \frac{P_3}{P_{rx} - \sum_{q=1}^{2} P_q} \quad \text{(coupling coefficient of the third element)} \tag{D.3}$$

In general,

$$\kappa_n = \frac{P_n}{P_{rx} - (P_1 + P_2 + \dots + P_{n-1})} = \frac{P_n}{P_{rx} - \sum_{q=1}^{n-1} P_q} \text{ (coupling coefficient of the } n^{\text{th}} \text{ element)} \quad (D.4)$$

For a lossless network shown in Figure D.1, the total power fed into the circuit will be equal to the sum of the powers delivered to the individual antenna elements and to the terminating load. If the same array acts as a receiver, then the total power received should be equal to the sum of the power received by the individual antenna elements (assuming that zero power was delivered to the load while transmission). Thus,

$$P_{rx} = P_1 + P_2 + \dots + P_n + \dots + P_{N-1} + P_N = \sum_{p=1}^{N} P_p \quad (D.5)$$

Expressing power in terms of current and impedance, one gets

$$P_{rx} = z_1 i_1^2 + z_2 i_2^2 + \dots + z_n i_n^2 + \dots + z_{N-1} i_{N-1}^2 + z_N i_N^2 = \sum_{p=1}^{N} z_p i_p^2 \quad (D.6)$$

where
z_p is the net impedance of the pth antenna element
i_p is the current at the terminals of pth antenna element
P_{rx} is the total received power

Here, z_p is given by $z_p = \sum_{q=1}^{N} z_{p,q} \frac{I_q}{I_p}$ with p and q being, respectively, the row and column numbers of the impedance matrix.

$$z_{p,q} = \begin{pmatrix} z_{1,1} & z_{1,2} & \cdots & z_{1,N} \\ z_{2,1} & z_{2,2} & \cdots & z_{2,N} \\ \vdots & \vdots & \ddots & \vdots \\ z_{N,1} & z_{N,2} & \cdots & z_{N,N} \end{pmatrix} \quad (D.7)$$

Thus, the coupling coefficient of couplers can be written as

$$\kappa_n = \frac{z_n i_n^2}{\sum_{p=1}^{N} z_p i_p^2 - \sum_{q=1}^{n-1} z_q i_q^2} \quad (D.8)$$

and the transmission coefficient is given by

$$\tau_{c_n}^2 = 1 - \kappa_n^2 \quad (D.9)$$

List of Symbols

Symbols listed are defined as per first occurrence in the text of this book. A few symbols which are defined locally within the chapters are not included below.

Lower case

a	Radius of dipole wire
a_n	Amplitude distribution
b_o	Detected symbol
c	Speed of light
d	Inter-element spacing
$\vec{d}_{mn}$	Position vector of mn^{th} element
$\vec{d}_n$	Distance vector
d_r	Relative distance between element pair
d_x	Element spacing along x-axis
d_y	Element spacing along y-axis
e	Error vector
e_i	Eigenvectors
f	Frequency
f_o	Centre frequency
f_i	Residual response values
f_j	Pattern response of jth element to an incoming signal
f_{ep}	Electric plasma frequency
f_m	Damping frequency
f_{mo}	Magnetic resonant frequency
f_{mp}	Magnetic plasma frequency
g	Gain vector
g_i	Eigenvector beam
$\vec{h}$	Effective height of an antenna
i_n	Current at the terminals of nth antenna element
i	Interfering signal
$\hat{i}_\theta$	Unit vector in the elevation direction
$\hat{i}_\phi$	Unit vector in the azimuth direction
j	$\sqrt{-1}$
k	Propagation constant
l	Length of the antenna element
n	Refractive index
n_k	Quiescent receiver noise voltage at k^{th} element
p_d	Polarisation of the desired signal
p_{ik}	Polarisation of the kth interfering signal
r_1	Distance from the top of Antenna 1 to the z height of Antenna 2
r_2	Distance from the base of image of Antenna 1 to the z height of Antenna 2
r_c	Reflection coefficient of coupler
r_o	Distance from the base of Antenna 1 to the antenna element height
s	Distance
t	Time
w	Weights
w_a, w_b	Weights used in DF–GSC
w_{opt}	Complex optimum weight vector
w_m	Equaliser tap weight
w_q	Complex quiescent weight vector of the beamformer

x Received signal vector
x_d Desired received signal vector
x_n Element excitation
$y(n)$ Signal at the output of the beamformer at time n
z Height of Antenna 2
$z_{a_{x,y}}$ Mutual impedance between x^{th} and y^{th} dipoles
z_p Net impedance at p^{th} antenna element

Upper case

A Amplitude
A_n Real excitation
A_e Effective elemental area of the antenna array
A_{em} Maximum effective area of a single element
A_p Physical area of antenna array
A_d Amplitude of desired signal
A_{ik} Amplitude of interfering signal
A_r Amplitude of reference signal
B Blocking matrix
C_i Cosine integral
C Euler's constant
D Diffraction coefficient
$\widehat{D}'$ Radiation pattern of an array element
E Electric field
E_t Tangential component of electric field
E_{Direct} Field associated with the direct ray
E_i Incident field
E_T Transmitted field
E_b Scattered field due to backward travelling wave
E_f Scattered field due to forward traveling wave
E'_Δ Reflected signal emerging from difference arm of magic tee
$\left(E_\Delta\right)_{mn}$ Combined signal at the difference arm of the magic tee
$E^s_{\Delta_q}$ Total scattered field at the difference arm of magic tee considering q-level of couplers
E'_Σ Reflected signal emerging from sum arm of magic tee
$\left(E_\Sigma\right)_{mn}$ Combined signal at the sum arm of the magic tee
$E^s_{\Sigma_q}$ Total scattered field at the sum arm of magic tee considering q-level of couplers
$(E_1)_{mn}$ Signal incident on port 2 of magic tee
$\left(E'_1\right)_{mn}$ Reflected signal arriving back at the input port (port 2) of magic tee
$\left(E'_1\right)_{m'n}$ Reflected signal arriving back at the input port (port 3) of magic tee
E_{in} Scattered field due to input load reflected wave
E^s Scattered field
E^{self} Scattered field due to self-reflected wave
$E_{Z_{21}}$ Parallel component of electric field along the axis of Antenna 2
$\vec{E}^r_{c_n}$ Scattered field due to n^{th} coupler
$\vec{E}^r_n$ Total scattered field
$\vec{E}^r_{p_n}$ Scattered field due to n^{th} phase-shifter
$\vec{E}^r_{r_n}$ Scattered field due to n^{th} dipole
$\vec{E}^r_{s_n}$ Total scattered field due to the scattering beyond the coupling port of couplers

$\vec{E}^s$ Total scattered field
F_{norm} Normalised element scattering pattern
G Antenna gain
H Magnetic field
H_t Tangential component of magnetic field
I Identity matrix
I_{1m} Current induced at Antenna 1
I_{2m} Current induced at Antenna 2
I_n Current at the feed terminals of nth antenna element
J Cost function
J_{ex} Excess MSE
J_{min} Minimum output power
L_n Length of delay lines
M Number of directions considered in the sidelobe envelope
M_i Individual source matrix
M_q Quiescent receiver noise matrix
M_r Covariance matrix of rth interfering signal
M_{rl} Covariance matrix of lth spectral line of rth interfering signal
N Number of antenna elements
N_x Number of elements along x-axis
N_y Number of elements along y-axis
P Projection operator
P_n Power received at the nth coupler
P_{rx} Total received power
P_{in} Total input power
P_L Power into the load
P_d Output desired signal power
P_{ik} Output interfering signal power
P_s Desired signal power
P_r Power ratio
R_x Array correlation matrix
R Distance between the target and the observation point
R_a Antenna resistance
R_d Ohmic resistance
R_r Radiation resistance
$R_\perp$ Reflection coefficient for perpendicular polarisation
$R_\parallel$ Reflection coefficient for parallel polarisation
R_{21} Mutual resistance between elements of Antenna 1 and Antenna 2
R_{mutual} Mutual resistance
R_n Normalised covariance matrix of undesired signals and thermal noise
R_{self_n} Self-coupling radiation resistance of nth dipole
S Sidelobe envelope
S_i Sine integral
T Transmission coefficient
U Steering vector
U_d Desired signal vector
U_{ik} Signal vector of kth interfering signal
U_s Steering vector
V Element output voltage
V_{21} Open circuit voltage at Antenna 2 produced by current in Antenna 1
V_g Open circuit voltage due to generator
V_j Input terminal voltage at j^{th} element
V_o Open circuit voltage at the antenna terminals
V_{oj} Open circuit voltage at the jth antenna element
V^s Source voltage
W Weight vector
$X(\phi)$ Element excitation vector
$X(t)$ Input signal
X_{21} Mutual inductance between Antenna 1 and Antenna 2
X_d Desired signal vector
X_{ik} Interfering signal vector
X_{mutual} Mutual inductance
X_n Noise vector
X_{self} Self-inductance

X_a Antenna reactance

Y Output signal

Z Impedance matrix

Z_a Antenna impedance

Z_{21} Mutual impedance between elements of Antenna 1 and Antenna 2

Z_g Internal impedance of generator

Z_{ii} Self-impedance

Z_{ij} Mutual impedance between i^{th} and j^{th} elements

Z_L Load impedance

Z_{mutual} Mutual impedance

Z_o Characteristic impedance

Z_{p_n} Terminal impedance of nth phase-shifter

Greek

α Inter-element space delay along x-direction

α_i z component of the wave vector in the ith layer

α_s Inter-element phase to scan the antenna beam along x-direction

β Incident wave inter-element space delay

β_i Eigenvalue of a source

β_o Quiescent eigenvalue

β_s Inter-element phase to scan the antenna beam

γ Conversion factor gain constant

γ_e Electrical damping factor

γ_m Magnetic damping factor

δ_{ij} Kronekar delta

$\delta_{m'n}$ Phase angle of the wave incident on port 2 of the magic tee

$\delta_{m'n}$ Phase angle of the wave incident on port 3 of the magic tee

ε Permittivity of the medium

ε_o Free space permittivity

ε_r Relative permittivity

ε'_r Real part of complex relative permittivity

ε''_r Imaginary part of complex relative permittivity

ζ_i Message contained in the signal

η Intrinsic impedance

η_o Impedance of free space

θ Angle of incidence

θ_d Angle of elevation of desired signal

θ_i Incident angle of the source

θ_{ik} Angle of elevation of interfering signal

θ_o Look direction angle

(θ, ϕ) Direction of incident signal

(θ_s, ϕ_s) Beam scan angle

κ_n Coupling coefficient of nth coupler

λ Wavelength

μ_o Free space permeability

μ_r Relative permeability

μ Permeability of the medium

μ_a Step size for adaptive weight w_a

μ_b Step size for adaptive weight w_b

$v(\theta)$ Steering vector in the direction θ

v_d Derivative of the steering vector in the look direction

v_s Steering vector in the look direction

ξ_d Ratio of desired signal power to the thermal noise power

ξ_{ik} Ratio of interfering signal power to the thermal noise power

ρ Sidelobe envelope

ρ_{dj} Desired signal phase at j^{th} element with respect to coordinate origin

ρ_{ikj} Interfering signal phase at j^{th} element with respect to coordinate origin

ρ_p Reflection coefficient of phase shifter

ρ_r Reflection coefficient of radiating element

ρ_{c_n} Reflection coefficient at the coupling port of the nth coupler
ρ_{in} Reflection coefficient at the receive port of the array
ρ_{l_n} Reflection coefficient of the load terminating isolation port of nth coupler
ρ_{p_n} Reflection coefficient of nth phase shifter
ρ_{r_n} Reflection coefficient of nth radiating element
σ Radar cross section of target
σ_n^2 Variance of uncorrelated noise on each antenna element
σ_c RCS of phased array due to couplers
σ_p RCS of phased array due to phase shifters
σ_r RCS of phased array due to radiating elements
σ_s RCS of phased array due to scattering
σ_Δ RCS of phased array due to scattered field at the difference arm of magic tee
σ_Σ RCS of phased array due to scattered field at the sum-arm of magic tee
$\sigma_{\Sigma\Delta}$ RCS of phased array due to the scattered field at sum and difference arm of magic tee
σ^2 Thermal noise power
τ Time delay of the element with respect to a spatial reference point
τ_c Transmission coefficient of coupler
τ_{c_n} Transmission coefficient of nth coupler
τ_p Transmission coefficient of phase shifter
τ_r Transmission coefficient of radiating element
τ_{p_n} Transmission coefficient of nth phase shifter
τ_{r_n} Transmission coefficient of nth radiating element
ϕ Azimuth angle
ϕ_d Azimuth angle of desired signal
ϕ_{ik} Azimuth angle of interfering signal
ψ Electrical length between couplers
ω Angular frequency
ω_o Carrier frequency
Γ_o Reflection coefficient
Γ_n Fraction of the signal entering the nth element; returning to the aperture for re-radiation
Γ_n^r Total reflected signal returning to n^{th} element
Δ Sampling interval
Δf Offset frequency
Λ Eigenvalue
Φ Signal covariance matrix
Ψ_d Carrier phase of the desired signal at the coordinate origin
Ψ_{ik} Carrier phase of the interfering signal at the coordinate origin

Suggestions for Further Reading

Phased array analysis

Balanis, C. A. 2012. *Advanced Engineering Electromagnetics.* Second edition. Hoboken, NJ: John Wiley & Sons, 1018.

Bernardi, G., M. Felaco, M. D. Urso, L. Thimmoneri, A. Ferina, and E. F. Meliado. 2011. 'A simple strategy to tackle mutual coupling and platform effects in surveillance systems.' *Progress in Electromagnetics Research C.* 20: 1–15.

Bhattacharyya, A. K. 2006. *Phased Array Antennas: Floquet Analysis, Synthesis, BFNs and Active Array Systems.* NY, USA: Wiley-Interscience, 496.

Compton, Jr., R. T. 1982. 'A method of choosing element patterns in an adaptive array.' *IEEE Transactions on Antennas and Propagation* 30: 489–93.

Elliott, R. S. 2005. *Antenna Theory and Design.* Singapore: John Wiley & Sons, 594.

Elliott, R. S. and G. J. Stern. 1981a. 'The design of microstrip dipole arrays including mutual coupling, Part I: Theory.' *IEEE Transactions on Antennas and Propagation* 29: 757–60.

Elliott, R. S. and G. J. Stern. 1981b. 'The design of microstrip dipole arrays including mutual coupling, Part II: Experiment.' *IEEE Transactions on Antennas and Propagation* 29: 761–65.

Fenn, A. J., D. H. Temme, W. P. Delaney, and W. E. Courtney. 2000. 'The development of phased-array radar technology.' *Lincoln Laboratory Journal* 12(2): 321–40.

Fourikis, N. 1997. *Phased Array-Based Systems and Applications.* New York, USA: Wiley-Interscience, 426.

Hansen, R. C. 2009. *Phased Array Antennas.* Second edition. Hoboken, NJ: Wiley-Interscience, 547.

Jha, R. M., V. Sudhakar, and N. Balakrishnan. 1987. 'Ray analysis of mutual coupling between antennas on a general paraboloid of revolution (GPOR).' *Electronics Letters.* 23: 583–84.

Kraus, J. D. and R. J. Marhefka. 2001. *Antennas for All Applications.* Third edition. UK: McGraw-Hill Higher Education, 962.

Kumar, B. P. and G. R. Branner. 1999. 'Design of unequally spaced arrays for performance improvement.' *IEEE Transactions on Antennas and Propagation* 47: 511–23.

Obelleiro, F., L. Landesa, J. M. Taboada, and J. L. Rodriguez. 2001. 'Synthesis of onboard array antennas including interaction with the mounting platform and mutual coupling effects.' *IEEE Antennas and Propagation Magazine* 43(2): 76–82.

Oraizi, H. and M. Fallahpour. 2011. 'Sum, difference and shaped beam pattern synthesis by non-uniform spacing and phase control.' *IEEE Transactions of Antennas and Propagation* 59: 4505–11.

Pathak, P. H. and N. Wang. 1981. 'Ray analysis of mutual coupling between antennas on a convex surface.' *IEEE Transactions on Antennas and Propagation* 29: 911–22.

Pesavento, M., A. B. Gershman, and Z. -Q. Luo. 2002. 'Robust array interpolation using second-order cone programming.' *IEEE Signal Processing Letters* 9: 8–11.

Singh, H., Sneha H. L., and R. M. Jha. 2013. 'Mutual coupling effect in phased arrays: A review.' *International Journal of Antennas and Propagation*. USA: Hindawi Publishers, 23.

Skobelev, S. P. 2011. *Phased Array Antennas with Optimised Element Patterns.* Norwood, MA: Artech House Publishers, 261.

Stern, G. J. and R. S. Elliott. 1981. 'The design of microstrip dipole arrays including mutual coupling, Part II: Experiment.' *IEEE Transactions on Antennas and Propagation* 29: 761–65.

Stutzman, W. L. and G. A. Thiele, 2013. *Antenna Theory and Design.* Third edition. Hoboken, NJ: John Wiley & Sons, 822.

Wirth W. -D. 2013. *Radar Techniques using Array Antennas.* Second edition. London, UK: IET Press, 560.

Conformal array analysis

Corcoles, J., M. A. Gonzalez de Aza, and J. Zapata. 2014. 'Full-wave analysis of finite periodic cylindrical conformal arrays with Floquet spherical modes and hybrid element-generalised scattering matrix method.' *Journal of Electromagnetic Waves and Applications* 28: 102–11.

Gerini, G. and L. Zappelli. 2005. 'Multilayer array antennas with integrated frequency selective surfaces conformal to a circular cylindrical surface.' *IEEE Transactions on Antennas and Propagation* 53: 2020–30.

He, Q. Q. and B. -Z. Wang. 2007. 'Radiation pattern synthesis for a conformal dipole antenna array.' *Progress In Electromagnetics Research, PIER* 76: 327–40.

Josefsson, L. and P. Persson. 2006. *Conformal Array Antenna: Theory and Design.* Hoboken, NJ: John Wiley & Sons, 472.

Kumar, B. P. and G. R. Branner. 2005. 'Generalized analytical technique for the synthesis of unequally spaced arrays with linear, planar, cylindrical or spherical geometry.' *IEEE Transactions on Antennas and Propagation* 53: 621–34.

Morton, T. E. and K. M. Pasala. 2006. 'Performance analysis of conformal conical arrays for airborne vehicles.' *IEEE Transactions on Aerospace and Electronic Systems* 42: 876–90.

Wan, L. -T., L. -T. Liu, W. -J. Si, and Z. -X. Tian. 2013. 'Joint estimation of 2D-DOA and frequency based on space-time matrix and conformal array.' *The Scientific World Journal*, 10.

Yang, K., Z. Zhao, Z. Nie, J. Ouyang, and Q. H. Liu. 2011. 'Synthesis of conformal phased arrays with embedded element pattern decomposition.' *IEEE Transactions on Antennas and Propagation* 59: 2882–88.

RCS estimation techniques

Adana, F. S. D., I. Gonzalez, O. Gutierrez, and M. F. Catedra. 2003. 'Asymptotic method for analysis of RCS of arbitrary targets composed by dielectric and/or magnetic materials.' *IEE Proceedings on Radar, Sonar and Navigation* 150: 375–78.

Adana, F. S. D., S. Nieves, E. Garcia, I. Gonzalez, O. Gutierrez, and M. F. Catedra. 2003. 'Calculation of RCS from the double reflection between planar facets and curved surfaces.' *IEEE Transactions on Antennas and Propagation* 51: 2509–12.

Bao, G. and W. Zhang. 2005. 'An improved mode-matching method for large cavities.' *IEEE Antennas and Wireless Propagation Letters* 4: 393–96.

Burkholder, R. J. and T. Lundin. 2005. 'Forward-backward iterative physical optics algorithm for computing the RCS of open-ended cavities.' *IEEE Transactions on Antennas and Propagation* 53: 793–99.

Hemon, R., P. Pouliguen, H. He, J. Saillard, and J. F. Damiens. 2008. 'Computation of EM field scattered by an open-ended cavity and by a cavity under radome using the iterative physical optics.' *Progress In Electromagnetics Research* 80: 77–105.

Jha, R. M. and W. Wiesbeck. 1995. 'The geodesic constant method: A novel approach to analytical surface-ray tracing on convex conducting bodies.' *IEEE Antennas and Propagation Magazine* 37(2): 28–38.

Jung, Y. -H., S. -M. Hong, and S. Choi. 2010. 'Estimation of radar cross section of a target under track.' *EURASIP Journal on Advances in Signal Processing*, 6.

Ling, H., R. -C. Chou and S. W. Lee. 1989. 'Shooting and bouncing rays: Calculating the RCS of an arbitrarily shaped cavity.' *IEEE Transactions on Antennas and Propagation* 37: 194–205.

Ling, H., S. W. Lee and R. -C. Chou. 1989. 'High-frequency RCS of open cavities with rectangular and circular cross sections.' *IEEE Transactions on Antennas and Propagation* 37: 648–54.

Lu, B., S. X. Gong, Y. J. Zhang, S. Zhang, and J. Ling. 2011. 'Analysis and synthesis of radar cross section of array antennas.' *Progress in Electromagnetics Research M* 16: 73–84.

Odendaal, J. W. and D. Grygier. 2000. 'RCS measurements and results of an engine-inlet system design optimization.' *IEEE Antennas and Propagation Magazine* 42(6): 16–23.

Thors, B., L. Josefsson, and R. G. Rojas. 2004. 'The RCS of a cylindrical array antenna coated with a dielectric layer.' *IEEE Transactions on Antennas and Propagation* 52: 1851–58.

Visser, H. J. 2005. *Array and Phased Array Antenna Basics*. West Sussex, England: John Wiley & Sons, 359.

Wang, W. T., Y. Liu, S. X. Gong, Y. J. Zhang, and X. Wang. 2010. 'Calculation of antenna mode scattering based on method of moments.' *Progress in Electromagnetics Research Letters* 15: 117–26.

Wentworth, S. M. 2006. *Fundamentals of Electromagnetics with Engineering Applications*. New York: John Wiley & Sons, 588.

Yan, Y., W. -X. Zhao, X. -W. Zhao, Y. Zhang, and C. -H. Liang. 2012. 'Parallel computation of RCS of electrically large platform with coatings modeled with NURBS surfaces.' *International Journal of Antennas and Propagation*, 10.

Zhang, S., S. X. Gong, Y. Guan, and B. Lu. 2011. 'Optimized element positions for prescribed radar cross section pattern of linear dipole arrays.' *International Journal of RF and Microwave Computer-Aided Engineering* 21(6): 622–27.

Active RCS reduction

Clarkson, P. M. 1993. *Optimal and Adaptive Signal Processing*. Boca Raton, Florida: CRC Press, 529.

Gupta, I. J., J. A. Ulrey, and E. H. Newman. 2005. 'Effects of antenna element bandwidth on adaptive array performance.' *IEEE Transactions on Antennas and Propagation* 53: 2332–36.

Haykin, S. and K. J. R. Liu. 2010. *Handbook on Array Processing and Sensor Networks*. Hoboken, NJ: Wiley-IEEE Press, 904.

Hui, H. T. 2004. 'A practical approach to compensate for the mutual coupling effect of an adaptive dipole array.' *IEEE Transactions on Antennas and Propagation* 52: 1262–69.

Hui, H.T. 2002. 'Reducing the mutual coupling effect in adaptive nulling using a re-defined mutual impedance.' *IEEE Microwave and Wireless Components Letter* 12: 178–80.

Kim, K., T. K. Sarkar, and M. S. Palma. 2002. 'Adaptive processing using a single snapshot for a nonuniformly spaced array in the presence of mutual coupling and near-field scatterers.' *IEEE Transactions on Antennas and Propagation* 50: 582–90.

Liu, H. 2011. *Smart Antenna Beam Pattern Synthesis via System-based Convex Optimisation Framework*. ProQuest, UMI Dissertation Publishing, 84.

Lu, J., H. Xiao, Z. Xi, and M. Zhang. 2013. 'Cued search algorithm with uncertain detection performance for phased array radars.' *Journal of Systems Engineering and Electronics* 24(6): 938–45.

Monzingo, R. A. and T. W. Miller. 2004. *Introduction to Adaptive Arrays.* Rayleigh NC: Scitech Publishing, 543.

Mouhamadou, M., P. Vaudon, and M. Rammal. 2006. 'Smart antenna array patterns synthesis null steering and multi user beam forming by phase control.' *Progress In Electromagnetics Research* 60: 95–106.

Parhizgar, N., M. A. M. Shirazi, A. Alighanbari, and A. Sheikhi. 2013. 'Adaptive nulling of a linear dipole array in the presence of mutual coupling.' *International Journal of RF and Microwave Computer Aided Engineering* 24(1): 30–38.

Shahbazpanahi, S., A. B. Gershman, Z. -Q. Luo, and K. M. Wong. 2003. 'Robust adaptive beamforming for general-rank signal models.' *IEEE Transactions on Signal Processing* 51: 2257–69.

Svendsen, A. S. C. and I. J. Gupta. 2012. 'The effect of mutual coupling on the nulling performance of adaptive antennas.' *IEEE Antennas and Propagation Magazine* 54(3): 17–38.

Tennant, A., M. M. Dawoud, and A. P. Anderson. 1994. 'Array pattern nulling by element position perturbations using a genetic algorithm.' *Electronics Letters* 30: 174–76.

Tsoulos, G. V. 2001. *Adaptive Antennas for Wireless Communications*. New York: Wiley-IEEE Press, 776.

Vorobyov, S. A., A. B. Gershman, and Z. -Q. Luo. 2003. 'Robust adaptive beamforming using worst-case performance optimization: A solution to the signal mismatch problem.' *IEEE Transactions on Signal Processing* 51: 313–24.

Xu, S., Y. M. Xu, and J. Huang. 2012. 'The research of active cancellation stealth system design.' *Applied Mechanics and Materials* 192: 390–96.

Yuan, Q., Q. Chen, and K. Sawaya. 2006. 'Performance of adaptive array antenna with arbitrary geometry in the presence of mutual coupling.' *IEEE Transactions on Antennas and Propagation* 54: 1991–96.

RCS reduction and control

Abdelaziz, A. A. 2008. 'Improving the performance of an antenna array by using radar absorbing cover.' *Progress in Electromagnetics Reasearch Letters* 1: 129–38.

Galarregui, J. C. I., A. Tellechea, J. L. M. D. Falcon, I. Ederra, R.G. Garcia, and P. D. Maagt. 2013. 'Broadband radar cross-section reduction using AMC technology.' *IEEE Transactions on Antennas and Propagation* 61: 6136–43.

He, X. X., S. S. Huang, T. Chen, and Y. Yang. 2013. 'Ladder arrangement method for stealth design of vivaldi antenna array.' *International Journal of Antennas and Propagation*, 7.

Lu, B., S.-X. Gong, J. Zhang, and J. Ling. 2011. 'Analysis and synthesis of radar cross section of array antennas.' *Progress In Electromagnetics Research*, 16: 73–84.

Moore, F. W. 2002. 'Radar cross-section reduction via route planning and intelligent control.' *IEEE Transactions on Control Systems Technology* 10: 696–700.

Shang, Y. P., S. Q. Xiao, J. L. Li, and B. Z. Wang. 2012. 'An electronically controllable method for radar cross section reduction for a microstrip antenna.' *Progress In Electromagnetics Research* 127: 15–30.

Zhu, X., W. Shao, J. L. Li, and Y. Dong 2012. 'Design and optimization of low RCS patch antennas based on a genetic algorithm.' *Progress In Electromagnetics Research* 122: 327–39.

RCS reduction using metamaterials

Alu, A. and N. Engheta. 2005. 'Achieving with plasmonic and metamaterial coatings.' *Physical Review E* 72: 9.

Alu, A. and N. Engheta. 2006. 'Erratum: Achieving with plasmonic and metamaterial coatings.' *Physical Review E* 73.

Caloz, C. and T. Itoh. 2006. *Electromagnetic Metamaterials: Transmission Line Theory and Microwave Applications: The Engineering Approach*. Hoboken, NJ: John Wiley & Sons, 360.

Cui, T. J., D. R. Smith, and R. Liu. 2010. *Metamaterials: Theory, Design, and Applications*. New York: Springer, 367.

Donzelli, G., A. Vallecchi, F. Capolino, and A. Schuchinsky. 2009. 'Anisotropic metamaterials made of paired conductors: Particle resonances, phenomena and properties.' *Metamaterials* 3(1): 10–27.

Galarregui, J. C. I., A. T. Pereda, J. L. M. de Falcón, I. Ederra, R. Gonzalo, and P. de Maagt. 2013. 'Broadband radar cross-section reduction using AMC technology.' *IEEE Transactions on Antennas and Propagation* 61: 6136–43.

Haghparast, M. and M. S. Abrishamian. 2011. 'Reducing backscattering cross section of an electrically large sphere with metamaterial coating.' *Applied Physics A: Materials Science & Processing* 103: 587–90.

Komeylian, S., and F. H. Kashani. 2012. 'Reducing radar cross section by investigation electromagnetic materials.' *Advanced Electromagnetics* 1(3): 111–15.

Munk, B. A. 2009. *Metamaterials: Critique and Alternatives*. Hoboken, NJ: John Wiley & Sons, 189.

Said, Z., A. Sihvola, and A. P. Vinogradov. 2009. *Metamaterials and Plasmonics: Fundamentals, Modelling, Applications*. Springer, 306.

Solymar, L. and E. Shamonina. 2009. *Waves in Metamaterials*. Great Britain, UK: Oxford University Press, 400.

Vallecchi, A. and F. Capolino. 2009. 'Metamaterials based on pairs of tightly-coupled scatterers.' *Theory and Phenomena of Metamaterials*. Boca Raton, FL: CRC Press.

Zhang, J., J. Wang, M. Chen, and Z. Zhang. 2012. 'RCS reduction of patch array antenna by electromagnetic band-gap structure.' *IEEE Antennas and Wireless Propagation Letters* 11: 1048–51.

RCS reduction using frequency selective surfaces

Costa, F., S. Genovesi, and A. Monorchio. 2012. 'A frequency selective absorbing ground plane for low-RCS microstrip antenna arrays.' *Progress in Electromagnetics Research* 126: 317–32.

Genovesi S., F. Costa, and A. Monorchio. 2012. 'Low-profile array with reduced radar cross section by using hybrid frequency selective surfaces.' *IEEE Transactions on Antennas and Propagation* 60: 2327–35.

Jiang, W., T. Hong, and S. -X. Gong. 2013. 'Application of bionics in frequency selective surface design and antenna radar cross section reduction.' *Progress in Electromagnetics Research C* 42: 29–38.

Kern, D. J., D. H. Werner, A. Monorchio, L. Lanuzza, and M. J. Wilhelm. 2005. 'The design synthesis of multiband artificial magnetic conductors using high impedance frequency selective surfaces.' *IEEE Transactions on Antennas and Propagation* 53: 8–17.

Khosravi M. and M. S. Abrishamian. 2007. 'Reduction of Monostatic RCS by switchable FSS elements.' *PIERS Online* 3(6): 770–73.

Mingyun, L., H. Minjie, and W. Zhe. 2009. 'Design of multi-band frequency selective surfaces using multi-periodicity combined elements.' *Journal of Systems Engineering and Electronics* 20(4): 675–80.

Sarabandi, K. and N. Behdad. 2007. 'A frequency selective surface with miniaturised elements.' *IEEE Transactions on Antennas and Propagation* 55: 1239–45.

Turpin, J., P. E. Sieber, and D. H. Werner. 2013. 'Absorbing ground planes for reducing planar antenna radar cross-section based on frequency selective surfaces.' *IEEE Antennas and Wireless Propagation Letters* 12: 1456–59.

Yang, G. H., S. Ma, E. C. Wang, Y. Qin, and X. Gu. 2013. 'Design of active frequency selective surfaces for the RCS reduction.' *International Journal of Security and Its Applications* 7(5): 407–14.

Author Index

Subject Index